Water and Energy Management in India

Mrinmoy Majumder · Ganesh D. Kale
Editors

Water and Energy Management in India

Artificial Neural Networks and Multi-Criteria Decision Making Approaches

Springer

Editors
Mrinmoy Majumder
Hydro-Informatics Engineering (under Department of Civil Engineering)
National Institute of Technology Agartala
Jirania, Agartala, India

Ganesh D. Kale
Department of Civil Engineering
Sardar Vallabhbhai National Institute of Technology
Surat, Gujarat, India

ISBN 978-3-030-66682-8 ISBN 978-3-030-66683-5 (eBook)
https://doi.org/10.1007/978-3-030-66683-5

This Springer imprint is published by the registered company Springer Nature Switzerland AG
The registered company address is: Gewerbestrasse 11, 6330 Cham, Switzerland

Preface

The uncontrolled extraction and the uneven distribution of the water resources worldwide have resulted in scarcity of water and now in many places of the Earth many people do not have regular supply of drinking water.

The burgeoning population added with the advancement of technology has also imbibed high demand for energy resources. GAIA has finite amount of fossil fuels. If the demand get increased compared to the available fossil-based energy resources, then deficit will occur and if the current rate of consumption continues then also scarcity will take place and soon most of the region will be under energy stress.

Now both water and energy resources are under stress and the solution lies in their interdependencies. Optimal management of water in energy systems and vice versa can reduce the stress on both the resources.

Management of Water and Energy Interdependencies in World

Globally, billions of citizens lack access to the clean drinking water and cleanliness and more than 3/4th of the wastewater is released untreated. Energy is a critical part of the solution. According to a study by International Energy Agency (IEA), it was found that attaining comprehensive access to clean water and sanitation (SDG 6) would increase less than 1% to global energy demand in the Sustainable Development Scenario by 2030. In rural areas, nearly 66% of those who lack access to electricity also lack access to the clean drinking water. Therefore, optimization of water supply and sanitation measures can minimize the demand for electricity (IEA 2020).

The physical, economic and environmental viability of future projects of electricity generation will need to consider the impact of water availability. Increased water stress has material as well as economic impact on the cooling technologies deployed across China's coal-fired power fleet. Moreover, in the same study conducted by IEA, it was found that hydropower in the Africa will emphasize the significance of policy measures on management of the water available to enhance the resilience of hydropower (IEA 2020).

The reduced freshwater reserves can lead to a larger dependence on energy-intensive sources of water supply like desalination, which is evident in case of the Middle East where share of total final energy consumption for desalination is predicted to increase from 5% today to almost 15% by 2040 (Moniz 2014).

The operation of some power plants and other energy production activities has become constrained when an acute drought affected more than 30% of the United States in 2012 due to the lack of adequate supply of water. Hurricane Sandy is another example, when because of no power, vital water infrastructures failed to function (Moniz 2014).

'The water-energy nexus—the concept refers to the relationship between the water used for energy production including both electricity and sources of fuel such as oil and natural gas, and the energy consumed to extract, purify, deliver, heat/cool, treat and dispose of water (and wastewater) sometimes referred to as the energy intensity (EI)' (Spang et al. 2014). All forms of energy production require some input of water making the relationship complicated and any water infrastructure requires energy to operate.

In India also the need to inculcate the concept of the water-energy nexus has become significant as the scarcity of water and energy is intense in this subcontinent due to the size of the population and their capacity to use modern technology.

Management of Water and Energy Interdependencies in India

The unruly extraction of water and energy resources is one of the major reasons why India is facing acute shortage of these resources. For example, overdrafting of groundwater for irrigation in the state of Punjab has made 79% of the groundwater wells 'overexploited' and 'critical' with extraction exceeding the supply as per a study conducted by Central Ground Water Board in the year of 2010 (CGWB 2010). Punjab is also contending with the significances of uncontrolled application of chemical fertilizers and pesticides, water for irrigation and power subsidies, in the form of drying up of aquifers, degraded soil quality, polluted groundwater and rivers. Cohesive evaluations associated with a nexus approach can become important tools of natural resource management and policy frameworks requiring the need to understand these resources and their use.

'In India, groundwater irrigation has witnessed a recognizable increase since 1970s and water tables are diminishing rapidly in many parts of the country'. Today, over half of India's geographical area has high to extremely high water stress. 'However, tweaking the existing preference for rice and wheat using policy has remained elusive'. As a result, continuous power supply was provided to the farmers resulting in the trip of the northern and eastern grids in 2012 and 2014 (Ganguly et al. 2018). The state electricity boards require heavy subsidies to stay afloat and the political will to subvert the massive agriculture power subsidy is growing only slowly.

Energy Policy of Indian Government has also given stress on the use of high-efficiency electrical devices in agriculture, which will ensure reduction of stress on available energy content of the region (Prajapati 2018).

In this aspect, the present monograph was initiated to provide a platform to the researchers who are working in the field of water and energy nexus in India for sustenance of mankind and their livelihood in the face of climate change and other 'man-made' natural disasters.

The soft computation and different data science techniques are inevitably used to solve the various problems of the nexus in between water and energy and to select the best solution available or to evaluate a novel system which may optimize the nexus. The solutions depicted in this monograph have utilized various soft computation and data science techniques like Artificial Intelligence (AI), Internet of Things-based Real Time Modules, Multi-criteria Decision-Making (MCDM)-aided decision support systems, and also various optimization techniques which follow nature are utilized to find the best solution available.

That is why, Chapter "A Review of Multiple Criteria Decision-Making Methods in Reference to Water Resources and Climate Science Applications" introduced the different MCDM techniques, which are now used to solve various water resources and climate-change-related problems in real-time or real-life scenario. Chapter "Development of Spatial Cognitive Model for Estimation of Ungauged Runoff for Mesoscale Rivers" depicts the development of a new model, which can predict runoff for ungauged catchment. This investigation applied the analytical hierarchy process (AHP), which is a popular MCDM technique to develop the model.

Chapter "Indicator Based Impact Analysis of Urbanization with Respect to Evapo-Transpiration" proposed an indicator, which can represent the vulnerability of an urban city due to rapid urbanization with the help of potential evapotranspiration-based indicator. Here, also, the MCDM techniques are used to develop the indicator and to automate the approximation process an AI-based technique was utilized. This indicator can be applied universally for any city of the world for evaluation of the impact of urbanization on local climate.

In Chapter "Trend Analyses in Groundwater Levels of the Bikaner District, Rajasthan", the trend in ground water levels was analysed for an important region of the state of Rajasthan. In this aspect, many trend evaluation metrics were utilized, and a specific trend was possible to be proposed for the selected blocks.

Virtual Water can be defined as the water used as a raw material to produce an item in an industry. This concept was proposed by John Anthony Allan in the year of 1993 (Allan 1993). It is an important indicator where it measures the level of impact on the available water due to the industry for which the Virtual Water Availability is being conducted. Chapter "Climate Change Impact on Virtual Water Availability: A Categorized Polynomial Neural Network Approach" tries to analyse the impact of climate change on the Virtual Water Availability of the watershed. Instead of specific industries, the study tries to evaluate the VWA of a watershed, which have the cumulative impact of all the industries on the available water in the context of impacts of climatic abnormalities.

In Chapter "Development of ANN Model for Simulation of the Runoff as Affected by Climatic Factors on the Jamuna River, Assam, India", Artificial Neural Network (ANN) is used for estimation of runoff in the face of climate change. The smart use of the technique for approximation of runoff for the watershed of a river in the Assam State of North East India depicts that the adaptiveness of the AI can be used for water resource management with optimal accuracy. Chapter "Modelling of Reference Evapotranspiration for Semi-arid Climates Using Artificial Neural Network" applied ANN for estimation of reference evapotranspiration (RET) of a river under semi-arid climate and situated in the state of Telangana in southern part of India. This investigation used ANN to optimize an existing empirical model available for prediction of RET.

Chapter "Verifying Storm Water Drainage System Capacity for Vadodara Airport" depicts the application of Storm Water Management Model for the evaluation of preparedness of the present drainage system for a 1- and 2-year return period floods of a soon to be international airport in the city of Vadodara located in the state of Gujrat in the western part of India.

In Chapter "Optimal Trade-Off Between the Energy—Economy of a Hydropower Plant for Better Management of the Renewable Energy Resources", the optimal trade-off between the energy production and financial liability of a hydropower plant located in state of Tripura of North East India was identified with the help of novel MCDM techniques and polynomial neural network-based mathematical frameworks, which will be later on embedded in real-time monitoring systems for monitoring the performance as well as financial liability of the power plant. Chapter "Impact Analysis of Water, Energy, and Climatic Variables on Performance of Surface Water Treatment Plants" utilizes the same techniques but applied a novel modified MCDM rule for evaluation of performance with respect to water, energy and climatic parameters of Surface Water Treatment Plant (SWTP). Chapter "Power Allocation in an Educational Institute in India: A Fuzzy-GMDH Approach" uses MCDM techniques to analyse the energy allocation methods of an educational institute. This study also uses a modified analytical hierarchy process where fuzzy ratings were applied to find the pairwise comparison ratio of the alternatives. The same study also used the polynomial neural network for optimization of the allocation process.

The last two chapters deal with system failure. Chapter "Application of New Convergent Point Decision Making Method in Estimation of Vulnerability Index for Hydro Power Reservoirs" aims to analyse the performance of hydropower reservoir failure and safety factor with new MCDM techniques whereas Chapter "Recognition of Fatigue Failure in Wave Energy Converter Using Statistical Control Chart, Multi-criteria Decision Making Tools and Polynomial Neural Network Model" tries to evaluate the fatigue failure of wave energy converter systems. This study also used MCDM and PNN for their performance analysis.

All these 13 chapters provide a solution to the problems faced in the water and energy-based systems like watershed, reservoirs, SWTPs and wave energy converters. Although none of the study encases the nexus between the two most used natural resources (water and energy), the data and information collected from the studies can be easily used to create such interdependencies in a more optimal and feasible manner such that maximum utilization under minimum liability can be achieved.

Jirania, India
Surat, India

Dr. Mrinmoy Majumder
Asst. Professor
Dr. Ganesh D. Kale
Asst. Professor

References

Allan, J. A., (1993). "Fortunately There are Substitutes for Water, otherwise our Hydro-Political Futures Would be Impossible." In *Priorities for Water Resources Allocation and Management* (pp. 13–26). London: Overseas Development Administration.

CGWB (Central Groundwater Board) (2010). *State profile. Groundwater scenario for Punjab.* Available at http://cgwb.gov.in/gw_profiles/St_Punjab.htm.

Ganguly, A. R., Bhatia, U., Flynn, S. E., (2018). *Critical infrastructures resilience: Policy and engineering principles Routledge.* ISBN 9781498758635.

IEA (International Energy Agency) (2020). *Introduction to the water-energy nexus*, IEA, Paris, Retrieved from https://www.iea.org/articles/introduction-to-the-water-energy-nexus.

Moniz, (2014). *Ensuring the Resiliency of Our Future Water and Energy Systems, Energy.gov, USA*, Retrieved from https://www.energy.gov/articles/ensuring-resiliency-our-future-water-and-energy-systems.

Prajapati, P., (2018). *Water-food-energy nexus in India, India*, Retrieved from https://www.teriin.org/article/water-food-energy-nexus-india.

Spang, E. S., Moomaw, W. R., Gallagher, K. S., Kirshen, P. H., Marks, D. H. 2014, "The water consumption of energy production: an international comparison." *Environmental Research Letters*, 9(10), 105002.

Acknowledgements

The editors would like to acknowledge the help of all the people involved in this project and, more specifically, to the authors and reviewers that took part in the review process. Without their support, this book would not have become a reality.

First, the editors would like to thank each one of the authors for their contributions. Our sincere gratitude goes to the chapter's authors who contributed their time and expertise to this book.

Second, the editors wish to acknowledge the valuable contributions of the reviewers regarding the improvement of quality, coherence and content presentation of chapters. Most of the authors also served as referees; we highly appreciate their double task.

We will also like to thank our Director Prof. H. K. Sharma and Prof. S. R. Gandhi, respectively, for providing the support and encouragement we required for the completion of the book in the era of Covid 19 and during an extremely busy academic schedule.

We would also like to acknowledge the assistance we received from our publishers during the preparation of the book; from proof to press their cooperation was exemplary and unparallel.

In addition, we would also like to recognize with gratitude to the numerous publishers, government agencies and authors who have allowed our chapter authors to use their works for betterment of their synthesis.

Although we have taken utmost care and, with sincerity, have tried to make this piece error free still few discrepancies may creep in. We will be highly indebted to our readers if they kindly point out those such that in the next edition we can make the book really error free.

Lastly, we would like to thank our family members for their undoubted and unconditional support during the preparation of the manuscripts and, of course, we would like to thank our readers. If any one of them can upgrade their research effort from the contributions in the collected chapters, then we will think that our endeavour and time that we have spent in the preparation of this edited book is a success.

Tripura, India
Gujarat, India

Dr. Mrinmoy Majumder
Asst. Professor
Dr. Ganesh D. Kale
Asst. Professor

Introduction

Introduction

Human beings have become a compelling force in manipulating environmental resources as per their convenience and requirements. The land use is affected by the agricultural activities, and the industrialization is playing a vital role in changing the atmosphere and surrounding ecosystem. Due to the increasing population and extraction of natural resources to satisfy the need and greed of the people, presently, Earth faces scarcity and stress in supplying the two fundamental requirement for sustenance of livelihood: water and energy. On the one hand, the change in the climatic pattern due to global warming has resulted in disbalance and extremity in the distribution of water resources. The technological advancements and luxurious lifestyle, on the other hand, have an enormous amount of stress on the available natural resources. As a result, the need to innovate and cultivate, either alternative sources or to prepare for an optimal utilization plan for the 'smart-use' of water and energy, is required urgently to stop the degradation and reverse the trend to prevent complete decay of humanity.

As discussed in the introductory paragraph, there are two solutions to the present crisis:

a) Inventing new sources and
b) Managing the existing resources *optimally*

1.1 Present Scenario of Water Resources

Even though water is an essential part of our lives, its deficiency and deterioration in quality seem to be an irrelevant and overlooked issue for humankind. Excessive pumping is not only lowering the groundwater table but also depleting its quality by letting intrusion of saltwater. Climate change poses an even more significant challenge against water resource planning as, on the one hand, it is making dry places drier resulting in water scarcity. On the other hand, it causes rise in the sea level, which could be dangerous for the people living in the coastal areas. By the end of 2020, around 30–40% of the world population will have to face the wrath of water scarcity. As per the report of the World Water Council, the global average annual per capita access of renewable water resources is estimated to fall from 6,600 m^3 to 4,800 m by 2025 owing to the excessive population growth and their demands. Water requirement is rapidly increasing with the rising global population, which is estimated to be around 10 billion by 2050 (Reid 2020).

Water pollution has a direct influence on human health. The most contributing factor in contaminating the water is the draining of the untreated domestic sewage into the surface water bodies. Apart from this, discharging hazardous materials, wastes and toxic chemicals from industries and agricultural activity is also polluting the water remarkably. As a result, a large section of people who are utilizing the surface water directly or indirectly is getting prone to chronic to severe health-related diseases. Study shows that every year around 10–20 million people are dying because of water-borne diseases. According to World Health Organization and UNICEF Joint Monitoring Program Report (2017), 785 million people across the world are deprived of potable water, and every 2 min, a child dies because of water- and hygiene-related diseases.

Chandrasekharam et al. (2020), in their study, had used enhanced geothermal sources owing to its low cost for desalination technologies to meet the annual freshwater and food deficit of the people of Egypt and ensure sustainable development for the country. The authors claimed that electricity from E.G.S. could create 58,400 × 109 m^3/year of detoxified water as compared to 659 × 109 m^3/year volume generated from hydrothermal sources. Hall et al. (2019) had proposed an evaluation methodology consisting of system simulation modelling that presents a settlement between risk and cost. A small set of low-cost plans were developed with an acceptable level of risk to achieve optimal water management plan balancing various environmental constraints.

1.2 Present Scenario of Energy Resources

Energy resources are crucial pillars for the socio-economic development of any country, prosperity of humanity and the decline of poverty. Access to these resources is becoming a growing and compelling challenge for global advancement (Ritchie et al. 2020). In the last two decades, energy demand and supply have witnessed a compounded growth of about 50%, which implies power sector inflation and enhanced energy efficiency. The study depicts that the global energy requirement is assumed to hike from 2014 to 2040 by around 25 percent (Cordeiro 2020). Presently, conventional energy resources like coal, petroleum and natural gas, which are dominating the world, contain hydrocarbons; crude oil being the most common source for transportation fuels (Bozkurt 2010).

As the consumption of conventional energy resources (oil and gas) is increasing, the ejaculation of carbon from these resources is also spreading, causing severe environmental problems like global warming and the greenhouse effect. Thus, there is a great necessity in the world to replace fossil-based energy resources with that of renewable resources such as solar and wind energies, or else the ecological balance will get destroyed. World Energy Outlook 2019 report states that renewables are the biggest challengers to coal in the Asian power and heat sector market. Still, the demand for natural gas is also rising at a breakneck pace as a fuel in the industries and other developing Asian countries. Moreover, the report says that electricity growth is more than double the average energy demand globally, which is set to overtake oil in the final energy consumption by 2040.

1.3 Climate Change

Climate change is a vast range of global phenomena comprising of an eloquent alteration of average weather conditions, which is exhibiting a similar increased temperature trend over the antecedent decades. Even though factors like sun's intensity and volcanic outburst contribute to the increased temperature, it is the human activity which is majorly responsible for rapid climatic changes globally. Enormous ejaculation of carbon dioxide gas assurgent from the constant burning of fossil fuels and deforestation along with other greenhouse gases like nitrous oxide and methane in the atmosphere is making the situation worse for the world with every passing day. In 2018, there was an emission of around 36.57 billion metric tonnes of carbon dioxide in the world.

As per the National Oceanic and Atmospheric Administration 2019 global climate report, integrated land and ocean temperature are increasing at an average rate of 0.81 °C per decade since 1981. The average air temperature of Arctic region increased by 5 °C within the last 100 years as per the report of the Intergovernmental Panel on Climate Change. The report further says that around 30% of the wildlife across the world can be on the verge of extinction with an additional 1.5 °C average rise. The aquatic life is also in danger due to increased temperature and absorption of an excessive amount of carbon dioxide and other harmful gases, which is making the water acidic. IPCC has projected a devolution of around 70–90% in the number of coral reefs at 1.5 °C. Countries worldwide are adopting energy-efficient technologies for the decrementation of greenhouse gas exhalation.

1.3.1 Climate Change Effect on Water Resources

The significant climate change impact is found to be in the change in the spatial and temporal patterns of precipitation and water cycle. The disbalance in this cycle will eventually accelerate two extreme phenomena, namely, droughts and floods. Increased temperature and less rainfall will lead to loss of water in soil and plants, which in turn causes more drought-prone regions. Contrarily, hike in sultriness for a more extended time holds more moisture in the sub-stratosphere, which might increase the volume of rainfall whenever it pours in the ground culminating in colossal floods. The faltering allotment of precipitation also leads to distinguished divergency in water availability between different territories, provoking the situation concerning water management (Pouget et al. 2016).

The intrusion of saltwater into freshwater boundaries due to ascent in sea level is also an alarming consequence of thermal augmentation and it can possess a severe threat to fisheries and soil conditions (Yuryev 2019). World Water Development Report 2020 states that about 52% of the world's population may have to face the problem of water scarcity by the year 2050 owing to accelerated global warming. In this global crisis, the use of climate-smart management tools and integrated regional water management techniques, along with other environmentally sound policies, should be adopted to protect people and the surrounding ecosystem.

Naschen et al. (2019) used the Water and Soil Appraisal tool to inspect the consequences of climate change on the water resources in the Kilombero Catchment in Tanzania. They had adopted Coordinated Regional Downscaling Experiment to look after the spatial and temporal variations in the climatic patterns in which the authors had recommended to utilize enunciated seasonality because of shifting rainfall patterns. Erler et al. (2018) had used the surface-subsurface model named Hydro Geo Sphere in analysing the eventuality of climate change on a catchment within the Laurentian Great Lakes region. Rasifaghihi et al. (2019) had adopted Bayesian statistical methods to predict the consumption of water. The authors showed that long-term forecasting technique is beneficial for the outlining of water supply and supports feasible urban development.

1.3.2 Impact of Climate Change on Energy Resources

The impact of climate change can be observed across the world through the whole of the energy system. As the temperature is surging, significant changes in the balance of heating and cooling demand patterns are taking place. As a result of this rising air temperature, peak demand for electricity is going to increase by 20% in the upcoming few years in many regions of the world. Variation in the rainfall patterns causes an alteration in the runoff, river flows and corresponding water level, which in turn affects the design and efficiency of hydropower plants. Reduction in hydropower energy generation is an alarming situation as it supplies 2.3% of the world's total energy. The scarcity of available water due to drying weather can adversely affect the cleaning and maintenance system needed for solar power and nuclear power generation mechanics.

An increase in air temperature leads to enhanced operational costs and less efficiency of the machine or system (Solaun et al. 2019). A minimal amount of rainfall owing to global climate change has resulted in the decline of wind power potential by about 13% in the last four decades as annual energy production from wind relies mostly on monsoon (Gopal 2019). Infrastructural shifts can mitigate the challenges of climate change. Some examples are being incorporated, such as renewable sources, namely, biofuels, wind, solar and hydropower, which are nowadays used enormously to generate electricity rather than coal-fired ones. The application of clean energy is a better alternative to deal with climate change. Nowadays, trends like the Internet of things, solar photovoltaic systems, automatic sensors and adopting green policies are gaining momentum towards feasible and efficient energy use and reducing the negative impacts of climate change.

Fan et al. (2020) had formulated an econometric model for regional hydropower propagation to investigate the consequences of climatic factors on hydropower generation in different regions of China in which the experimental results showed that rainfall is a pivotal factor in hydropower generation. A surge in temperature can increase electricity demand and improve hydropower output. De Jong et al. (2019), in their study, assessed the impact of variations in climatic patterns on Brazil's solar and wind energy resources using three downscaled global climate models. The results inferred that in the northeast and southeast regions of Brazil, solar radiation potential could increase marginally, and there might be a substantial increase of around 40% in wind energy generation in most of the regions of the country because of climate change.

Thus, there is a need to prepare novel but innovative management and conservation plan for optimal allocation of both these resources such that mutually beneficial nexus between water and energy can be achieved. In this aspect, recent advancements in soft computation and decision-making techniques can be utilized. Sections 1.4 and 1.5 discuss such methods that can be used for finding feasible, optimal and real-time solutions.

1.4 Soft Computation Techniques

Soft Computing is an assembly of computational techniques that provide imprecise but usable and cost-effective solutions to very complex problems, which is either unsolvable or time-consuming with the current hardware. This approach enhances the power of human beings to perform a logical assessment of a complicated issue that generally cannot be solved using mathematical methodologies. Soft computing has full applications not only in engineering areas such as communication grid, inverters, electric power arrangement, etc. but also applicable in our workaday life, for instance, speech and vision recognition systems, data compression, manufacturing systems, identification of handwriting and so on. Soft computing techniques such as artificial neural network, fuzzy logic, genetic algorithm, expert systems, ant colony optimization and particle swarm optimization are more like complementary to each other instead of being competitive. They can be applied together to solve the concerned real-world problem (Ibrahim 2016).

Aryafar et al. (2019), in their study, have approached a hybrid PSO-based ANN model for modelling metallic and dye pollutant adsorption process from 54 industrial wastewater samples based on which performance indices are evaluated. Statistical Package for the Social Sciences software has been used to perform multivariate regression analysis, which shows that contact time and solution concentration are the two most influential factors in the adsorption mechanism. These factors are controlled and optimized by the PSO-ANN technique, which is found to be a useful tool in assessing the adsorption of industrial pollutants. Najafzadeh et al. (2019) employed four soft computing models, namely, feedforward backpropagation neural network, radial basis function neural network, adaptive neuro-fuzzy inference system and support vector machine in estimating the daily flow rates for designing Kerman wastewater treatment plant. The results of the assessment of training and testing stages had shown SVM model as the most efficient technique with a convincing degree of accuracy for the flow rate prediction. Thus, the authors concluded that these soft computing techniques could be effectively applied as robust tools in the designing process of wastewater treatment plants and other environmental engineering works. Nguyen et al. (2020) proposed the assessment of groundwater potential and identification of suitable areas of recharge in Dal Lak Province, Vietnam, by employing four ensemble soft computing models. The models are logistic regression mingled with the dagging, bagging, random subspace and cascade generalization techniques which were constructed to optimize the concerned training datasets. The models were validated using statistical measures, and G.I.S. software was used to display the groundwater potential maps. The authors found that dagging logistic regression was the most computationally accurate model in assessing the potential groundwater zones.

Some of the prominent soft computing techniques are described below:

1.4.1 Artificial Neural Networks

Artificial neural network is a prevalent computing system comprising of anastomotic artificial neurons that accomplishes optimization objectives by simulating different aspects related to the behaviour and capacity of the human brain, such as comprehending information processing, high level of analogousness and aptitude of learning

(Gallo 2015). ANN is known to be concordant with situations, flexible with data and competent enough for prognosticating any kind of complex problems. If a component of the neural network fails, the system can still forge ahead without any problem by their parallel constitution, but high processing time is required for large neural networks (Katkar 2012).

This study emphasized the application of artificial neural network in modelling a nonlinear system to predict the groundwater level of southwestern Nigeria. They used the geoelectric parameters such as aquifer resistivity, aquifer thickness, overburden resistivity, overburden thickness and coefficient of anisotropy as an input to the ANN model (Adiat et al. 2019). Moreover, these geoelectric parameters were generated by the interpretation of vertical electrical soundings data, which were obtained by adopting Schlumberger array configuration. Developing and validating the ANN-based model have shown satisfactory and reliable results that can be used in groundwater planning and management. Gadekar et al. (2019), in their study, had adopted a combined response surface methodology and artificial neural network approach to model and optimize colour removal of disperse dye by batch adsorption technique. They used poly-aluminium chloride-based water treatment residuals. The optimal conditions for colour removal were initially achieved using R.S.M. Then, the experimental results were used to train the neural network with the final pH parameter to predict the quantity of colour removal. The authors concluded that at optimum conditions, colour removal of $52.6 \pm 2.0\%$ was obtained. Thus, the adsorption could be used as a primary treatment for dye removal reducing the cost of treatment. A genetic algorithm-enhanced artificial neural network hybrid model was developed to forecast the monthly water production in the drinking water treatment plants in China. The water quality parameters like temperature and chemical oxygen demand and operational parameters like electricity consumption and chemical consumption were selected as the input variables. The production of drinking water was considered as the output. The training algorithm was optimized by minimizing the mean squared error (MSE) and fitness until the optimum architecture for the hybrid model was reached. The experiment results showed that the combined GA-ANN-based model could efficiently predict variation in water production based on parameter variations through scenario analysis (Zhang et al. 2019). The findings can help the decision-makers to take proper steps in advance in terms of regulatory changes and market demand.

1.4.2 Fuzzy Logic

Dr. Lotfi A. Zadeh first advanced the concept of fuzzy logic in the 1960s. It is a method of reasoning based on degrees of truth that work on the level of possibilities. It simulates human behaviour in decision-making considering digital outputs 0 and 1 as the extreme cases of fact. The fuzzy logic computing system deals with uncertainties that are usually developed in almost every professional sector to make effective decisions and solve real-world problems. The fuzzy logic can be implemented in systems such as micro-controllers, workstation-based or large network-based systems for achieving the correct output.

Reva Nagi et al. (2020) studied on detecting infection in the crop plants, classifying the disease post-infection and appraising the severity of the disease with the application of fuzzy logic in which the authors proposed that the approach will help the farmers to identify the disease class at an early stage. Al-Dmour et al. (2019) studied the delineation and execution of an analogous alarming process in the form of a warning score that has adopted the fuzzy logic approach to typecast patients' status and disease's austerity. The study inferred that the applied approach imparts satisfactory results that are affiliated with the present Modified Early Warning Score system. Andrew et al. (2020), in their work, emphasized the utilization of fuzzy logic in decision-making for sustenance planning of the manufacturing equipment by using main inputs. The average time between failures, the average time to repair, availability of spares and the age of the material, which will ultimately help the maintenance engineers and plant engineers to plan their activities productively, were evaluated for their potential to be considered as primary inputs.

1.4.3 Genetic Algorithm

Genetic algorithms are the machine learning search techniques applied to solve constrained and unconstrained optimization problems that simulate the process of the general biological evolution of species. This soft computing approach excogitates towards an ideal solution by customizing the best fit among the population of individual solutions over successive generations. Unlike other statistical models which are solved using derivative knowledge or Hessian matrix, this algorithm uses logic based on natural genetics and natural selection. The general steps in this technique start with creating an initial population that comprises randomly generated solutions. The most precise solutions as per the desired objective are selected further for reproducing new solutions by altering some genes randomly, crossover, mutation and the new offspring solutions are better and more optimal than their parent solutions.

Sakshi et al. (2019) proposed a novel method for the optimization of artificial neural network weights integrating pruning and genetic algorithm. They have observed that ANNs trained with GA-optimized weights manifest higher convergence with lower execution time and higher success rate. Abbasi et al. (2020) study the scheduling of the flow updates utilizing a hybrid genetic algorithm flow scheduler for attenuating the finishing time of the network upgrade by searching the solution space in software-defined networking. Moreover, the authors observed that the GA flow scheduler improves network performance when combined with other existing flow scheduling methods. Hanh et al. (2019) in their study proposed a genetic algorithm metaheuristic approach to compute the fitness function such that an ideal placement scheme is obtained to maximize the range of wireless sensor nodes. The results showed that the method provides solution quality and long stability with less computational time.

1.4.4 Particle Swarm Optimization

Particle swarm optimization is an evolutionary soft computational method whose performance is based on the social behaviour of bird flocking or fish schooling (Aydoan et al. 2019). In this technique, the random search is accoutered with swarm intelligence to attain an optimal solution. Every particle in the swarm signifies a candidate solution to the optimization problem. Particle progressively amends its position and velocity in the search space as per the required proximity to achieve its personal best location and also the global best place among its neighbours so that a more precise and accurate optimal solution can be obtained (Hossain et al. 2019). Particle swarm optimization technique has a wide range of applications in engineering as well as unconstrained global optimization problems owing to its fast convergence rate and easy implementation (Parsopoulos et al. 2002).

Tharwat et al. (2019) have proposed a chaotic particle swarm optimization to optimize the control points of the Bézier curve, with which the optimum smooth path that minimizes the distance between starting and ending points is found out. From the study, they have found that the algorithm can find the optimal way. Sengupta et al. (2019) made a study on historical and recent developments with hybridization perspectives making use of particle swarm optimization. In their research, they put the particular emphasis on deployment, development and improvements of the basic and its most modern state-of-the-art implementation. Deng et al. (2019) have introduced the improved particle swarm optimization to diagnose the faults of the motor bearing, which the existing fault diagnosis methods could not effectively recognize. They found that the improved PSO algorithm can productively improve the classification accuracy of least-square support vector machine, and the proposed fault diagnosis method performs better than other methods.

1.5 Multi-criteria Decision-Making Techniques

Multi-criteria decision analysis is an esteemed mechanism that is exercised to many complex decisions. These approaches are most associable in solving problems that are characterized as a choice among a set of candidate alternatives concerning multiple criteria (Hafezalkotob et al. 2018). Many times, in solving real-world problems, the decision-maker faces confusion and uncertainty when there are numerous criteria, both of which are important but constraint such as quality and cost. In such cases, the multi-criteria decision analysis tool is handy in dealing with diversiform criteria and prioritizing the alternatives or the projects which can accordingly aid the decision-makers to take appropriate judgement (Janse 2018). MCDM can be utilized and employed in all quarters right from our everyday problems to convoluted scientific affairs; specifically, in research areas of software engineering, networking, robotics and graphics (Singh et al. 2014).

Various MCDM methods such as analytical hierarchy process, analytical network process, TOPSIS and data development analysis are extensively used in the real-world problems by the researchers nowadays as they are highly competent in accomplishing decisions under ambiguity. Two of the most prominent MCDM techniques are discussed below:

1.5.1 Analytical Hierarchy Process

Analytic hierarchy process is a multi-criteria decision-making method where a stratified framework is developed. This method attributes a definite value delineating the preference degree for a concerned alternative over the additional options. With the help of pairwise comparison, judgments are adapted to multiple criteria, which ultimately propagate the overall priority for ranking the alternatives (Maletic et al. 2016).

Zhang et al. (2020) study the application of combined fuzzy inclusive evaluation and analytic hierarchy process in determining risk assessment of large-scale seawater desalination projects based on a comprehensive evaluation of an incorporated fuzzy and analytic hierarchy process method in which two levels of risk indicators are recognized. The analysed reports suggest that the overall risks of all the concerned projects are at the 'Very low' level. Karakuş (2019) studied the utilization of water quality index and GIS-based analytic hierarchy process in evaluating groundwater quality in the wet and dry seasons of Sivas Province (Turkey). The findings showed that 91.66% of groundwater samples collected during the wet season and around 77% of samples collected during the dry season are apt for consumption purposes.

1.5.2 Analytic Network Process

The analytic network process is a decision finding approach which is also an abstract of the analytic hierarchy process. Nodes represent criteria, sub-criteria and alternatives. Each of these nodes is correlated with each other in case of ANP. Ranking of alternatives and criteria relies upon each other. This technique enforces detailed definitions of nodes and interconnections, which require arrant thinking of the problem (Tuzkaya et al. 2008). As a result of this, ANP technique is extensively used in decision-making, which comprises diversiform criteria.

Mokarram et al. (2019) evaluated the groundwater quality in Northern Fars Province, Iran, with the use of GIS-based ANP and AHP in which fuzzy charts are created for each layer using a trapezoidal membership function. The outcomes indicate that ANP generates higher preciseness than fuzzy-AHP (0.954 in comparison to 0.845). Moreover, the findings suggest that calcium, chlorine and sodium content along with high electrical conductivity has adversely affected groundwater quality conditions of Northern Fars Province. Dano et al. (2019), in their study, used a unified technique that consolidates GIS, ANP and remote sensing for assessing and delineating susceptibility of flood in which ANP mathematical model was used to estimate the relative weighted significance of the several influencing factors.

The soft computing techniques like MCDM and ANN and its variants can be implemented to identify optimal policy for managing both water and energy resources. The present monograph aims to showcase some exemplary and novel methods which are applied to find an innovative solution for managing the natural resources maintaining the demand of the ever-growing population and their hunger for luxury.

Deep Jyoti Deb
Hydro-Informatics Engineering
under Department of Civil Engineering
National Institute of Technology Agartala
Agartala, Tripura, India

Prachi Kharbanda
Hydro-Informatics Engineering
under Department of Civil Engineering
National Institute of Technology Agartala
Agartala, Tripura, India

Khakchan Tripura
Hydro-Informatics Engineering
under Department of Civil Engineering
National Institute of Technology Agartala
Agartala, Tripura, India

References

Abbasi M Guleria A Devi M. (January 2020). “A genetic algorithm-based flow update scheduler for software-defined networks.” *International journal of Communication System. 33*(2), 25.

Adiat, K. A. N., Ajayi, O. F., Akinlalu, A. A. et al. (2020). Prediction of groundwater level in basement complex terrain using artificial neural network: a case of Ijebu-Jesa, southwestern Nigeria. *Applied Water Science* 10, 8.

Al-Dmour, Jumanah A, Assim Sagahyroon, AR Al-Ali, Salah Abusnana. (September 2019). “A Fuzzy Logic–Based Warning System for Patients Classification.” *Health Informatics Journal, 25*(3): 1004–24.

Andrew, A., Kumanan, S. (2020). Development of an intelligent decision-making tool for maintenance planning using fuzzy logic and dynamic scheduling. *International Journal of Information Technology, 12*, 27–36.

Arnell, N. (1999). Climate change and global water resources. *Global Environmental Change, 9*, S31–S49. doi:10.1016/s0959-3780(99)00017-5

Aryafar, A., Mikaeil, R., Doulati Ardejani, F., Shaffiee Haghshenas, S., Jafarpour, A. (2019). Application of nonlinear regression and soft computing techniques for modeling process of pollutant adsorption from industrial wastewaters. Journal of Mining and Environment, *10*(2), 327-337.

Aydoğan, E. K., Delice, Y., Özcan, U. et al. (2019). Balancing stochastic U-lines using particle swarm optimization. *Journal of Intelligent Manufacturing, 30*, 97–111.

Bozkurt, I. (2010). Energy Resources and Their Effects on Environment. WSEAS TRANSACTIONS on ENVIRONMENT and DEVELOPMENT. Issue 5, Volume 6, May 2010. ISSN: 1790-5079

Chandrasekharam, D., Lashin, A., Nassir Al Arifi, Abdulaziz M. Al-Bassam Varun, C. (2020). Geothermal energy for sustainable water resources management. *International Journal of Green Energy, 17*:1, 1–12.

De Jong, P., Barreto, T. B., Tanajura, C. A. S., Kouloukoui, D., Oliveira-Esquerre, K. P., Kiperstok, A., Torres, E. A. (2019). Estimating the impact of climate change

Deng, W., Yao, R., Zhao, H. et al. (2019). A novel intelligent diagnosis method using optimal LS-SVM with improved PSO algorithm. *Soft Computing, 23*, 2445–2462.

Erler, A. R., Frey, S. K., Khader, O., d' Orgeville, M., Park, Y.-J., Hwang, H.-T., … Sudicky, E. A. (2018). Simulating Climate Change Impacts on Surface Water Resources within a Lake Affected Region using Regional Climate Projections. *Water Resources Research.*

Fan, J., Hu, J., Zhang, X., Kong, L., Li, F., Mi, Z. (2018). Impacts of climate change on hydropower generation in China, *Mathematics and Computers in Simulation.*

Gadekar, M. R., Ahammed, M. M. (2019). Modelling dye removal by adsorption onto water treatment residuals using combined response surface methodology-artificial neural network approach. *Journal of Environmental Management, 231*, 241–248.

Gallo, C., "Artificial Neural Networks Tutorial." In Encyclopedia of Information Science and Technology, Third Edition. edited by Mehdi Khosrow-Pour, D.B.A., 6369–6378. Hershey, PA: I.G.I. Global, 2015.

Gopal S. (2019). Study shows climate change impacts wind energy industry. https://india.mongabay.com/.

Guermazi, E., Milano, M., Reynard, E., Zairi, M. (2018). *Impact of climate change and anthropogenic pressure on the groundwater resources in arid environment. Mitigation and Adaptation Strategies for Global Change.*

Hafezalkotob, A., Hafezalkotob, A., Liao, H., Herrera, F. (2018). An overview of MULTIMOORA for multi-criteria decision-making: Theory, developments, applications, and challenges. *Information Fusion.*

Hall, J. W., Mortazavi-Naeini, M., Borgomeo, E., Baker, B., Gavin, H., Gough, M., … Watts, G. (2019). Risk-based water resources planning in practice: a blueprint for the water industry in England. *Water and Environment Journal.*

Hanh, N. T., Binh, H. T. T., Hoai, N. X., Palaniswami, M. S. (2019). An Efficient Genetic Algorithm for Maximizing Area Coverage in Wireless Sensor Networks. *Information Sciences.*

Hossain, S., Imran, k., Akhand, M. A. H., Shuvo, M. I. R.,Siddique, N., Adeli, H. (2019). Optimization of University Course Scheduling Problem using Particle Swarm Optimization with Selective Search. *Expert Systems with Applications, 127*, 9–24.

Ibrahim, D. (2016). An Overview of Soft Computing. *Procedia Computer Science, 102*, Pages 34–38.

Janse, B. (2018). Multiple Criteria Decision Analysis (MCDA). Retrieved [insert date] from ToolsHero: https://www.toolshero.com/decision-making/multiple-criteria-decision-analysis-mcda.

Katkar, G. S. (2012). The Assembly of Neural Network in Soft Computing. *International Journal of Advanced Research in Computer Engineering & Technology, 1*(3), ISSN: 2278–1323.

Maletič, D., Lasrado, F., Maletič, M., Gomišček, B. (2016). Analytic Hierarchy Process Application in Different Organisational Settings. Applications and Theory of Analytic Hierarchy Process - Decision Making for Strategic Decisions.

Nagi, R., Sanjaya, S. T. (2020). "Application of Fuzzy Logic in Plant Disease Management." In *Fuzzy Expert Systems and Applications in Agricultural Diagnosis*, ed. A. V. Senthil Kumar, M. Kalpana, 261–302, Accessed February 26, 2020.

Najafzadeh, M., Zeinolabedini, M. (2019). *Prognostication of Waste Water Treatment Plant Performance Using Efficient Soft Computing Models: An Environmental Evaluation. Measurement.*

Näschen, K.; Diekkrüger, B.; Leemhuis, C.; Seregina, L.S.; van der Linden, R. (2019). Impact of Climate Change on Water Resources in the Kilombero Catchment in Tanzania. Water, 11, 859.

Nguyen, P. T., Ha, D. H., Avand, M., Jaafari, A., Nguyen, H. D., Al-Ansari, N., et al., Soft Computing Ensemble Models Based on Logistic Regression for Groundwater Potential Mapping. Appl. Sci. 2020, 10, 2469.

Parsopoulos, K., Vrahatis, M., (2002). Particle swarm optimization method in multiobjective problems. *Proceedings of the A.C.M. Symposium on Applied Computing.*

Reid, K. (2020, March 19). Global water crisis: Facts, F.A.Q.s, and how to help. World Vision. https://www.worldvision.org/clean-water-news-stories/global-water-crisis-facts.

Ritchie, H. Roser, M. (2020) - "Energy." Published online at OurWorldInData.org. Retrieved from: 'https://ourworldindata.org/energy' [Online Resource].

Sakshi S., Ravi K. (2019). A Neuro-Genetic Technique for Pruning and Optimization of ANN Weights. *Applied Artificial Intelligence, 33*(1), 1–26.

Sengupta, S.; Basak, S.; Peters, R.A., (2019). Particle Swarm Optimization: A Survey of Historical and Recent Developments with Hybridization Perspectives. *Machine Learning and Knowledge Extraction,* 1, 157–191.

Singh, A. Malik, K. S. (2014). Major MCDM Techniques and their application-A Review. *IOSR Journal of Engineering*, ISSN (p): 2278-8719, Vol. 04, Issue 05, PP 15–25.

Singh, A., Saha, D., Tyagi, A. C. (2019). Emerging Issues in Water Resources Management: Challenges and Prospects. *Water Governance: Challenges and Prospects*, 1–23. doi:10.1007/978-981-13-2700-1.

Solaun, K., Cerdá, E. (2019). Climate change impacts on renewable energy generation. A review of quantitative projections. *Renewable and Sustainable Energy Reviews, 116*, 109415.

Tharwat, A., Elhoseny, M., Hassanien, A.E. et al. (2019). "Intelligent Bézier curve-based path planning model using Chaotic Particle Swarm Optimization algorithm." *Cluster Computing, 22*, 4745–4766.

Tundisi, J. G. (2008). Recursos hídricos no futuro: problemas e soluções. *Estudos Avançados, 22*(63), 7–16.

Tuzkaya G, Önüt S, Tuzkaya Umut R., Gülsün B. (2008). An analytic network process approach for locating undesirable facilities: An example from Istanbul, Turkey. *Journal of Environmental Management, 88*(4), 970–983, ISSN 0301-4797.

Versini, P.-A, Pouget, L., Mcennis, S., Custodio, E., Escaler, I. (2016). Climate change impact on water resources availability – Case study of the Llobregat River basin (Spain). *Hydrological Sciences Journal*, Taylor & Francis.

Yuryev, A (2019). Impact of Climate Change on Water Resources. https://www.wateronline.com/doc/impact-of-climate-change-on-water-resources-0001.

Zhang, Y., Gao, X., Smith, K., Inial, G., Liu, S., Conil, L. B., Pan, B. (2019). Integrating water quality and operation into prediction of water production in drinking water treatment plants by genetic algorithm enhanced artificial neural network. *Water Research*, 114888.

Contents

Energy Management

Water Management

A Review of Multiple Criteria Decision-Making Methods in Reference to Water Resources and Climate Science Applications

Hiteshri Shastri, Kaustubh Salvi, Shashikanth Kulkarni, and Saptarshi Misra

Abstract This chapter provides a review of the literature on multiple criteria decision-making (MCDM) applications to water resources and climate science problems. Inputs from these fields are extremely important for effective policy formulations, especially for agro-economy of India. However, characterized by uncertainties originating from different sources and complex governing physics, the outcomes of these fields are required to be applied to a constrained based system. Hence, a robust decision making tool is required to augment the policy formations against the systems with conflicting constraints. MCDM techniques are devised mainly to evaluate such problems with conflicting constraints and provide solutions. These methods reveal the credibility to deal with such situations. Seventy references describing eleven different MCDM approaches have been evaluated in this paper. This provides a unified source of references that could be useful for researchers and practitioners. Based on the review, some recent trends and future research directions are also highlighted.

Keywords MCDM · Climate change · Decision-making

H. Shastri
Department of Civil Engineering, Charotar University of Science and Technology, Anand, Gujarat, India
e-mail: hiteshrishastri.cv@charusat.ac.in

K. Salvi (✉)
Alabama Transportation Institute, University of Alabama, Tuscaloosa, USA
e-mail: kaustubh.iitm@gmail.com

S. Kulkarni
Department of Civil Engineering, Osmania University, Hyderabad, Telangana, India
e-mail: kulkarni.shashikanth@gmail.com

S. Misra
Department of Computer Science and Engineering, IIT Kharagpur, Kharagpur, West Bengal, India

M. Majumder and G. D. Kale (eds.), *Water and Energy Management in India*,
https://doi.org/10.1007/978-3-030-66683-5_1

1 Introduction

Climate science is the current center of the research universe. Characterized by colossal complexity and unparalleled research scope, climate sciences have captivated the attraction of scientists, across all over the world. In a true sense, the climate system is an interdisciplinary research field. It is filled with research opportunities for every possible scientific and social domain. Scientists working in various fields of basic sciences, all engineers, medical professionals, and even economists can find a plethora of research problems, pertaining to their respective domains. However, the roles played by different professionals are different. While scientists work on unraveling the governing physics and understand the system, engineers use that knowledge to generate the data products, which can be used for formulating effective planning policies. This can be elaborated with a simple example. Effect of El Niño–Southern Oscillation on Indian summer monsoon has been studied by many scientists and a lot of striking results were established based on this research. However, a common man will get satisfied if he gets accurate information on today's forecast. These two situations are linked. However, it requires a bridge to transfer the information from the synoptic scale to a much smaller perspective. Downscaling techniques such as (1) statistical, (2) dynamical form the bridge between the two domains for generating the data for impact-relevant variables. The high-resolution data generated by deploying downscaling techniques is an important step towards decision making.

Governing bodies most often use the data as one of the inputs for formulating decision-making policies. The implications of changing climatic conditions over the hydrologic cycle, food and agriculture, and extremes are already well established. Hence, along with changing climatic conditions, the governing bodies are required to deal with multiple contrasting constraints. Further many resources and currant resource allocation systems will be at risk under changing climate. For example, the elevated sea level due to global warming and melting of ice glaciers affects the coastal ecosystems; the occurrences of extreme events endanger water-related infrastructure. Considering the consequences of such impacts is highly important for policymakers in the process of regional planning to effectively negotiate the related dangers (Yin 2001). The tussle between the contrasting constraints makes the decision-making complicated. Under such circumstances, state-of-the-art decision-making techniques are deployed for better decision-making. In this chapter, we intend to discuss one of the credible decision-making frameworks named 'multi-criteria decision-making (MCDM)', which has been extensively adopted. This framework is an umbrella with a lot of decision-making techniques, especially with contrasting constrains. MCDM methods are broadly classified into two modules namely, Multi-Objective Decision Making (MODM) and Multi-Attribute Decision Making (MADM) (Zimmermann 1991). MODM involves decision problems, where decision space is incessant, whereas, MADM focuses on the problems where decision space is disconnected (Triantaphyllou et al. 1998).

2 Multiple Criteria Decision Making

MCDM (Khosravi et al. 2019; Everest et al. 2020; Liao et al. 2020) is regarded as a confluence of three fundamental areas (Bonissone et al. 2009) as represented in Fig. 1.

'Solution generation via search' involves the search over multi-dimensional space to identify a non-dominated feasible solution. This search is characterized by the presence of challenges such as non-convex solution spaces, multi-dimensional objective space, and complicated objective-constraints nexus. Hence, the possible solutions are mostly nonlinear and probabilistic in nature. 'Solution Generation via preference aggregation and trade-off (PA and T) involves combining the preferences (that might be ill-defined and time-variant) using simple or linear combination otherwise complex or knowledge-driven models.

Interactive visualization involves the process to augment the cognitive model of the problem and enables progressive decision making. This also helps the decision-makers in the process of selection loop and solution refinement.

The earth's climate system is highly nonlinear in nature. Climate change has been evident around the globe and the vulnerability of different components of ecological cycles to climate hazard is no exception. While this element is widely acknowledged, the indicator-based MCDM methods offer a theoretically-sound approach to identify various causes of incommensurability and nonlinearity vulnerability assessments related to climate change. MCDM is widely used for selection, ranking,

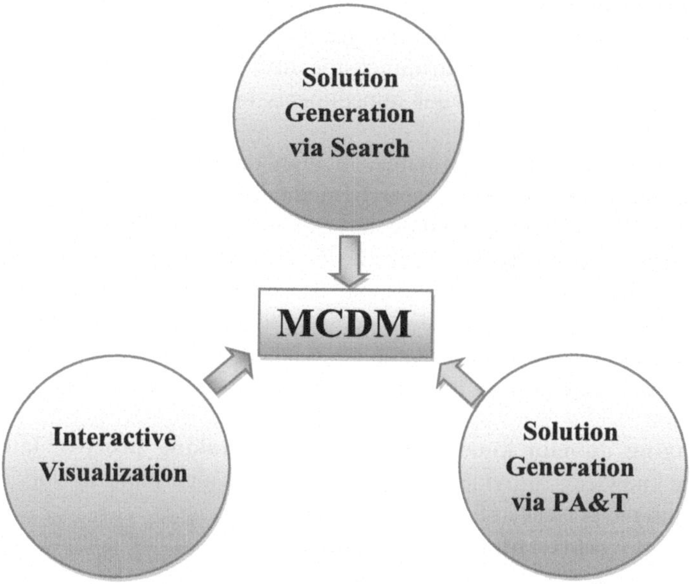

Fig. 1 MCMD overview (Bonissone et al. 2009)

and evaluation in the fields of management, education, agriculture, planning, transportation, construction, manufacturing, medical, control, and logistics to name a few. This chapter presents a review of some useful methods predominantly applied to solve decision problems pertaining to climate. Although most of the examples included here are drawn from water resource selection, these methodologies are largely independent of the particular application involved and can be applied to a wide range of climate applications that range from an energy assessment, environmental performance, etc.

3 MCDM Framework

MCDM focused on the interactive visualization, preference tradeoff, and interaction of search. Its framework consists of different attributes right from the deployment requirements to decision making (Bonissone et al. 2009). These attributes are discussed here to portray the orchestration of the MCDM framework.

a. **Deployment requirements**: MCDM workload can be deployed into two modes viz. batch and online. Batch mode involves deployment in groups, while online deployment is real-time.
b. **Deployment architecture**: The decision-making framework shows the flexibility in terms of the deployment architecture. It may be centralized or distributed in nature.
c. **Response Evaluation**: As stated earlier, the solution of MCDM is not necessarily deterministic. It may be uncertain, vague, or imprecise.
d. **Search Complexity**: Multi-dimensionality and complicated objective-constraints nexus are the major impediments in the path of successful search algorithms. As typical in design problems many evolutionary search algorithms do not preserve their performance in high dimensional spaces. This poses a strong need to use hybrid search methodologies to leverage the domain knowledge and interactively guide the search.
e. **Objectives and constraints complexity**: Complicated connections between objective function and constraints and non-convex regions prevent the use of aggregations or fast searches. It is necessary to take care of such complexities.
f. **Uncertainty management**: Uncertainty is an integral part of the climate system. Usually, it is represented in terms of vagueness or imprecision in solutions evaluation.
g. **Leveraging domain knowledge in decision-making process**: Internal and external knowledge provides an edge to the search algorithms to work in a better manner in the presence of conflicting constraints.
h. **Preferences representation/aggregation**: This involves complete or partial ordering, linear or nonlinear aggregations of preferences.
i. **Decision-making requirements and methods**, e.g., automated decisions for progressive and interactive decision-making and real-time batch applications.

j. **Update requirements for solution fidelity**, e.g., updating or retraining the deployed knowledge or data-driven evaluators.

The next section describes a few important MCDM methodologies which are largely applied to climate science problems. Here an emphasis is upon (a) applications with respect to the restrictions imposed by the formulation of a MCDM problem, (b) how to use MCDM systems to deal with uncertain and incomplete information, and (c) the decision-making process of the particular MCDM systems.

4 MCDM Methods

Expert Systems are becoming important tools in the management and planning of environmental systems (Durkin 1994; Huang and Chang 2003). However, to identify the best suitable MCDM technique for a given problem is very challenging. Sun and Li (2010) revealed that to simplify the decision procedure, there are more than 70 MCDM methods been introduced. A comparative analysis of MCDM methods is presented in several publications (Escobar and Moreno-Jimenez 2002; Opricovic and Tzeng 2007; Triantaphyllou 2000; Velasquez and Hester 2013, Zavadskas et al. 2014). Voogd (1983) revealed that for 40% of cases, the application of each MCDM method generated a different outcome from the other method. As suggested by Zanakis et al. (1998), theses dissimilarities arise due to the following reasons:

(1) Different MCDM methods employ weights differently.
(2) Different MCDM techniques vary by the way they select the best option.
(3) Most MCDM methods try to comprehend the decision factors or objectives differently.
(4) A number of MCDM methods use supplementary parameters that manipulate the solution.

Following Velasquez and Hester (2013), this section briefly describes these methods with a summary and brief outline of the overall approach with the assessment of the advantages and limitations of each methodology. Velasquez and Hester (2013) identify the following MCDM methods through a review of the literature: 1. Multi-Attribute Utility Theory (MAUT), 2. Analytic Hierarchy Process (AHP), 3. Fuzzy Set Theory, 4. Case-based Reasoning (CBR), 5. Data Envelopment Analysis (DEA), 6. Simple Multi-Attribute Rating Technique (SMART), 7. Goal Programming, 8. Elimination and choice translating reality (ELECTRE), 9. Preference Ranking Organization Method for Enrichment Evaluations (PROMETHEE), 10. Simple Additive Weighting (SAW), and 11. A technique for Order of Preference by Similarity to Ideal Solution (TOPSIS).

4.1 Multi-attribute Utility Theory (MAUT)

Multi-Attribute Utility Theory (Fishburn 1967; Keeney and Fishburn 1974; Keeney 1977) is the very frequently applied MCDM method. MAUT is fundamentally an addition to the Multi-Attribute Value Theory (MAVT) proposed by Keeney and Fishburn (1974). This methodology provides a higher precision to incorporate uncertainty and enhances the capability for evaluation of preferences for the performance and associated uncertainty (Loken 2007). MAUT helps in calculating the best possible utility to conclude upon the most suitable line of action for a given problem by assigning a utility to every possible consequence (Konidari and Mavrakis 2007). The ability to assign a utility to every uncertainty is considered as the major advantage of MAUT. This special quality is not accounted for with many MCDM methods. An extensive summary of the application of outranking methods in different decision-making problems as well as MAUT is provided by Siskos et al. (1984). MAUT is largely utilized in problems related to natural resource management with an approach cantered around societal risk preferences, survey-based attributes. The ability to take into account uncertainty is the key strength of this methodology. Ananda and Herath (2005) applied MAUT to analyze forest land-use risk preferences in Australia. Kailiponi (2010) applied MAUT methodology in support of evacuation decisions to address storm surge scenarios and identify levels, where evacuation actions are required. In the most recent applications, the combination of two or more MCDM methods is used to overcome shortcomings in any one particular method. Konidari and Mavrakis (2007) utilized a number of methodologies to estimate climate change mitigation. This study, in addition to the simple Multi-Attribute Ranking Technique (SMART), applied Analytic Hierarchy Process (AHP) to allocate grades to the instruments.

Raju and Vasan (2007) illustrated the application of MAUT to rank subsystems of Mahi Bajaj Sagar Irrigation Project in the state of Rajasthan, India. MAUT is employed to rank seven performance evaluation criteria, namely land development works, timely supply of inputs, economic impact, farmers participation in an irrigation system, crop yield, conjunctive use of water resources, and preservation of the environment. The irrigation subsystems are classified with the help of Kohonen Artificial Neural Networks (KANN), which can be utilized for further ranking by MAUT.

4.2 Analytic Hierarchy Process (AHP)

The Analytic Hierarchy Process (Saaty 1980) applies pair-wise evaluations to judge the alternatives with reverence to the different criteria and to estimate weights to be allocated to different criteria (Loken 2007). AHP is linear and hierarchical in nature with the objective at top and different alternatives at lower levels (Wang et al. 2012). This methodology is developed independently of other decision theories. AHP is one of the widely accepted methods of MCDM having various advantages and disadvantages. The ease of use is the first and foremost advantage of AHP with the approach of providing comparison in pairs. This method allows decision-makers to compare alternatives and weight coefficients with relative ease. The methodology is scalable, though it demands abundant data to perform pairwise comparisons in a proper fashion; the execution is less data-intensive than MAUT. The major disadvantage of this method is that it suffers interdependence problems between alternatives and criteria. Apart from this, AHP at the same time can be subjected to inconsistencies in ranking criteria and judgment through not allowing to grade one apparatus in isolation, but in evaluation with the rest, without identifying its strength and weaknesses (Konidari and Mavrakis 2007). A major limitation of AHP is that the common form of AHP is subjected to rank reversal. AHP is largely applied to resolve resource management issues, political strategy planning, public policymaking, and other planning problems. The AHP applications with the problem of resource management overcome the disadvantage of reversal of rank by having a smaller limited number of alternatives to start with.

Analytic Network Process (ANP) is the common form of AHP (Saaty 2006). ANP is nonlinear, as in contrast to AHP, and possesses a higher apprehension with network structure. It also possesses the ability to prioritize clusters of elements and handle them with a higher interdependence than AHP. It supports multifarious networked decision-making with different intangible criteria (Tsai et al. 2010). The main drawback, of ANP, is that it ignores the different effects among clusters (Wang et al. 2012). ANP has seen a larger field of application, particularly in combination with other different MCDM methods. ANP is frequently utilized in product planning, supply chain management, project selection, and optimal scheduling.

4.3 Fuzzy Theory

Fuzzy Theory (Zadeh 1965) is a widely utilized MCDM method. The fuzzy set theory developed based on the classical set theory considers incomplete information and with the help of available knowledge solves the problems. This is highly useful in problems related to imprecise data (Balmat et al. 2011). This methodology permits imprecise input and may work based on a few rules to encompass highly complex problems. The fuzzy systems, however, can be difficult to develop sometimes, as the development of a robust solution that can be applied to the real world demands numerous simulations. Fuzzy set theory is recognized and has been used in applications such as resources and supply chain management, environment and climate problems, engineering problems, etc. Khadam and Kaluarachchi (2003) demonstrated an application fuzzy ranking procedure for a cost–benefit analysis to support decision making in environmental projects addressing groundwater contamination.

Teegavarapu (2010) addressed the modeling of climate change uncertainty with an application of a fuzzy set theory for climate-sensitive administration of hydro systems. A fuzzy MCDM methodology aided by the BIM variable was developed by

Chen et al. (2016) for selecting low carbon building (LCB) measures, particularly in densely populated subtropical urban areas. The selection of optimal locations for the generation of usable energy from solar energy using solar panels is an important issue, as it will impact the economic growth and development of a region. Zoghi et al. (2017) developed a method for optimal site selection based on a combination of fuzzy logic, weighted linear combination (WLC), and MCDM processes. This method showed considerable accuracy in locating the most favorable sites for solar power generation.

4.4 Case-Based Reasoning (CBR)

The case-based reasoning (CBR) methodology is proposed by Li and Sun (2008) with an application for predicting economic distress in companies with a lead time over a year before its actual occurrence. For a given problem CBR retrieves all the similar cases from an existing record of cases and recommends decision-making based on the case having the highest similarity (Daengdej et al. 1999). The first advantage of this methodology is that it requires modest effort for acquiring supplementary data. The methodology also requires lower maintenance for the database to be previously available and demands little upkeep. Another important advantage of this approach over most of the MCDM methods is that as higher numbers of cases are added to the database, the CBR can improve over time. A CBR with many cases can also adapt to changes in the environment. The major weakness of this method is its sensitivity to data inconsistency. (Daengdej et al. 1999). For example, in a given problem, the cases existing in the database could be invalid leading to provide invalid answers. This is most common for applications that include comparisons of engineering designs, medicine, and businesses. CBR has also been utilized in understanding various aspects of climate variables from the application point of view. De Caro et al. (2017) used a combination of adaptive CBR models and cardinality reduction techniques to solve wind forecasting problems avoiding complex computations and get realistic forecasts of wind power in generators. Liu et al. (2016) used remote sensing data and principles of CBR to propose a method to remove the shortcomings of existing lightning-caused forest fire risk rating assessments.

4.5 Data Envelopment Analysis (DEA)

DEA applies a linear programming approach to evaluate the relative efficiencies amongst the available alternatives (Thanassoulis et al. 2012). This methodology rates the efficiencies of available alternatives against each other. DEA is skilled to handle multiple inputs and outputs with effective efficiency quantification. DEA shows the credibility to reveal relationships that may remain hidden with other methods. A significant shortcoming of this method is that it does not deal with inaccurate data and functions on the basic assumption that all the output and input and output variables are exactly known. While dealing with any actual problem, this assumption, however, may not hold well (Wang et al. 2005). It can be very well realized that the results are highly sensitive to the model inputs. DEA is very commonly applied for efficiency comparison. This is commonly used in agriculture, economics, utilities, medical, road safety, business, and retail problems mainly because of the availability of accurate facts that could be utilized for input, which overcomes one of the major deficiencies of this method. DEA is the most useful technique to aid the efficient management of limited water resources. Rodriguez et al. (2004) applied the DEA approach in Andalusia districts to study irrigation efficiency. In recent times, DEA approaches have been dynamically used in the field of sustainable energy. A complete discussion on origin, advancement, and prospects of DEA application in the field of sustainable solutions is thoroughly discussed by Zhou et al. (2018). Martin-Gamboa et al. (2018) demonstrated a novel combined Life Cycle Assessment and dynamic DEA approach to estimate the environmental impact and associated efficiencies of power plants in Spain. Hosseinzadeh-Bandbafha et al. (2018) recognized the efficiency of DEA for the optimization of energy use and the reduction of greenhouse gas emissions in the peanut production of Iran. The study demonstrates how the reduction of GHG emissions may be achieved with appropriate management measures, resulting in of applied DEA approach.

4.6 Simple Multi-attribute Rating Technique (SMART)

SMART (Edwards 1971, 1977) is one of the simplest forms of multi-attribute utility theory (MAUT). It offers a straight forward way to execute the principles of MAUT. The decision methodology is conceptualized on the basis that decisions depend on subjective quantities namely probabilities and values. An inaccuracy in output maybe because of error in modeling or due to model application with simplified assumptions. The two basic assumptions inherent with the SMART are preferential independence and utility independence (Chen et al. 2010). This method suitably converts significance weights into definite numbers. The major advantages of SMART include the simplicity of the application as it allows utilization of different weight assignment techniques such as absolute, relative, etc. SMART also handles data well under each criterion and considerably reduces the efforts required by decision-makers as compared to MAUT. A major disadvantage of this method similar to MAUT is the inconvenience in the methodology for determining work in view of the complex framework (Konidari and Mavrakis 2007). SMART's common applications are in transportation, manufacturing, construction, environmental, logistics, and military problems. The simplicity of the application facilitates situations, where a reasonable amount of information is easily accessible to decision-makers. Mi et al. (2017) used the SMART approach to develop a climate change mitigation index to assess regional efforts on climate change mitigation at the sub-national level and applied to provincial China. It helped to establish mitigation performance over different regions in

China. Seghezzo et al. (2017) applied a SMART methodology to increase the authenticity and sustainability of land use processes in Argentina. The method is applied to resolve the conflicts over ownership and land use. Read et al. (2017) applied the SMART procedure as a five-step method for the choice of energy project having larger uncertainty. On the other hand, farming sustainability has been calculated and assessed by Talukder et al. (2017) with a similar methodology.

4.7 Goal Programming

Goal Programming is a pragmatic programming methodology that is capable to select from a very large number of options. The capability to hold large-scale problems is a major advantage of this method. The ability to produce infinite alternatives for the given set of conditions provides a considerable advantage over some other methods. A key drawback of this methodology is its lack of ability to weight coefficients. The proper weight of coefficients in such applications is determined by applying other methods such as AHP. The application of goal programming is very well demonstrated in the field of manufacturing, planning and scheduling, health care, portfolio selection, delivery system proposal, energy scheduling, reservoir management, scheduling of timber harvest, and wildlife administration problems. These studies mainly demonstrate the application of goal programming in combination with other methods to have an appropriate weight assignment. Guo et al. (2016) illustrate the application of Goal Programming approach for decision making under uncertainty. Pal et al. (2012) demonstrate the Fuzzy Goal Programming (FGP) method for resolving and modeling agricultural planning problems for the best possible production of several crops in the Bardhaman District of West Bengal in India. The proper distribution of irrigation water in different seasons of a year is estimated based on FGP based models. The methodology achieves the highest value of the membership goals based on the desires of the decision-maker.

4.8 Elimination and Choice Translating Reality (ELECTRE)

ELECTRE is a highly popular MCDM methodology for ranking of alternatives pairwise comparison. The discordance and concordance measures are combined to generate a fair amount of outranking between each pair of alternatives. This is an outranking method that applies several iterations, based on concordance analysis. Its major advantage is that it takes into account the uncertainty and ambiguity of the data. The advantage of the ELECTRE is that the tradeoffs between various attributes are compensatory and the information enclosed in the judgment matrix is completely utilized. The diverse versions of ELECTRE such as ELECTRE-2, ELECTRE-3, and ELECTRETRI are popular and widely applied (Karagiannidis and Mounssiopoulos 1997). One of the major disadvantages of this methodology is that its progression and results may be difficult to explicate with non-technical terminology. At the same time integrated, the lowest performances for the provided criteria are not displayed. Therefore the weakness, strengths, impacts, and results of given alternatives cannot be directly recognized, verified (Konidari and Mavrakis 2007). The application of ELECTRE has been demonstrated in the field of water management, energy, economics, and transportation, and environmental problems. Similar to the other MCDM methods, this method also takes uncertainty into account. This is an important requirement for many of the applications. Chitsaz and Banihabib (2015) demonstrated the applications of the ELECTRE technique for flood risk problems. This case study demonstrates the application of non-parametric stochastic tests, sensitivity analysis, and aggregation models to identify the best applicable ranking model. ELECTRE is identified as the best model based on the outcomes of the mentioned tools.

4.9 Preference Ranking Organization Method for Enrichment Evaluations (PROMETHEE)

PROMETHEE similar to ELECTRE is an outranking method that has several iterations. The ease of application of PROMETHEE has made it a common method. It has been widely applied for many decades with many improved iterations. Different methodologies of PROMETHEE include PROMETHEE I that provides a fractional ranking of the alternatives and PROMETHEE II that provides a complete ranking of the alternatives. These methodologies were developed in the year 1982. Other advanced versions of PROMETHEE methods, such as PROMETHEE III that provides a ranking based on an interval, PROMETHEE IV that provides a complete or partial ranking of the alternatives for a continuous set of viable solutions, PROMETHEE V that solves problems with constraints and segmentation, and PROMETHEE VI that is widely applied as human brain representation are important (Behzadian et al. 2010).

One of the major limitations of this method is that it does not present an apparent methodology for the assignment of the weights. The application of PROMETHEE is well demonstrated in the decision-making problems in the field of environmental and water management, hydrology, financial and business management, logistics and transportation, chemistry, manufacturing and assembly, energy management, and agriculture. An overview of the drawbacks of different PROMETHEE methods is presented by Keyser and Peeters (1996).

Vulevic and Dragovic (2017) applied the PROMETHEE technique for ranking the watersheds and provide valuable information for watershed management. Anand et al. (2017) employed PROMETHEE-2 methodology based on five performance indicators to rank the Regional Climate Models (RCM) to evaluate the impact of climate change on the Ganges. Markl-Hummel and Geldermann (2014) applied the outranking method PROMETHEE to obtain preference from available alternatives for action implementation for the local-level for climate protection decision aid. This study demonstrates that PROMETHEE provides greater transparency by representing the different independent preferences of the decision-makers and weighing each alternative against all other preferences.

4.10 Simple Additive Weighting (SAW)

SAW method is a standard form of multi-attribute value methodology. As suggested by the name itself, this methodology is established based on a straightforward addition of scores that represent goal accomplishment under each principle, multiplied by particular weights (Qin et al. 2008). This methodology possesses an ability to balance the provided criterion and remains intuitive to decision-makers. The major advantage of this methodology is the simplicity of calculation that may be performed even without the help of complex computer programs. At the same time, particular disadvantages of this methodology include: (1) all values of the criteria in the given problem should be transformed to maximizing ones, (2) the result obtained may not be logical because the estimates yielded by SAW do not always reflect the real situation with values of one particular criterion largely differing from the of others (Podvezko 2011).

SAW has had applications in water management, business, and financial management. Mendas and Delali (2012) applied SAW to identify suitable agriculture lands for a particular type of cultivation. The study used GIS and SAW techniques to establish agricultural land suitability map.

4.11 Technique Ordered Preference by Similarity to the Ideal Solution (TOPSIS)

TOPSIS methodology identifies an alternative, which is nearest to the best solution and farthest away to the negative ideal solution in a multi-dimensional computing space (Deng et al. 2000; Qin et al. 2008). TOPSIS methodology involves a simple procedure to apply and can be easily programmed. The number of steps to arrive at a decision with TOPSIS remains unchanged regardless of the number of attributes (Ic 2012). TOPSIS has been used in environmental management, supply chain management, and logistics design, engineering and manufacturing systems, business and marketing management, human resources management, and water resources management. The success of this approach lies in its simplicity and ability to maintain the same number of steps regardless of problem size. The major shortcoming of this method is that it applies Euclidean Distance and does not consider the association of different attributes with each other. It is complicated to weight attributes and keeps the uniformity of judgment, particularly with additional attributes. Behzadian et al.

(1998) presents the state of the art survey of different applications of TOPSIS algorithm in different fields. Simonovic and Verma (2008) presented an application of multicriteria decision making under uncertainty using a fuzzy TOPSIS algorithm. Chung et al. (2014) successfully applied the TOPSIS technique to determine the optimal solutions for water resource vulnerability characteristics under the climate change impact. The weights for each of the vulnerability indicators are established based on both Shannon's entropy and Delphi technique, to reduce uncertainty in the procedure of determining the weights.

The above approaches may provide comparisons based on both qualitative and quantitative material for the climate change impacts/vulnerabilities assessment. They are effective for evaluating alternative adaptation policies based on a multitude of evaluation criteria. A plethora of other approaches are also available to handle the multi-criteria decision-making problems. Over the past decades, extensive studies of expert-system applications were reported in many environmental fields such as groundwater remediation (He et al. 2006; Qin et al. 2006), air pollution control (Zhou et al. 2004), solid waste management (Marianne 1996), and water resources management (Jamieson and Fedra 1996). The multi-criteria approaches have shown a lot of promise in solving environmental problems (Agrell et al. 1998; Fukuyama et al. 1994; Haimes et al. 1975; Marttunen and Hämäläinen 1995; Ridgley et al. 1997). However, here it is important to mention that in many of these studies the framework is derived with an approach, where the true decision-makers are involved. As a result of the same, specific decision support tools remain suitable within the framework and depends on the case. The framework becomes highly complicated as the environmental problems are very often of international nature. At the same time, the applications to climate change impact assessment and adaptation planning remain very few. For example, Li et al. (2003) developed a decision support system for managing pesticide losses in agricultural watersheds under changing climate; Huang et al. (2005) developed an expert system for assessing climate-change impacts within the petroleum sector to the formulation of the relevant adaptation policies. These studies demonstrated that the knowledge of climate change impacts and the relevant adaptation policies could be obtained interactively and dynamically (Huang et al. 2002). Connell et al. (2000) suggest MCDM as one of the approaches for developing sustainable water resources management, application-oriented papers (Duckstein et al. 1994; Bender and Simonovic 2000) illustrate the applications to bridge the gap between theory and practice. Raju et al. (2000) highlights the requirement of a larger number of studies to demonstrate the application of the MCDM approach to real engineering planning and design problems involving conflicting objectives.

The MCDM approaches have potential applications in water resource management and climate science. All climate-related applications have inherently uncertain, MCDM techniques provide a better assessment of the uncertainty and provide meaningful solutions to water resources related problems and issues. There is a lot of scope for furthering the applications of the same in climate science. Moreover, the advantage of these methods is simplicity and easy to apply. The different MCDM methodologies described above assist users to process large amounts of integrated assessment of the data to guide suitable decisions. The climate change decision experts can develop the MCDM approaches for integrated development of the region of interest to manage conflicting resources efficiently. There is remarkable scope for applications of MCDM in the climate and water development sector.

5 Summary and Conclusion

The earth's climate system is highly nonlinear. Climate change has been evident around the globe and the vulnerability of different components of ecological cycles to climate hazard is no exception. The indicator-based MCDM methods offer a theoretically-sound approach to identify different sources of nonlinearity in climate vulnerability assessments. MCDM is widely used for selection, ranking, and evaluation in the fields of planning, education, management, manufacturing, transportation, construction, logistic, medical, control, and agriculture. This chapter presents a review of some of the useful methods, predominantly pertaining to the decision-making problems in climate science. The examples are carefully drawn from a wide spectrum of climate applications that ranges from the different aspects of climate studies such as water resource management, agricultural planning, climate hazards, and vulnerability assessment, energy assessment, environmental performance selection, etc. Here is important to mention that these methodologies are not domain-specific and can be applied to a wide spectrum of climate-related problems.

Many different approaches are proposed to facilitate the multicriteria decision-making problems. In recent times, owing to advancing technologies and ease of usage by, combining different methods MCDM has become commonplace in MCDA. The combination of many different methods addresses the shortcomings of a particular method. The different MCDM methods applied in combination, along with the methods in their original forms, can be highly successful in their applications. However, this is achieved only through a proper assessment of their strengths and weaknesses in relationship to the given problem. For example, a particular problem could easily employ a methodology that may not be best suitable to answer it. This chapter is conceptualized to provide a broad assessment of MCDM-based procedure based on individual qualities of different MCDM techniques for climate and water resources application point of view. Briefly, the contribution of the present effort is twofold:

(1) The study provides an extensive review of the characteristics of MCDM methodologies in general practice as well as under the climate application paradigm.
(2) The study provides a broad review of the selection of the appropriate MCDM method for the common climate problem.

References

Ananda, J., & Herath, G. (2005). Evaluating public risk preferences in forest land-use choices using multi-attribute utility theory. *Ecological Economics, 55*(3), 408–419.

Anand, J., Devak, M., Gosain, A. K., Khosa, R., & Dhanya, C. (2017, April). Applicability of ranked Regional Climate Models (RCM) to assess the impact of climate change on Ganges: A case study. In: *EGU General Assembly Conference Abstracts* (Vol. 19, p. 780).

Balmat, J., Lafont, F., Maifret, R., & Pessel, N. (2011). A decision-making system to maritime risk assessment. *Ocean Engineering, 38*(1), 171–176.

Behzadian, M., Agrell, P. J., Lence, B. J., & Stam, A. (1998). An Interactive multicriteria decision model for multipurpose reservoir management: The Shellmouth reservoir. *Journal of Multi-Criteria Decision Analysis, 7*, 61–86.

Behzadian, M., Kazemzadeh, R., Albadvi, A., & Aghdasi, M. (2010). PROMETHEE: A comprehensive literature review on methodologies and applications. *European Journal of Operational Research, 200*(1), 198–215.

Bender, M. J., & Simonovic, S. P. (2000). A fuzzy compromise approach to water resource systems planning under uncertainty. *Fuzzy Sets and Systems, 115*, 35–44.

Bonissone, P. P., Subbu, R., & Lizzi, J. (2009). Multicriteria decision making (MCDM): A framework for research and applications. *IEEE Computational Intelligence Magazine, 4*(3), 48–61.

Chen, L., & Pan, W. (2016). BIM-aided variable fuzzy multi-criteria decision making of low-carbon building measures selection. *Sustainable Cities and Society, 27*, 222–232.

Chen, Y., Okudan, G., & Riley, D. (2010). Decision support for construction method selection in concrete buildings: Prefabrication adoption and optimization. *Automation in Construction, 19*(6), 665–675.

Chitsaz, N., & Banihabib, M. E. (2015). Comparison of different multi criteria decision-making models in prioritizing flood management alternatives. *Water Resource Management*.

Connell, E. O., Bathurst, J., Kilsby, C., Parkin, G., Quinn, P., Younger, P., Anderton, S., & Riley, M. (2000). Integrating mesoscale catchment experiments with modelling: The potential for sustainable water resources management. In: *Fifth IHP/IAHS George Kovacs Colloquium, HELP, International Hydrological Programme*, UNESCO, Paris.

Daengdej, J., Lukose, D., & Murison, R. (1999). Using statistical models and case-based reasoning in claims prediction: Experience from a real-world problem. *Knowledge-Based Systems, 12*(5–6), 239–245.

De Caro, F., Vacaro, A., & Villacci, D. (2017). Spatial and temporal wind power forecasting by case-based reasoning using big-data. *Energies, 10*(2), 252. https://doi.org/10.3390/en10020252.

De Keyser, W., & Peeters, P. (1996). A note on the use of PROMETHEE multicriteria methods. *European Journal of Operational Research, 89*(3), 457–461.

Deng, H., Robert, J. W., & Yeh, C. H. (2000). Inter company comparison using modified TOPSIS with objective weight. *Computer & Operation Research, 27*(10), 963–973.

Duckstein, L., Treichel, W., & Magnouni, S. E. (1994). Ranking ground-water management alternatives by multicriterion analysis. *Journal of Water Resources Planning Management, ASCE, 120*, 546–565.

Durkin, J. (1994). *Expert systems: Design and development*. New York: Macmillan.

Edwards, W. (1971). Social utilities. In: *Engineering Economist Summer Symposium Series* (Vol. 6, pp. 119–129).

Edwards, W. (1977). How to use multi attribute utility measurement for social decision making. *IEEE Transactions on Systems, Man, and Cybernetics, 7*(5), 326–340.

Escobar, M. T., & Moreno-Jimenez, J. M. (2002). A linkage between the analytic hierarchy process and the compromise programming models. *Omega, 30*, 359–365.

Everest, T., Sungur, A., & Özcan, H. (2020). Determination of agricultural land suitability with a multiple-criteria decision-making method in Northwestern Turkey. *International Journal of Environmental Science and Technology, 2020*, 1–16.

Fishburn. (1967). Conjoint measurement in utility theory with incomplete product sets. *Journal of Mathematical Psychology, 4*(1), 104–119.

Fukuyama, K., Kilgour, D. M., & Hipel, K. W. (1994). Systematic policy development to ensure compliance to environmental regulations. *IEEE Transactions on Systems, Man and Cybernetics, 24*(9), 1289–1305.

Guo et al. (2016). Dependent-chance goal programming for water resources management under uncertainty. *Scientific Programming*, 1747425.

Haimes, Y. Y., Hall, W. A., & Freedman, H. T. (1975). *Multi-objective optimization in water resources systems*. New York: Elsevier Scientific Publishing.

He, L., Chan, C. W., Huang, G. H., & Zeng, G. M. (2006). A probabilistic reasoning-based decision support system for selection of remediation technologies for petroleum-contaminated sites. *Expert Systems with Applications (Elsevier Science), 30*(4), 783–795.

Hosseinzadeh-Bandbafha, H., Nabavi-Pelesaraei, A., Khanali, M., Ghahderijani, M., & Chau, K. (2018) Application of data envelopment analysis approach for optimization of energy use and reduction of greenhouse gas emission in peanut production of Iran. *Journal of Cleaner Production, 172*, 1327–1335. https://doi.org/10.1016/j.jclepro.2017.10.282. ISSN 0959-6526.

Huang & Chang. (2003). The perspectives of environmental informatics and systems analysis. *Journal of Environmental Informatics, 1*(1), 1–7.

Huang, G. H., Chen, Z., Liu, L., Huang, Y. F., & Li, J. B. (2002). *Development of an information system for supporting climate change impact and adaptation strategies studies within the prairie's petroleum industries*. Final report, Prairie Adaptation Research Collaborative, Canadian Climate Impacts and Adaptation Research Network.

Huang, Y. F., Huang, G. H., Hu, Z. Y., Maqsood, I., & Chakma, A. (2005). Development of an expert system for tackling the public's perception to climate-change impacts on petroleum industry. *Expert Systems with Applications, 29*(4), 817–829.

Ic, Y. (2012). An experimental design approach using TOPSIS method for the selection of computer-integrated manufacturing technologies. *Robotics and Computer-Integrated Manufacturing, 28*(2), 245–256.

Jamieson, D. G., & Fedra, K. (1996). The 'WaterWare' decision-support system for river-basin planning. 1. Conceptual design. *Journal of Hydrology, 177*(3–4), 163–175.

Kailiponi, P. (2010). Analyzing evacuation decisions using multi-attribute utility theory (MAUT). *International Conference on Evacuation Modeling and Management (ICEM), 3*, 163–174.

Karagiannidis, A., & Mounssiopoulos, N. (1997). Application of ELECTRE 3 for the integrated management in the Greater Athens Area. *European Journal of Operational Research, 97*, 439–449.

Keeney, R. (1977). The art of assessing multiattribute utility functions. *Organizational Behavior and Human Performance, 19*(2), 267–310.

Keeney, R., & Fishburn, P. (1974). Seven independence concepts and continuous multiattribute utility functions. *Journal of Mathematical Psychology, 11*(3), 294–327.

Khadam, I., & Kaluarachchi, J. (2003). Multi-criteria decision analysis with probabilistic risk assessment for the management of contaminated ground water. *Environmental Impact Assessment Review, 23*(6), 683–721.

Khosravi, K., Shahabi, H., et al. (2019). A comparative assessment of flood susceptibility modeling using multi-criteria decision-making analysis and machine learning methods. *Journal of Hydrology, 573*, 311–323.

Konidari, P., & Mavrakis, D. (2007). A multi-criteria evaluation method for climate change mitigation policy instruments. *Energy Policy, 35*(12), 6235–6257.

Liao, H., Gou, X., Xu, Z., Zeng, X., & Herrera, F. (2020). Hesitancy degree-based correlation measures for hesitant fuzzy linguistic term sets and their applications in multiple criteria decision making. *Information Sciences, 508*(2020), 275–292.

Li, H., & Sun, J. (2008). Ranking-order case-based reasoning for financial distress prediction. *Knowledge-Based Systems, 21*(8), 868–878.

Liu, W., Wang, S., Zhou, Y., et al. (2016). *Natural Hazards, 81,* 347. https://doi.org/10.1007/s11069-015-2083-1.

Li, Y. F., Li, Y. R., Huang, G. H., Wang, X. Z., & Chen, B. (2003). Development of a decision support system for managing pesticide losses in agricultural watersheds. *International Journal of Sediment Research, 18*(1), 60–73.

Loken, E. (2007). Use of multi-criteria decision analysis methods for energy planning problems. *Renewable and Sustainable Energy Reviews, 11*(7), 1584–1595.

Marianne, L. M. (1996). A multi-attribute spatial decision support system for solid waste planning. *Computers Environment and Urban Systems, 20*(1), 1–17.

Markl-Hummel, L., & Geldermann, J. (2014). A local-level, multiple criteria decision aid for climate protection. *EURO Journal on Decision Processes, 2*(1–2), 121–152.

Martín-Gamboa, M., Iribarren, D., & Dufour, J. (2018). Environmental impact efficiency of natural gas combined cycle power plants: A combined life cycle assessment and dynamic data envelopment analysis approach. *Science of the Total Environment, 615,* 29–37. https://doi.org/10.1016/j.scitotenv.2017.09.243. ISSN 0048-9697.

Marttunen, M., & Hämäläinen, R. P. (1995). Decision analysis interviews in environmental impact assessment. *European Journal of Operational Research, 87,* 551–563.

Mendas, A., & Delali, A. (2012). Support system based on GIS and weighted sum method for drawing up of land suitability map for agriculture—Application to durum wheat cultivation in the region of Tiaret in Algeria. *WSEAS Transactions on Environment and Development, 8*(2). E-ISSN: 2224–3496.

Mi, Z. F., Wei, Y. M., He, C. Q., et al. (2017). *Mitigation and Adaptation Strategies for Global Change, 22,* 45. https://doi.org/10.1007/s11027-015-9660-1.

Opricovic, S., & Tzeng, G. H. (2007). Extended VIKOR method in comparison with outranking methods. *European Journal of Operational Research, 178,* 514–529.

Pal, B. B., Goswami, S. B., Sen, S., & Banerjee D. (2012). Using fuzzy goal programming for long-term water resource allocation planning in agricultural system: A case study. In: P. Balasubramaniam & R. Uthayakumar (Eds.), *Mathematical modelling and scientific computation. Communications in computer and information science* (Vol. 283). Berlin, Heidelberg: Springer.

Podvezko, V. (2011). The comparative analysis of MCDA methods SAW and COPRAS. *Inzinerine-Ekonomika-Engineering Economics, 22*(2), 134–146.

Qin, X. S., Huang, G. H., Zeng, G. M., Chakma, A., & LiNRSRM, J. B. (2006). a decision support system and visualization software for the management of petroleum-contaminated sites. *Energy Sources Part a (Taylor & Francis), 28,* 199–220.

Qin, X., Huang, G., Chakma, A., Nie, X., & Lin, Q. (2008). A MCDM-based expert system for climate-change impact assessment and adaptation planning—A case study for the Georgia Basin Canada. *Expert Systems with Applications, 34*(3), 2164–2179.

Raju, S., & Vasan, A. (2007). Multi attribute utility theory for irrigation system evaluation. *Water Resources Management, 21*(4), 717–728. https://doi.org/10.1007/s11269-006-9060-0.

Raju, K. S., Duckstein, L., & Arondel, C. (2000). Multicriterion analysis for sustainable water resources planning: A case study in Spain. *Water Resources Management, 14*(6), 435–456.

Read, L., Madani, K., Mokhtari, S., & Hanks, C. (2017, 15 January). Stakeholder-driven multi-attribute analysis for energy project selection under uncertainty. *Energy, 119,* 744–753. https://doi.org/10.1016/j.energy.2016.11.030.

Rodriguez et al. (2004). Application of data envelopment analysis to studies of irrigation efficiency in Andalusia. *Journal of Irrigation and Drainage Engineering, 130*(3). https://doi.org/10.1061/(ASCE)0733-9437(2004)130:3(175).

Ridgley, M. A., Penn, D. C., & Tran, L. (1997). Multicriterion decision support for a conflict over stream diversion and land-water reallocation in Hawaii. *Applied Mathematics and Computation, 83,* 153–172.

Saaty, T. L. (1980). *The analytic hierarchy process.* New York: McGraw-Hill.

Saaty, T. L. (2006). Rank from comparisons and from ratings in the analytical hierarchy/network processes. *European Journal of Operational Research, 16*(2), 270–271.

Seghezzo, L., Venencia, C., Buliubasich, E. C., Iribarnegaray, M. A., & Volante, J. N. (2017). Participatory, multi-criteria evaluation methods as a means to increase the legitimacy and sustainability of land use planning processes. The case of the Chaco Region in Salta, Argentina. *Environmental Management, 59*(2), 307–324. https://doi.org/10.1007/s00267-016-0779-y.

Simonovic, S. P., & Verma, R. (2008). A new methodology for water resources multicriteria decision making under uncertainty. *Physics and Chemistry of the Earth, Parts A/B/C, 33*(5), 322–329.

Siskos, J., Wascher, G., & Winkels, H. (1984). Outranking approaches versus MAUT in MCDM. *European Journal of Operational Research, 16*(2), 270–271.

Sun, X., & Li, Y. (2010). An intelligent multi-criteria decision support system for systems design. In: *10th AIAA Aviation Technology, Integration, and Operations (ATIO) Conference, 2010* (pp. 1–11). https://doi.org/10.2514/6.2010-9222.

Talukder, B., Blay-Palmer, A., Hipel, K. W., & van Loon, G. W. (2017). Elimination method of multi-criteria decision analysis (MCDA): A simple methodological approach for assessing agricultural sustainability. *Sustainability, 9*(2), 287. https://doi.org/10.3390/su9020287.

Teegavarapu, R. V. (2010). Modeling climate change uncertainties in water resources management models. *Environmental Modeling Software, 25*(10), 1261–1265. https://doi.org/10.1016/j.envsoft.2010.03.025.

Thanassoulis, E., Kortelainen, M., & Allen, R. (2012). Improving envelopment in data envelopment analysis under variable returns to scale. *European Journal of Operational Research, 218*(1), 175–185.

Triantaphyllou, E. (2000). Multi-criteria decision making methods. In: *Multi-criteria decision making methods: A comparative study. Applied optimization* (Vol. 44). Springer, Boston, MA.

Triantaphyllou, E., Shu, B., Sanchez, S. N., Ray, T. (1998). Multi-criteria, decision making: An operations research approach. In: J. G. Webster (Ed.), *Encyclopedia of Electrical and Electronic Engineering* (Vol. 15, pp. 175–186). New York, NY: Wiley.

Tsai, W., Leu, J., Liu, J., Lin, S., & Shaw, M. (2010). A MCDM approach for sourcing strategy mix decision in IT projects. *Expert Systems with Applications, 37*(5), 3870–3886.

Velasquez, M., & Hester, P. T. (2013). An analysis of multi-criteria decision making methods. *IJOR, 10*(2), 56–66.

Voogd, H. (1983). *Multicriteria evaluation for urban and regional planning.* Pion Ltd.

Vulevic, T., & Dragovic, N. (2017). Multi-criteria decision analysis for sub-watersheds ranking via the PROMETHEE method. *International Soil and Water Conservation Research, 5*(1), 50–55. https://doi.org/10.1016/j.iswcr.2017.01.003.

Wang, Y., Greatbanks, R., & Yang, B. (2005). Interval efficiency assessment using data envelopment analysis. *Fuzzy Sets and Systems, 153*(3), 347–370.

Wang, X., Li, Y., & Cui, W. (2012). Design and implementation of a collaboration-based AHP evaluation system. In: *Proceedings of the 2012 IEEE 16th International Conference on Supported Cooperative Work in Design (CSCWD).*

Yin, Y. (2001). *Designing an integrated approach for evaluating adaptation options to reduce climate change vulnerability in the Georgia Basin.* Final report, Climate Change Action Fund, Adaptation Liaison Office.

Zadeh, L. (1965). Fuzzy sets. *Information and Control, 8*(3), 338–353.

Zanakis, S. H., Solomon, A., Wishart, N., & Dublish, S. (1998). Multi-attribute decision making: A simulation comparison of select methods. *European Journal of Operational Research, 107*(3), 507–529.

Zavadskas, E. K., Turskis, Z., & Kildienė, S. (2014). State of art surveys of overviews on MCDM/MADM methods. *Technological and Economic Development of Economy, 20*, 165–179.

Zhou, Q., Huang, G. H., & Chan, C. W. (2004). Development of an intelligent decision support system for air pollution control at coal-fired power plants. *Expert Systems with Applications, 26*(3), 335–356.

Zhou, H., Yang, Y., Chen, Y., & Zhu, J. (2018). Data envelopment analysis application in sustainability: The origins, development and future directions. *European Journal of Operational Research, 264*(1), 1–16. ISSN 0377-2217.

Zimmermann, H. J. (1991). *Fuzzy set theory and its applications*. Boston: Kluwer Academic Publishers.

Zoghi, M., Ehsani, A. H., Sadat, M., Javad Amiri, M., & Karimi, S. (2017). Optimization solar site selection by fuzzy logic model and weighted linear combination method in arid and semi-arid region: A case study Isfahan-IRAN. *Renewable and Sustainable Energy Reviews, 68*, 986–996.

Development of Spatial Cognitive Model for Estimation of Ungauged Runoff for Mesoscale Rivers

Tilottama Chakraborty

Abstract The burgeoning population and uncontrolled extraction to satisfy the need of the growing population, has induced major changes in the land use and land cover which in turn has imbibed global warming and corresponding changes in the regular pattern of climate. As a result, the flow amount and frequency in a river also get affected and often fails to maintain the equilibrium required to sustain a healthy watershed. The estimation of stream-flow was a popular topic for a long span of time especially for the ungauged basins. The linearity and inability to represent the randomness of stream flow estimation by the empirical models yields the necessity of modern and advanced models which can respond to the uncertainty involved in prediction of flow amount or frequency. However, due to the data dependency and lack of data for the ungauged basins, different models were proposed to estimate stream-flow from lumped data set which ultimately generates an erroneous model. That is why, the present investigation aims to develop a model which tries to retrieve required data from changes in the land use and covers. As land features and uses can be captured remotely by satellites such a model where stream-flow is estimated as a function of changes in different land use and covers can reduce the need of primary or on-site data along with historical data sets for prediction of flow pattern. The present investigation aims to develop a model for estimation of flow amount which have the input variable depending on the change in land use features by the help of neural networks and multi criteria decision making models. An accuracy level of above 95% indicates the advancement of the model and its wide circulation for providing sustainable benefits to the local inhabitants.

Keywords Multi-criteria decision analysis · Land-use change · Runoff modeling · Cognitive indexes

---

T. Chakraborty (✉)
Hydro-Informatics Engineering, National Institute of Technology Agartala, Barjajala, Jirania, Agartala 799046, Tripura, India
e-mail: tilottama86@gmail.com

M. Majumder and G. D. Kale (eds.), *Water and Energy Management in India*,
https://doi.org/10.1007/978-3-030-66683-5_2

1 Introduction

Changes in runoff can induce uncertainties in the inter-relationships that exist between the various parameters within a watershed. Abnormalities like floods, water logging, droughts etc. are common due to changes in runoff. Changes in peak runoff can also expose design failures of different hydraulic structures, and can also cause an uneven distribution of water availability within a single watershed. It is for these reasons that predicting surface runoff for future scenarios is a very popular and important topic of research (Gardner 2009; Kling et al. 2012).

For example, functional multiple regression (FMR) is utilized to get day by day stream flow (Requena et al. 2018) estimation low duration curves (FDC) series at ungauged sites. A model has been created to go about as a helpful device in dealing with watershed management and the land use approaches for ungauged catchment (Cheamson et al. 2018). In the investigation of Jodar et al. (2018) release stream in an ungauged watershed is evaluated utilizing the hydrological factors of a HBV (Hydrologiska ByrånsVattenbalansavdelning) model. Amin et al. (2017) examinations the effects of environmental change on the hydrologic forms under future environmental change conditions over Muda and Dungun watersheds of Peninsular Malaysia. The examination can give future water management strategies and can be useful in the ID and evaluation of the likely risks because of the expected future outrageous occasions. Because of the absence of data, it was accepted in this investigation that the principal physical qualities (geology, landuse/spread, vegetation, water powered foundation and so on.) continue as before between the present and future time frames. Kundu et al. (2017) examine the change in landuse and its effect on the water balance of the investigation territory. Result shows the change in CN values is expanding with time which may have consequences for the water balance of the region in the future with decreased ET and groundwater. These impacts can be considered to improve the procedures in planning, management and development strategies for the future.

Stone et al. (2001) identified that greenhouse gases and the increment of atmospheric carbon dioxide (CO_2) will probably bring about a changed atmosphere and this changed atmosphere will influence water accessibility in the Missouri River Basin. Besides, Meenu et al. (2013) saw that environmental change would essentially influence water accessibility, overflow, and the stream in waterways given its impact on numerous hydrologic frameworks. Xu (2000) consider the effects of changed climate on water resources in central Sweden. Akhtar et al. (2008) expressed that environmental change and its fluctuation are probably going to influence water assets given the impact on horticulture. Gardner (2009) identified that, there is a change in mean annual runoff because of changes in hydrological parameters. A seasonal runoff and peak flow change was observed due to climate change in a study of Chang and Jung (2010).

In the course of recent years there have been huge changes in the geological degree of urban improvement in different places all through the world Wei et al. (2007) observed that runoff and erosion processes are influenced by numerous variables, but among those factors, rainfall and land use are two of the most important. Many social and economic benefits also depend upon land-use pattern changes, and water resources and water-quality degradation is one major direct environmental impact of such development (USEPA 2001). Vast areas of grassland, agriculture, forest etc. are constantly being converted to impervious land cover due to urbanization. Fox et al. (2012) revealed that Mediterranean environments have been liable to significant land-cover change since the end of the Second World War, and that because of urbanization which influences the runoff. The particular impact in such manner is reflected in an expansion in the volume and rate of surface runoff and decreases in base flow (Carter 1961; Anderson 1970) which has resulted in larger and increasingly visit occurrences of neighborhood flooding (Field et al. 1982; Hall 1984), decrease in water supply, and decreased base flow into stream channels during dry climate (Harbor 1994). Different effects of changes in release conduct because of urbanization incorporate expanded water levels in lakes and wetlands (Calder et al. 1995; Schueler and Holland 2000), adjusted watershed water balance (Fohrer et al. 2001), and increased erosion in river channel banks (Doyle et al. 2000). Changes from pervious to impervious surfaces are additionally liable for deterioration in the quality of storm-water runoff. In the Great Lakes region, Mao and Charkauer (2009) examined changes in land use and found that the most sensational changes were in the focal pieces of the investigation region where 5,000,000 hectares of deciduous woodland have been changed over to lush meadow and column crop agribusiness, bringing about a 5–15% abatement in evapotranspiration (ET) and a 10–30% expansion in absolute in total runoff.

Soil and Water Assessment Tool (SWAT) model has been applied to channelize the stream flow in an ungauged catchment in Stung Pursat catchment (Ang and Oeurng 2018). But this model is applicability for local conditions.

For the modeling of monthly stream flow Wavelet-linear genetic programming approach has been adopted in a study conducted by Ravansalar et al. (2017) to predict the monthly stream flow. The results of the study show that the model could be applied for the simulation of cumulative stream flow data prediction in one month ahead.

To find improved stream flow estimation, the Soil and Water Assessment Tool (SWAT) and Artificial Neural Network (ANN) models have been examined and compared (Jimeno-Saez et al. 2018). Two different catchments in terms of climatic condition have been taken for this study. Both the models show the good results. However, SWAT was found to be more reliable in case of lower flows, while ANNs were better at estimating higher flows in all cases.

Three different landuse scenarios i.e. High-Tech agriculture (HT), Agriculture for Nature (AN) and Market-Driven agriculture (MD) on water discharge has been assessed using SWAT (Molina-Navarro et al. 2018). The results show that the scenario-specific climate inputs were most important for the simulation of hydrological cycle.

Maintainable Yield Estimator has been used for describing vulnerability in day by day stream estimation at ungauged areas (Farmer and Levin 2018). The strategy has been changed to give suitably limit certainty spans that contain 95% of the watched stream streams in cross-approval. Multiday methods for day by day streamflow are additionally assessed to portray the vulnerability. The outcomes show that the method can be utilized for improved and feasible water resource management. MLR has also been applied in similar studies (Arsenault et al. 2019).

To predict the discharge of ungauged catchments a new statistical test of model performance has been developed (Prieto et al. 2019). To estimate the runoff the parameter transfer (PT) method and a method of area ratio (AR) were combined (Li et al. 2019a, b). Reddy et al. (2019) studies developed a rainfall-runoff relationship which can be utilized to estimate the flood discharge at ungauged catchments (Reddy et al. 2019).

A multi-target system named stream Prediction under Extreme Data-shortage (SPED) is proposed for stream expectation in ungauged catchments (Alipour 2019). Observational (SCS-CN), Artificial Neural Network (ANN) and Hydrological Model (HEC-HMS) has been used for modeling of Stream flow (Meresa 2019).

Global Scenario of Urbanization and land use status of India is depicted in sub Sects. 1.1 and 1.2. In Sect. 2 Survey of literature with respect to the objective of the study and the objective of the present study has been discussed. In Sect. 3 the location of the study has been depicted, respectively.

1.1 Global Scenario of Urbanization

Over half of the world's populations stay in urban communities, and the urban population is developing at an a lot quicker rate than the Earth's population. It is normal that 61% of the total population will live in urban settlements in 2030. The Total World Population on 2016 is 7,156,392,318 and the rate of change in population density from rural to urban region is 1.98%.

According to World Urbanization Prospect 2011 revision, a report published by UN, the maximum change in population density to urban area was observed in Burundi (6.8%) and Moldova (−1.5%).

In case of China, India and USA, the three countries having the highest number of population in the World, the rate of change to urban population is 2.7, 2.4 and 1.3% respectively.

According to population density, Bangladesh (1034 pop/km^2), Taiwan (646 pop/km^2) and South Korea (505 pop/km^2) is the three most populated countries and the rate of change in urban population is 3.5%, 0.3% and 0.6% respectively.

All the rate of change in population given above was calculated with respect to the urban population of 2014.

Various studies shows that, the world keeps on urbanizing in every one of the SSPs however results vary generally across them, with urbanization arriving at 60, 79, and 92% before the finish of century in SSP3, SSP2, and SSP1/SSP4/SSP5, individually (Jiang et al. 2018). In an investigation of urbanization impacts on vegetation spread in significant African urban areas, MODIS land spread and upgraded vegetation record (EVI) information were utilized (Yao et al. 2019). Wang et al. (2019) contemplated the effects of urbanization on biological system administrations from 2000 to 2010 in the Beijing-Tianjin-Hebei (BTH) urban mega region in China. The outcomes show the bits of knowledge for upgrading urban supportability in the BTH locale. In an investigation of urbanization impacts on vegetation spread in significant African urban areas, MODIS land spread and upgraded vegetation record (EVI) information were utilized (Yao et al. 2019). This examination proposed that urbanization and its impacts on VC in Africa should take more consideration. Miao et al. (2019) appears in their investigation that immediate spillover is essentially identified with urbanization and direct overflow increase profundity influenced by urbanization. Remote sensing raster order is used to evaluate the land use and related Curve Number (CN) variety and the SCS technique was utilized to assess the overflow volume in an investigation of Urbanization Growth Effect on Hydrological Parameters in Mega Cities (Helmi et al. 2019). Study shows that, there is an inordinate increment in overflow because of urban turn of events. Li et al. (2018) presents a structure for recognizing hotspots of overflow increment, which can give significant direction to urban directors in future green framework arranging.

1.2 Landuse Status of India

India has topographical zone of 328.7 million hectares. Land assets in India are fundamentally separated into horticultural land, woods land, land implied for field and nibbling, land for other non-agrarian use and waste land. Waste land incorporates rough, dry and desert regions. Land is likewise utilized for other non-rural purposes, for example, lodging, streets and industry. According to the GoI (Government of India), 2015, around 21% of the geological territory was involved by woodlands, 8% was used for non-rural purposes, 5% was infertile and unculturable and 7.5% stayed decrepit (Pandey and Ranganathan 2018). Figure 1 is showing the land use map of India.

2 Literature Review

Guzha et al. (2018) conducted an investigation in East Africa to assessing the effects of Land Use and Land Cover Changes (LULCC) on runoff, surface overflow, and low streams. To comprehend the present status of information on forest cover loss and LULCC impacts on catchment hydrology in the East Africa district, the investigation aggregated significant accessible examinations including catchment studies, modeling studies and pattern investigations. Because of the absence of data effects of forest cover loss is viewed as uniform.

To examine the hydrological effects of land use land spread (LULC) changes in the Andassa watershed for a time of 1985–2015 and to foresee the LULC change impact on the hydrological status in year 2045. The outcome can recommend the need of guideline of the LULC so as to keep up the hydrological balance.

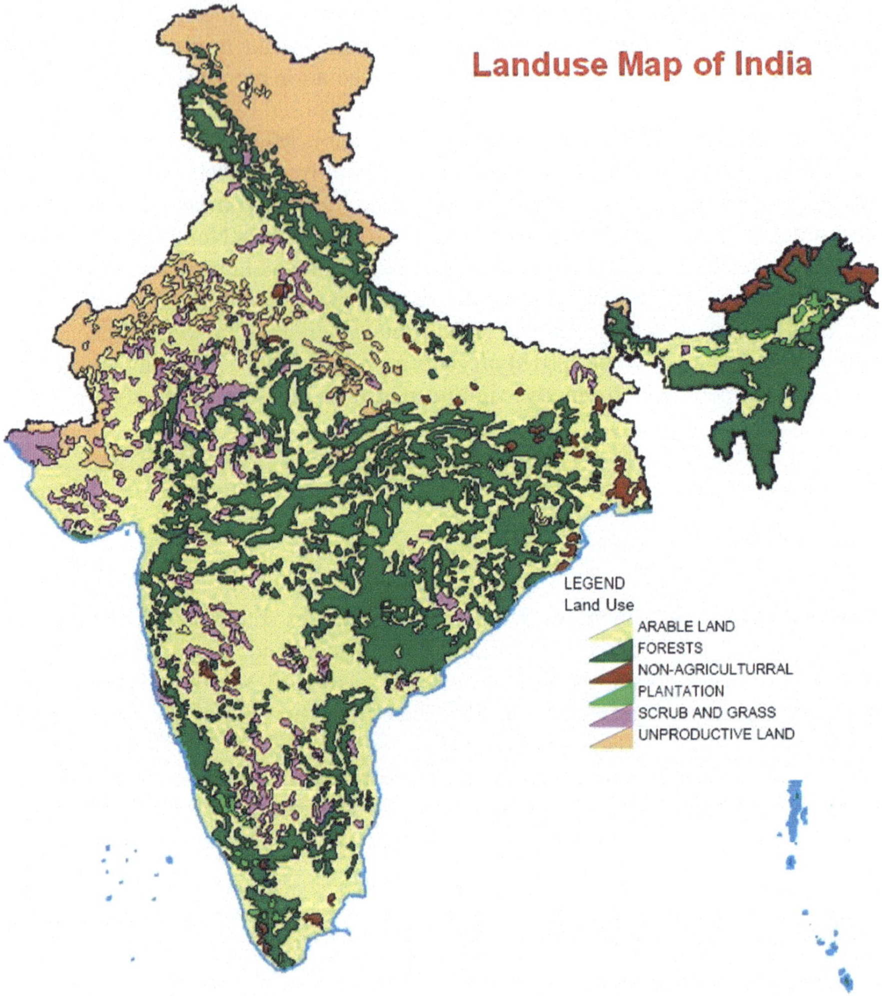

Fig. 1 Landuse status of India (*Source* https://www.indiawaterportal.org/articles/land-use-map-india-national-institutehydrology)

Bajracharya et al. (2018) examined the effect of environmental change on the water balance and hydrological system in Kaligandaki Basin. The examination can add to a successful administration and arranging of water flexibly and demand in the Kaligandaki basin considering the impact of environmental change. As future anticipated atmosphere factors precipitation and temperature were viewed as it were.

Anache et al. (2018) considered the effect of environmental change and land use on discharge and soil erosion at the hill slope scale in the Brazilian Cerrado. One of the most continuous reasons for death and property from natural hazards in

China is Flash floods. One of the fundamental difficulties is finding proper hydrological model boundary esteems for ungauged catchments. Ragettli et al. (2017) locate another model which can distinguish flood occasions in accepted ungauged catchments. The model is received for the pseudo-ungauged catchments essentially further model enhancements are prescribed to expand adjustment in catchments of the dryer Northern territories. A methodology with a mix of bivariate record flood approach with event type explicit engineered plan hydrograph development has been applied to configuration sets of hydrographs for ungauged catchments (Brunner et al. 2018). In an investigation of Li et al. (2019a, b), the effects of LUCC and precipitation change on surface runoff depth by considering the Soil Conservation Service-Curve Number has been assessed. (SCS-CN) method and the widely used Long-Term Hydrologic Impact Assessment (L-THIA) model.

Napoli et al. (2017) conducted the investigation to introduce the approach and the after effects of an examination for evaluating the hydrological reactions to land use and atmosphere changes by methods for ArcSWAT model. Such examination could help policymakers associated with the land use planning in considering land use change, since it can intensify the impacts of atmosphere and segment changes on bowl hydrodynamics. Berihun et al. (2019) break down discrete and joined impacts of Land use/land spread (LULC) change and atmosphere changeability in differentiating agro-natural situations.

Gumindoga et al. (2017) utilized the HEC-HMS model for simulation of runoff in the ten gauged and ungauged Upper Manyame sub catchments in Zimbabwe. This investigation will have a critical commitment for the future advancement of water resource programs in Upper Manyame Catchment. Nyaupane et al. (2018) has been shows the precipitation forecast in HEC-HMS Model for Irwin Creek, Charlotte, North Carolina.

Urbanization along with quick degradation of environments because of expanding requests from the increasing population have prompted huge scope changes in the current land use and climatology of numerous watersheds all through the world (Dietz and Clausenb 2008; Dewan and Yamaguchi. 2009; Lambin and Meyfroidt 2011).

One of the ongoing pushes in hydrologic modeling is assess the impacts of land use and land cover changes on water resources and floods (Yang et al. 2013), which are fundamental for arranging and activity. Akhtar et al. (2008), has expressed that environmental change and its inconstancy are probably going to influence on the water resources as it is influencing the farming. Gardner (2009) has seen that because of the change in hydrological boundary, for example, temperature and precipitation there is an adjustment in mean yearly runoff which may compromise the surface water supply and diminishes in soil moisture and hence have unfavorable effects on farming.

Zadeh (1965) presented the idea that, in zones of imprecision and vulnerability, fuzzy set theory permits better scientific modeling to occur. Burrough (1986) stated that studies show that the use of Boolean logic yields far inferior results. Indeed, To be sure, for designing and technical disciplines, fuzzy logic has proven to be efficiently applicable; and Burrough et al. (1992) later proceeded to show that, fuzzy classification can applicable in areas for farming. Barreto-Neto (2008) stated about the applicability of fuzzy logic. Fuzzy logic has also been applied in evaluating environmental impacts (Enea and Salemi 2001; Klingseisen et al. 2008). Zhang et al. (2019) utilized fuzzy linguistic rules to estimate the river runoff prediction problem. Azad et al. (2019) identified modified adaptive neuro fuzzy inference system for water quality parameters modelling of surface water.

Kontogianni et al. (2012) detailed in their investigations that Fuzzy Cognitive Mapping (FCM) is step by step developing as an elective procedure fit for helping scientists in the space of natural arrangement, proposing that the FCM approach can help explicitly through its application in multi-partner examination for chance evaluation, catching qualities, and situation development.

Lee et al. (2009) proposed an incorporated MCDM approach to organize the priority value of energy technology. Souissi et al. (2019) applied AHP MCDM to set up a flood risk vulnerability guide of the Gabes district.

Tan et al. (2018) built up a versatile center and long long-term runoff estimate model utilizing EEMD-ANN crossover approach. This model is applied to conjecture the 1-month ahead stream flow of three stations in China; however it isn't that much appropriate for dry season estimate.

Alizadeh et al. (2017) contemplated another methodology is introduced to forecast precipitation and runoff in Tolt River bowl. It is found from the investigation that a dependable forecast of the precipitation and runoffr process both for one and two months ahead can be accomplished when the new technique is applied. As indicated by this investigation, including forecasts can be a proficient method to expand the model exactness in multi time step.

Nasseri et al. (2008) expressed that artificial neural systems (ANNs), which play out a nonlinear planning among sources of info and yields for exceptionally non-straight, non-curved and dimensionalized procedures. Falah et al. (2019) applied ANN model for planning flood immersion regions of Emam-Ali town transport in Mashhad city, Khorasan Razavi Province, Iran. In an investigation of precipitation runoff modeling Asadi et al. (2019) applied ANN.

In most of the studies, stated above has shown that there is an impact of LULC on runoff. But the studies have not defined the link between these two parameters clearly. In some studies the forest cover loss has taken as constant and in some studies only precipitation, temperature etc. have considered as projected climate change data. In this regards the present study aims to find the interrelationship between these variables to cope with climate change.

2.1 *Objective of the Study*

The aim of the study is to establish a strong link between changes in landuse and changes in runoff; and given that change in runoff cause uncertainties in terms of the design and integrity of hydraulic structures, an objective tool that can detect the change in runoff as a function of landuse and land cover was developed. If the interrelationship between these variables can be mapped, design modifications to cope with climate change can be made also. In this regard, an idea was prepared with the help of MCDM and ANN techniques which can automatically detect the change in runoff if the values of change in landuse are fed to the model. Such index can help to detect the change in runoff due to change in landuse and land cover which can help the related personnel to develop mitigation measures in advance to percent damages to the hydraulic structure and other related issues. The index was validated by sensitivity analysis, Multi linear regression analysis (MLR) and case studies.

Stream-flow estimation models are commonly developed by various scientists all over the World. However for un-gauged catchment collection of relevant climatic data is complex as well as costly Again due to change in land use and cover the amount of stream-flow may change and if the stream-flow is predicted with the help of land use and cover then output can be estimated only from satellite imageries of the land use. Not only this, but the novelty of the study also lies in the fact that for the first time a cognitively weighted hydrologic model can be developed where requirement of data collection will be minimum.

3 Study Location

Dhalai district of Tripura is situated between two hill ranges of Longtharai and Atharamura. The district are bordered by Bangladesh and Unakoti district of Tripura on the north, by Bangladesh on the south, by North district of Tripura on the east and by Khowai and Gomati districts of Tripura on the west. The district is accessible with rest of Tripura including the capital city Agartala through National Highway 44. The Headquarter of the district is located at Ambassa, situated beside the National Highway 44 in the Ambassa Block. Dhalai district has four Sub-divisions namely-Ambassa, Kamalpur, Longtarai valley, and Gandacharra, comprising of five Blocks i.e. Ambassa, Salema, Manu, Chhawmanu and Dumburnagar. Dhalai, Manu, Khowai, Gomti are the major rivers of this district (Dhalai District Adminitration 2020) Location map of study area is showing in Fig. 2.

3.1 Geology

Geologically the hill tracks of Tripura in general and Dhalai district in particular consists of tertiary rocks, which has been later covered by the alluvial deposits. The force of the tectonic movement has caused the sub-stratum to raise up several parallel anticline ridges. This perhaps started during the upliftment of the Himalayas during the Eocene period (Geological Survey of India 2011). Barail, Surma and Tipam are the three sedimentary rock formations found here lying one above the other. The Barail series are composed mainly of moderate hard fine grained and yellow to pink sand stones. The Surma group overlies the Barail series and has two formations i.e. the lower Bhuban formation and the upper Bokabil formation.

Fig. 2 Location map of study area (*Source* Google Earth™)

From the seismological point of view, the entire district falls under the earthquake zone V, which point out that there is lack of stability in the arrangements of the rock stratum (Soil survey).

3.2 *Topography*

The topography of Dhalai district comprises mostly of rugged terrain with some undulating surface. About 75% of the district geographical areas are characterized by hilly terrain covered with dense forests and only about 25% are plains. Three hills range i.e. Atharamura, Longtarai and Sakhan range are found here. All these three ranges run almost parallel to one another in a north–south direction and in between are the valleys. In the northern part of the district, the height of the hills range starts decreasing and ultimately merges with the Sylhet plains of Bangladesh. The district can be divided into two divisions-the hill ranges, and the valley/plain areas physio graphically (District Survey Report 2018).

3.3 Drainage

Manu River and Dhalai River is the two-principal river of the district. These rivers are seasonal and non-perennial as well as not navigable (Chateerjee 1984). The Manu River is one of the longest river originates from the Kohoisib peak of the Sakhan range having a total length of about 167 km. This river has hydroelectric potentialities. The Dhalai River rises in the Dolajari peak of the Longtarai range having a total length of about 117 km (Chateerjee 1984). These two rivers are rain fed and transport large volume of water especially during the monsoon season causing floods in the lower portion of the valleys. During winter season the water level of these rivers falls and the rivers becomes almost stagnant, the cannel sometimes gets disconnected (Chatterjee 1984). Due to this the rivers water cannot be use for irrigational purpose.

3.4 Climate

Dhalai district has a monsoon type of climate. However, there is a difference of temperature between the hills and plains, which ranges between sub-tropical in the plains to temperate climatic conditions found in the hilly areas. The topographic features seem to have influenced the climatic condition of the Dhalai district, where the plains are hotter and humid in comparison to the hills, which have a salubrious climate. The four main seasons here are—Post Monsson, Moonsoon, Pre monsoon and Winter (Bhatt and Bhargava 2006). The winter season is marked by fall in temperature and heavy fog cover in both the plains and the hills. However, the hills experience lower temperature in comparison to the plains mainly due to its altitude. In the pre-monsoon season the region experiences cyclones with hailstorms (usually occurs in the month of April and May) causing damage to the crops as well as human property (Chatterjee 1984). The south-west monsoon strikes the district by the month of June bringing about heavy rainfall. The average annual rainfall of the district is around 2,220 mm and lasts for a period of about five months from May to September. The maximum and minimum temperatures during summer are 36 and 16.9 °C and during winter are 28 and 5.3 °C.

3.5 Soil

The soil found in the hilly tracks of Dhalai district in ranges between red literatic soil and sandy loam to silty clay soil. The soil of the river valleys is composed mostly of alluvium rich in humus content and range between sandy clay loam to clayey loam. The soils of the lowlands are extremely fertile and are enriched almost annually by the deposition from the hilly tracks. It is transported by the rivers with their small tributaries originating from the surrounding hills range. (District Survey Report 2018).

3.6 *Landuse and Land Cover*

In Dhalai district the hill ranges supports thick vegetation as rainfall are well distributed and the area receives sufficient sunlight. Varieties of bamboo species as well as other plants are found here. Cane and grasslands mainly cover the plains and the small hillocks of the lowlands. In Dhalai district, dense forest occupies major chunk, followed by open forest, agriculture and plantation. (Dhalai District Administration 2020).

3.7 *Water Quality*

Most of the rural areas in Dhalai district receive drinking water from deep tube wells, hand pumps, ring wells and few having overhead water tanks. In dry season many villages need drinking water supplied by tanker. The turbidity is very high in water is (Das et al. 2014).

Limited study has been done to understand about the runoff of the Dhalai district of Tripura. Unseemly land usage, informal cutting of slope side slants, urbanization, agrarian development, and abatement of vegetation spread in bumpy zones have prompted the adjustments in the overflow in Dhalai catchment. Dhalai River conveys enormous measure of sedimentation in the storm season. Because of the adjustment in the land use and change in the runoff, it impacts the residue transport of the Dhalai River. It might cause topped off of stream divert and flood in certain pieces of the Dhalai area. This issue shows that there is a requirement for evaluation of runoff in Dhalai catchment area.

Methodology of the study objective has been discussed briefly under the Sect. 3 which comprises Data collection in Sect. 3.1, Image Processing in Sect. 3.2, HEC-HMS for Runoff Prediction in Sect. 3.3, MCDM in Sect. 3.4, ANN in Sect. 3.5, Development of the indicator in Sect. 3.6, Validation of the model is in Sect. 3.7.

4 Methods Applied

In the present investigation the MCDM techniques like Analytical Hierarchy Process, Fuzzy, Weighted Sum Method and Weighted Product Method was used for feature selection and priority estimation of the input variables and Artificial Neural Networks was applied to approximate the runoff with the help of the selected input variables. Sections 4.1, 4.2, 4.3, 4.4 and 4.5 depicts a brief description of these techniques.

4.1 Analytic Hierarchy Process (AHP)

AHP, developed by Saaty (Venkat Rao 2007), addresses how to determine the relative importance of a set of activities in a multi-criteria decision problem. The process makes it possible to incorporate judgments on intangible qualitative criteria alongside tangible quantitative criteria. Analytic Hierarchy Processes a multiple criteria decision-making tool.

This is an Eigen value way to deal with the pair-wise examinations. It likewise gives an approach to align the numeric scale for the estimation of quantitative just as subjective exhibitions. The scale ranges from 1/9 for 'least esteemed than', to 1 for 'equivalent', and to 9 for 'totally more significant than' covering the whole range of the examination. The initial step, an unpredictable choice issue is organized as a chain of importance. The AHP technique depends on three standards: first, structure of the model; second, relative judgment of the other options and the rules; third, amalgamation of the needs (Venkat Rao 2007).

4.2 Fuzzy MCDM Method

The fuzzy scale of ratings consisted of eleven ratings of importance: Excessive High (ExH), Extremely High (EH), Very High (VH), Semi-High (SH), Neither High nor Low (NHNL), Semi-Low (SL), Low (L), Very Low (VL), Extremely Low (EL), Excessively Low (ExL); the crisp value of which was determined by the Theory of Maximization. The ratings were connected to their respective ranks of importance and the ranks of the ratings in each row were divided by the worst rank of the row. The results were averaged to determine the weightage of each parameter.

A 4 × 4 matrix of alternative is developed to find the weightage of the parameters. The comparative rating was given as per Fuzzy scale. If A is the alternative matrix and A is the alternative matrix then,

$$A = \{\mathrm{Di}, \mathrm{Di}\} \tag{1}$$

where Di = {change in forest (Δf), change in water body (Δw), Change in settlement (Δs), change in rainfall (Δp)}.

$$\text{So, } A = \{(\Delta\mathrm{f}, \Delta\mathrm{w}, \Delta\mathrm{s}, \Delta\mathrm{p})\} \tag{2}$$

4.3 Weighted Sum Model (WSM)

The weighted add model (WSM) is that the best glorious and multi-criteria higher cognitive process methodology for evaluating variety of alternatives in terms of variety of call criteria. In general, suppose that a given MCDA drawback is outlined on m alternatives and n call criteria. Moreover, allow us to assume that everyone the factors area units profit criteria, that is, the upper the values area unit, the higher it's. Next suppose that W_j denotes the relative weight of importance of the criterion C_j and a_{ij} is that the performance worth of different Ai once it's evaluated in terms of criterion Cj. Then, the overall importance of different Ai, denoted as A_i. WSM-score, is outlined as follows

$$A_i = \sum_{j=1}^{n} w_j a_{ij}, \; for i = 1, 2, 3, \ldots, m \tag{3}$$

4.4 Weighted Product Model (WPM)

The weighted product model (WPM) could be a well-liked multi-criteria decision analysis (MCDA)/multi-criteria decision method (MCDM) technique. It's just like the weighted sum model (WSM). The most distinction is that rather than addition within the main operation currently there's multiplication like all MCDA/MCDM ways, given could be a finite set of decision alternatives delineated in terms of variety of decision criteria. Every decision different is compared with the others by multiplying variety of ratios, one for every call criterion. Every quantitative relation is raised to the ability like the relative weight of the corresponding criterion. A number of the primary references to the present technique area unit. Suppose that a given MCDA downside is outlined on m alternatives and n call criteria. What is more, allow us to assume that everyone the standards area unit profits criteria, that is, the upper the values area unit, the higher it's. Next suppose that wj denotes the relative weight of importance of the criterion Cj and aij is that the performance worth of different Ai once it's evaluated in terms of criterion Cj. Then, if one desires to check the two alternatives AK and AL (where m $\geq$ K, L $\geq$ 1) then, the subsequent product needs to be calculated:

$$P(A_K/A_L) = \prod_{j}^{n} \left(a_{Kj}/a_{Lj}\right)^{wj}, for \; K, L = 1, 2, 3, \ldots m \tag{4}$$

4.5 *Artificial Neural Network Model (ANN)*

ANN is a computational model made out of many handling components associated by a variable weight. The systems have layers of equal components, known as neurons. The idea of ANN was first presented in 1943, when Warren McCulloch, a neurophysiologist, and a youthful mathematician, Walter Pitts, composed a paper on how neurons may function; they displayed a basic neural system with electrical circuits. As of late, ANNs have been commonly utilized in numerous territories, for example, control, information pressure, anticipating, improvement, design acknowledgment, characterization, discourse and vision (Saravanan and Balakrishnan 2013; Boushehr et al. 2014; Tjadera et al. 2014; Halwatura and Najim. 2013). The objective equation of the ANN model is described by Eqs. 5 and 6.

$$h_m = f\left(w_n x_n + b_j\right) \tag{5}$$

$$y = g(\in_n h_n + b_k) \tag{6}$$

where, W_n is the weight, X_n is the input, b_j and b_k are the bias for the input-to-hidden layer and the hidden-to-output layer, respectively, hn is the hidden layer, and $\in_n$ is the weight of the hidden layer. f and g are the activation functions applied in between the input and hidden layers and the hidden and output layers, respectively.

5 Methodology

First stage of the present study was collection of data for relevant parameters and then determination of their priority with respect to the objective of the decision-making problem. The priority of the parameters was assigned with the help of different MCDM methods which is described in detail at the Sect. 4. The predictive model was developed with the help of Artificial Neural Networks.

5.1 *Data Compilation*

The relevant data were collected and compiled as explained in the next paragraph:

(a) Literature survey for identification of input parameters: Various literatures are surveyed along with govt. and non govt. reports to find the input parameters which majorly influence the change in runoff in a watershed. The literature published in reputed journals of Elsevier and Springer were consulted before preparing a list of most important parameters which influences the change in the runoff. The lists of these parameters were placed to various hydrological along with agricultural experts for their opinion about the most important parameter in their aspects. After taking the views of the experts the list was placed to the stake holders for their opinions. The five most important parameters is selected by incorporating views of experts and stakeholders. The data about the same parameters are then collected in such a manner that the dynamism of the parameters can be reflected on change in runoff.

(b) Change in landuse and land cover data.
(c) Change in rainfall data.
(d) Change in runoff data.
(e) Meteorological data for the study location were obtained from the India Meteorological Department IMD, Agartala, and Tyndal distributed weather data (IWP 2014). Nearly 50 years of average monthly rainfall data were collected and preprocessed for application in the model.
(f) Satellite images for the last three decades were obtained from the Bhushampad satellite (Indian Remote Sensing; IRS) and Google Earth.

5.2 Image Processing

Satellite Images of Dhalai District, captured at an eye altitude of 20 km, were imported into image processing software Geomatica Freeview V10.3. The images were divided into 28 grids and classified into 'agricultural land', 'water body', 'barren land', 'forest cover' and 'settlement'. However, in this study, only 'water body', 'settlement' and 'forest cover' were selected; the remaining unidentified portion of the image was taken as 'barren land'. It has been found in various studies (e.g., Coutu and Vega 2007; Zhao et al. 2004) that barren land is not an important land use and can be ignored if the proportion is equal to or less than 20%. Further, agricultural land could be divided into two sub-parts: 'orange plantation' and 'pineapple shrub'. However, owing to the resolution of the image, the area of shrubland overlapped with the area of barren land, and orange plantation areas overlapped with 'forest cover'. For this reason, the area of agricultural land could not be delineated. Furthermore, most of the cultivation in these areas can be classified as jhum (shifting) cultivation, which does not depict a regular pattern in the same way that 'normal' agricultural practices do. Therefore, there may have been a certain proportion of (jhum) agricultural land not captured by the resolution of the images, and instead classified as 'forest cover' or perhaps 'barren land'. The images were also classified into pervious and impervious areas, with water bodies, forest cover and barren land considered as pervious and the rest classified as impervious. Figure 3 shows the methodology used for determining the percentage area of pervious and impervious land in the grids with the help of the image processing software.

Some part of the images may not be identifiable and can be clustered under unidentified categories. In the recent study the multi chromatic images of $5 \times 5\ \text{km}^2$ dimension was captured at an eye altitude of 20 km. The spectral resolution of the

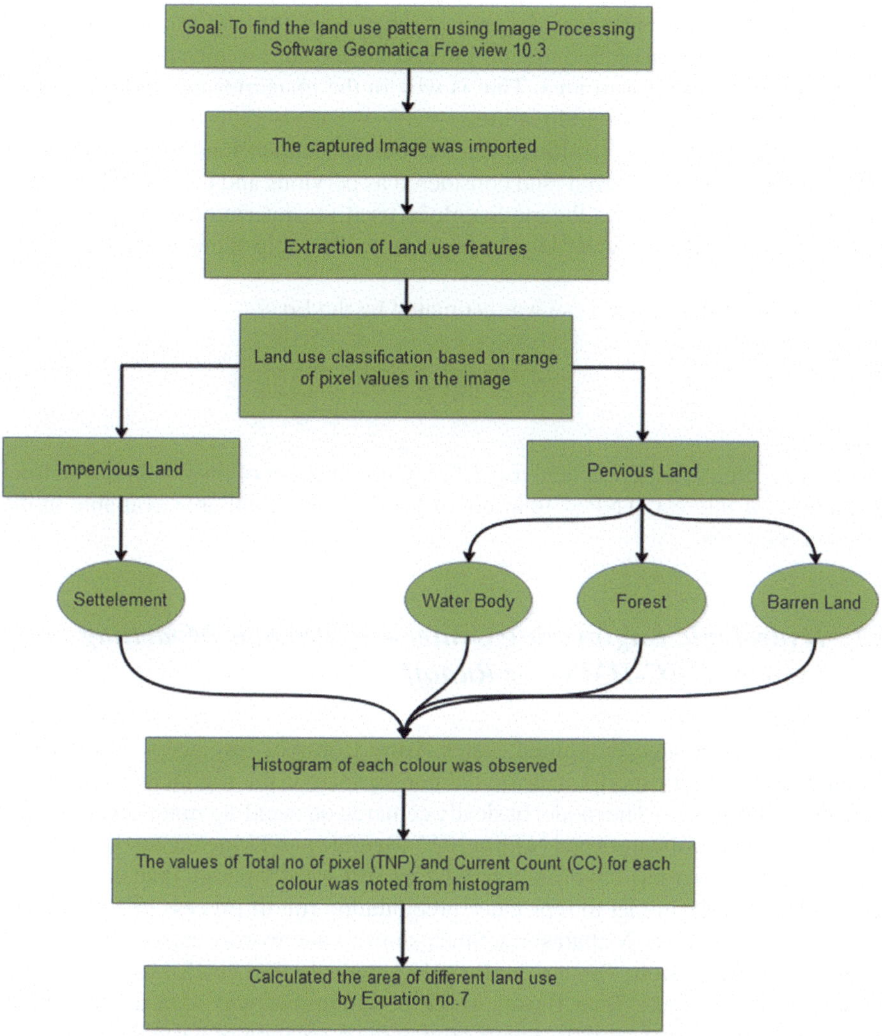

Fig. 3 Flow chart depicting the image processing methodology to determine the area of pervious and impervious land

scale of the image was 1:3930. The noise percentage of the imageries was found to be 2–19%.

In the present investigation after processing of the image mainly four kinds of land uses were identified viz. forestland, water body, settlement and barren land.

But as discussed in the section the study area also has some agricultural landuse where orange and pineapples are cultivated. As orange is a tree type plant while segments it is difficult to separate an orange plant from the forest area. Similarly pineapple is a shrub and seasonally cultivated. As within a decade only two images were captured for each season and then the area of different land use were averaged to had the resultant area of the same for the decade, area due to pineapple cultivation was incorporated into forest area. That is why in the imageries no land use can be separated under the agricultural features.

The images were also classified into pervious and impervious areas, with water bodies, forest cover and barren land considered as pervious and the rest classified as impervious. Figure 3 shows the methodology used for determining the percentage area of pervious and impervious land in the grids with the help of the image processing software.

Area of the different land use was estimated by the Eq. 7

$$A_n = \frac{CC}{TNP} \times A_w \tag{7}$$

where, A_n = area of different land use, CC = Current count of the pixel representing the landuse of interest, TNP = total no. of Pixel, A_w = Total area available in the image of the grids.

5.3 Hydrologic Engineering Center—Hydrologic Modelling System (HEC-HMS) for Runoff

HEC-HMS was created by United States Army Corps of Engineers (USACE). It is a lumped boundary model with spatial conveyance of the region of a watershed region into sub-basins. The model basically centered on deciding runoff hydrographs from sub-basins (Halwatura and Najim. 2013; Beighley and Moglen 2003). McColla et al. conjecture the runoff hydrograph utilizing HEC-HMS model. Patil et al. (2019) utilized HEC-HMS model to reproduce precipitation runoff process in Nashik area (Upper Godavari bowl), Maharashtra. Study shows a noteworthy increment in runoff because of urbanization. Tassew et al. (2019) applied HEC-HMS Model for Flow Simulation in the Lake Tana Basin, where bend number was seen as the touchy boundary. Rangari et al. (2020) centers around the recreation of basic tempest occasion for examining waste limit with regards to part of Hyderabad city utilizing HEC-HMS and HEC-Geo HMS. The mimicked model gives a peak discharge of 590.5 m^3/s (Hussain Sagar Lake) for the August 2008 precipitation occasion having 221.4 mm precipitation for the term of 34 h. Basic Rainfall for Small-Watershed Flash Floods has been determined utilizing the HEC-HMS hydrological model. The HEC-HMS model was utilized to reenact the precipitation runoff process and decide the early admonition pointers under various precipitation designs (Yuan et al. 2019). Derdour et al. (2018) recreate runoff in the semi-dry locale of Ain Sefra watershed applying the HEC-HMS. The peak releases got from the model for the 10, 50, 100 and multiyear storms are separately 425.8, 750.5, 904.3 and 1328.3 m^3/s. Jani et al. (2018) shows the utilization of a HEC-HMS model to evaluate the run-off to anticipate the flood possibility in ineffectively checked catchment. In an investigation of

Hadaf watershed, land use/land spread change has been recognized utilizing diverse change strategies utilizing remote detecting, GIS and HEC-HMS displaying method (Pampaniya et al. 2018). HEC-HMS was utilized to set up the semi-conveyed hydrological model of the slope bowl to ascertain the precipitation overflow and flood directing procedure in Xueye Lake bowl in Laicheng District. The outcomes show that, the relative blunders between top release and runoff profundity are under 20% and the model has great appropriateness to the flood reenactment of the Xueye Lake Basin (Wang et al. 2018).

5.4 *Determination of Priority Value Using Multi Criteria Decision Making (MCDM)*

MCDM is applicable for establishing choice and arranging issues including various standards investigation. These doesn't exist an interesting ideal answer for issues and it is important to utilize expert's opinion to find the solutions. MCDM might be considered as an unpredictable and dynamic procedure including one administrative level and one designing level. The approaches gave by MCDM are not simply some scientific models accumulating measures, perspectives, or qualities, yet moreover they are choice help situated. All things considered, support is a key idea in MCDM, inferring that the models are not created through a clear successive procedure where the leader's role is important. The MCDM approach can be applied in various situations whenever and wherever decision-making is required (Kabira et.al. 2013). Decision-making generally involves the following five steps:

- Factors selection
- Criteria and Alternative selection
- Identification of weighing methods to represent importance;
- Aggregation;
- Decision-making.

All the criteria or factors to be chosen must be rational with the choice goal, free of one another, quantifiable, and ideally spoke to on a similar scale. Likewise, the choice of choices must be doable, reasonable, accessible, and not a copy of different other options.

5.5 *Development of the Cognitive Index for Representing Runoff Probability*

The index for depiction of change in runoff the runoff probability index (RPI) was determined with the help of Eq. 8.

$$Index = \frac{\sum_{i=1}^{n}(w_n c_n)}{\sum_{i=1}^{n} w_n} \tag{8}$$

where, W_n is the weights of the different land use types, and C_n is the change in magnitude of the different input parameters.

The weight vector was estimated by the four different MCDM methods. The parameter of the index was used as input to the ANN and MLR model and index was used as output. The model was developed to avoid repetitive determination of weights when new places are introduced.

5.6 *Validation of the Model*

By adopting sensitivity analysis the influence of the parameter on the output decision has been verified. A statistical model MLR (Ramirez et al. 2017) also has been used to predict the index value for the validation of the developed model by ANN. The validation process of the model has been depicted in Sects. 5.6.1 and 5.6.2.

5.6.1 Sensitivity Analysis

A sensitivity analysis of the model was performed based on the change in 'forest cover', change in 'water body', change in 'settlement', and change in rainfall.

5.6.2 Comparing the Results with Multi-linear Regression Analysis (MLR)

This is a statistical analysis that permits analyzing how different free factors are identified with a variable. When this connection is distinguished, data pretty much the entirety of the autonomous factors can be utilized to make all the more impressive and precise expectations concerning why things are how they are. A MLR analysis software but non cognitive method, MaxStat Lite Software (Chakraborty 2017) has been used in the present study.

5.6.3 Case Study Analysis

For the case study analysis Dhalai district of Tripura, India has been chosen. The detail description of the study area has been depicted in Sect. 3.

Results from the above procedure were discussed in Sect. 6.

6 Results and Discussion

The results from the methodology followed as per the description depicted in Sect. 4 was divided into three parts: Result from Image Processing (Sect. 6.1), MCDM (Sect. 6.2) and validation (Sect. 6.3).

6.1 Results from Image Processing

The land use pattern of each decade was estimated with the help of image processing. Results of the captured Satellite Images of Dhalai District, after image processing are showing in Table 1. Table 1 depicts the land use pattern i.e. the percentage of forest, settlement, barren land and undefined land of each grid for the year 1999.

From the Table 1, it is found that in 28 nos. of grid most of the grids are covered by forest and water body. It shows less urbanization on that particular area.

6.2 Results from MCDM

The priority value of the selected indicators was estimated by four different methods, result of which is discussed in Sects. 6.2.1, 6.2.2, 6.2.3 and 6.2.4.

6.2.1 Results from Fuzzy MCDM

The pair-wise comparison matrix and the fuzzy grading (FG) matrix for the application of MCDM used in the study are given in Tables 2 and 3.

According to the fuzzy littoral ratings the relative importance of each of the input the relative importance of each of the input parameter was given in the Table 3. After converting the littoral ratings into numerical equivalent it was found that importance of rainfall is maximum and water body is minimum among the parameters compared.

6.2.2 Results from AHP

Weight Vector by AHP MCDM of each landuse type and rainfall is given in the Table 4.

As per the AHP MCDM method, Change in rainfall (C_4) was found to be the most influential and Change in Water Body (C_2) was found to be the least influential parameter as per the study objective.

Table 1 Grid wise land use percentage (1999)

Grid	Forest (%)	Settlement (%)	Water body (%)	Barren land (%)	Undefined (%)
1	30	5	50	Nil	15
2	65	15	5	5	10
3	55	Nil	15	9	21
4	75	10	Nil	3	12
5	45	25	20	5	5
6	50	Nil	30	Nil	20
8	30	40	Nil	10	20
9	50	30	10	5	5
10	60	15	20	Nil	5
11	40	25	10	7	18
12	30	30	30	Nil	10
13	90	Nil	10	Nil	0
14	80	Nil	20	Nil	0
15	80	Nil	20	Nil	0
16	95	Nil	Nil	Nil	5
17	95	Nil	Nil	Nil	5
18	75	5	10	5	5
19	55	Nil	15	9	21
20	75	10	Nil	3	12
21	45	25	20	5	5
22	50	Nil	30	10	10
23	60	15	20	Nil	5
24	40	10	20	7	23
25	30	30	30	Nil	10
26	90	Nil	10	Nil	0
27	10	40	30	5	15
28	40	10	20	7	23

6.2.3 Results from WSM

Weight vector by WSM of each landuse type and rainfall is given in Table 5.

As per the WSM MCDM method, Change in rainfall (C_4) was found to be the most influential and Change in Settlement (C_3) was found to be the least influential parameter as per the study objective.

Table 2 The pair-wise comparison matrix considered to find the cognitive index

	Change in forest cover (C_1)	Change in water body (C_2)	Change in settlement (C_3)	Change in rainfall (C_4)
Change in forest cover (C_1)	X	VH	H	L
Change in water body (C_2)	VL	X	L	VL
Change in settlement (C_3)	L	H	X	EL
Change in rainfall (C_4)	H	VH	EH	X

*VH = very high, VL = very low, H = high, L = low

Table 3 Fuzzy grading matrix utilized in the study

	C_1	C_2	C_3	C_4	Final weightage
C_1	0.00	0.88	0.77	0.42	0.52
C_2	0.29	0.00	0.42	0.29	0.25
C_3	0.42	0.77	0.00	0.05	0.31
C_4	0.77	0.88	1.00	0.00	0.66

Table 4 Weight vector by AHP

	Final weights value
C_1	0.218
C_2	0.071
C_3	0.031
C_4	0.304

Table 5 Weight vector by WSM

	Final weights value
C_1	0.517
C_2	0.246
C_3	0.306
C_4	0.656

6.2.4 Results from WPM

Weight Vector by WSM of each landuse type and rainfall is given in the Table 6.

As per the WSM MCDM method, Change in rainfall (C_4) was found to be the most influential and Change in Settlement (C_3) was found to be the least influential parameter as per the study objective.

Table 6 Weight vector by WPM

	Final weights value
C_1	0.437
C_2	0.221
C_3	0.301
C_4	0.515

Table 7 Average weight vector

Alternatives	AHP	Fuzzy	WSM	WPM	Ensemble	Rank
C_1	0.218	0.52	0.517	0.437	0.422	2
C_2	0.071	0.25	0.246	0.221	0.197	4
C_3	0.031	0.31	0.306	0.301	0.237	3
C_4	0.304	0.66	0.656	0.515	0.533	1

Table 8 Table showing the decadal change in runoff due to the changes in different types of landuse and cover

Decade	% change in 'forest cover'	% change in discharge	% change in 'water body'	% change in discharge	% change in 'settlement'	% change in discharge
1990s	6.088	7.585	5.000	5.074	5.500	7.115
1980s	5.250	7.037	4.000	4.746	4.750	6.262
1970s	4.611	5.900	1.000	3.140	3.00	4.412
1960s	3.855	4.627	1.000	2.682	2.500	2.762
1950s	2.627	3.844	0.350	2.164	1.250	2.607

As per the average weight vector (Table 7) Change in rainfall (C_4) and Change in Settlement (C_3) was found to be the most influential and least influential parameters as per the study objective. Further all the weights value as determined from the five MCDM methods was arranged to find the Index value.

In the scenario analysis (Table 8), it was observed that with an increase in 'forest cover' by 75%, there was a 49% decrease in runoff. Even for a 25% increase, the decrease in runoff was found to be 9%. Furthermore, with a decrease in 'water body' by 50%, a 25.5% increase in runoff was predicted by the model, which depicts the sensitivity of the model with respect to each of its inputs.

6.3 *Validatio of the Model Results*

The model results were validated by the sensitivity analysis and case study analysis.In the first type of validation (Sect. 6.3.1) the reliability was adjudged by similar priority value and sensitivity of the selected indicators. But in the second type of validation

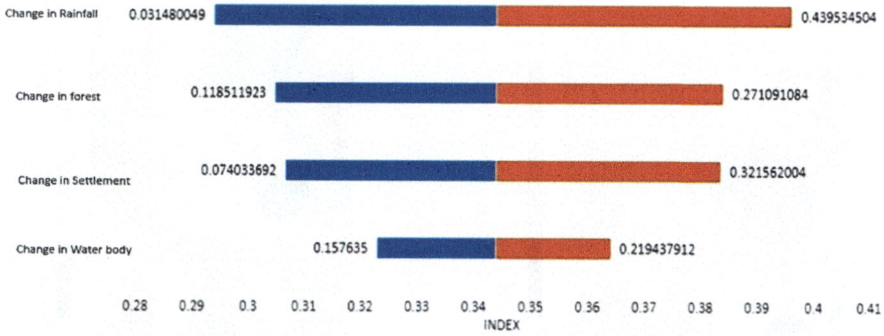

Fig. 4 Sensitivity analysis of the model

Table 9 Swing percentage correspondence to input variable

Input variable	Percent swing^2 (%)
Change in rainfall	42.90
Change in forest cover	26.00
Change in settlement	24.30
Change in water body	6.90

(Sect. 6.3.2) the model is tested in real life scenario and the results can be validated by the historical records of the location.

6.3.1 Sensitivity Analysis of the Model

Sensitivity analysis of the model was performed to understand the reliability of the model. Figure 4 is showing the reliability analysis of the model (Table 9).

As per the sensitivity analysis change in rainfall was found to be the most sensible parameter among all the four different parameters. It shows that sensitivity of indicators confirms the priority of the same parameters in estimation of runoff from land use change.

6.3.2 Results from the Case Study Analysis

The values of actual runoff were computed with the help of HEC-HMS hydrological model. An image was captured for the Dhalai district of the year 1998 and already gridded image was processed to estimate the area of different land use. The data were fed to the software to estimate the value of an index. It is shown that, the runoff is directly proportional to the runoff probability (Fig. 5). The change in runoff was found to be maximum in the grid no. 4 and 27 respectively.

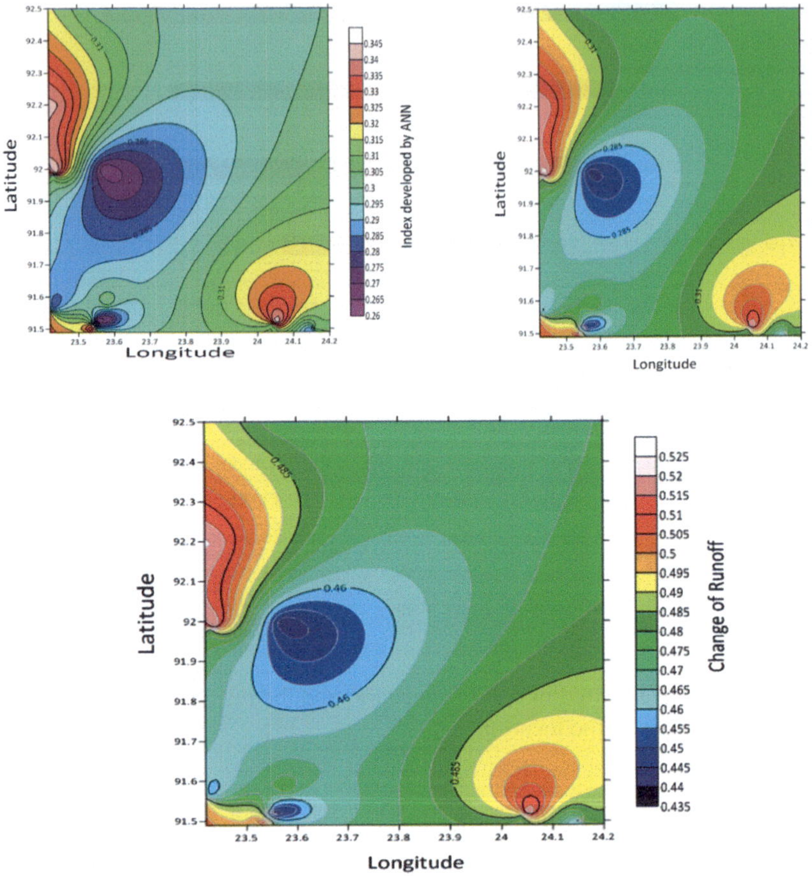

Fig. 5 Index value predicted by artificial neural network (ANN) with the actual runoff of each grid for the year 1999

As per the results it is showing that the index predicted by MLR (Fig. 5) and trend of Index with the Change in Runoff (Fig. 6) is very similar to the results predicted by the ANN model. It shows the validation of the model.

6.4 *SWOT Analysis of the Index*

This software can be used for predicting the runoff change. The impact of extreme events and rapid urbanization could be simulated. The output can be utilized to predict damages to hydraulic structure.

The index is still in experimental stage and was applied in one catchment only.

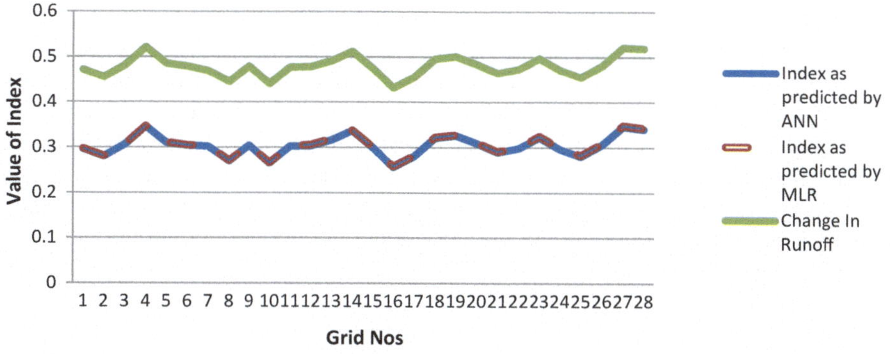

Fig. 6 Trend of index predicted by ANN and MLR with the change in runoff

The reviews of literature are followed to identify the parameters. Expert and stakeholders input can be included for a more accurate selection.

Early decision can be taken regarding adaptive measures.

Protective measures can be undertaken to avoid damages from extreme events.

Socio economic management measures of the stakeholders will become more plausible as the index is distributed and can detect the change both spatially and temporally.

7 Conclusion

An attempt was made to estimate the stream flow of un-gauged river as a function of change in land use and cover. In this aspect an objective and cognitive method was developed where the significance of each of the input parameters were determined by the MCDM methods and utilized to develop an index which represents the change in run off in the watershed. The advantage of the method is it can predict the amount of stream flow and also the frequency of the stream-flow by processing images of the water shed captured remotely. The change in runoff is estimated at a discrete domain indicated by the grids of uniform dimension. The advantage of Fuzzy MCDM is utilized and cog-nativity of ANN models were used to create a framework for estimation of stream flow in a discrete domain from the information of change in land use and cover of the watershed. As per the authors knowledge this is the first time when stream flow can be predicted from land use data and at a discrete domain. However as change in runoff is indicated by an index not in magnitude a fuzziness of the prediction will be eminent as the significance can be changed with change in MCDM methods or change in number of input parameters. This problem can be solved if various methods and different combinations of inputs and output were compared then method which estimate the significance most prominently and have minimum steps of calculation may be selected for determination of importance. Although this step is one time and may be completed at the initial stages only.

Impact of climate change on watersheds is the inherent aim of the present study for sustainable management of un-gauged watersheds. The present monograph tries to highlight such studies which show to manage the uncertainties that can be imposed by the change in climate on the functioning of watersheds. The present study also shows a procedure for intelligent decision making to solve the problems of climatic abnormalities.

References

Ai-Nasseri, B. (2008). Sevoflurane and propofol: original and generic. *Ann Fr Anesth Reanim, 27*(1), 120–122.

Akhtar, M., Ahmad, N., & Booij, M. J. (2008). The impact of climate change on the water resources of Hindukush–Karakorum–Himalaya region under different glacier coverage scenarios. *Journal of Hydrology, 355,* 148–163.

Alipour, M. (2019). Streamflow prediction in ungauged basins located within data-scarce regions.

Alizadeh, M. J., Kavianpour, M. R., Kisi, O., & Nourani, V. (2017). A new approach for simulating and forecasting the rainfall-runoff process within the next two months. *Journal of Hydrology, 548*, 588–597.

Amin, M. Z. M., Shaaban, A. J., Ercan, A., Ishida, K., Kavvas, M. L., Chen, Z. Q., & Jang, S. (2017). Future climate change impact assessment of watershed scale hydrologic processes in Peninsular Malaysia by a regional climate model coupled with a physically-based hydrology modelo. *Science of the Total Environment, 575,* 12–22.

Anache, J. A. A., Flanagan, D. C., Srivastava, A., & Wendland, E. C. (2018). Land use and climate change impacts on runoff and soil erosion at the hillslope scale in the Brazilian Cerrado. *Science of the Total Environment, 622*, 140–151.

Ang, R., & Oeurng, C. (2018). Simulating streamflow in an ungauged catchment of Tonlesap Lake Basin in Cambodia using Soil and Water Assessment Tool (SWAT) model. *Water Science.*

Arsenault, R., Breton-Dufour, M., Poulin, A., Dallaire, G., & Romero-Lopez, R. (2019). Streamflow prediction in ungauged basins: Analysis of regionalization methods in a hydrologically heterogeneous region of Mexico. *Hydrological Sciences Journal.*

Asadi, H., Shahedi, K., Jarihani, B., & Sidle, R. C. (2019). Rainfall-runoff modelling using hydrological connectivity index and artificial neural network approach. *Water, 11*(2), 212.

Azad, A., Karami, H., Farzin, S., Mousavi, S.-F., & Kisi, O. (2019). Modeling river water quality parameters using modified adaptive neuro fuzzy inference system. *Water Science and Engineering, 12*(1), 45–54.

Bajracharya, A. R., Bajracharya, S. R., Shrestha, A. B., & Maharjan, S. B. (2018). Climate change impact assessment on the hydrological regime of the Kaligandaki Basin, Nepal. *Science of the Total Environment, 625*, 837–848.

Barreto-Neto, A. A., & De Souza Filho, C. R. (2008). Application of fuzzy logic to the evaluation of runoff in a tropical watershed. *Environmental Modelling & Software, 23*, 244–253.

Berihun, M. L., Tsunekawa, A., Haregeweyn, N., Meshesha, D. T., Adgo, E., Tsubo, M., Masunaga, T., et al. (2019). Hydrological responses to land use/land cover change and climate variability in contrasting agro-ecological environments of the Upper Blue Nile basin, Ethiopia. *Science of the Total Environment.*

Bhatt, S. C., & Bhargava, G. K. (2006). *Land and people of Indian state and Union territories—In 36 volumes* (pp. 17–18). Delhi: Kalpaz Publications.

Boongaling, C. G. K., Faustino-Eslava, D. V., & Lansigan, F. P. (2018). Modeling land use change impacts on hydrology and the use of landscape metrics as tools for watershed management: The case of an ungauged catchment in the Philippines. *Land Use Policy, 72,* 116–128.

Boushehr, I., Hasanzadeha, M., & Danehkarb, A. (2014). Environmental site selection for oil jetty using the analytical network process method case study. *Ocean Engineering, 77*, 55–60.

Brunner, M. I., Seibert, J., & Favre, A.-C. (2018). Representative sets of design hydrographs for ungauged catchments: A regional approach using probabilistic region memberships. *Advances in Water Resources, 112*, 235–244.

Burrough, P. A., MacMillan, R. A., & VanDeursen, W. (1992). Fuzzy classification methods for determining land suitability from soil profile observations and topography. *Journal of Soil Science, 43,* 193–210.

Calder, I. R., Hall, R. L., Bastable, H. G., Gunston, H. M., Shela, O., Chirwa, A., & Kafundu, R. (1995). The impact of land use change on water resources in sub-Saharan Africa: A modelling study of Lake Malawi. *Journal of Hydrology, 170*(1–4), 123–135.

Carter, W. R. (1961). Magnitude and frequency of floods in suburban areas. *US geological survey professional, 424-B* (pp. B9–11).

Chakraborty, T., & Majumder, M. (2017). Application of AHP-DEMATEL and GMDH framework to develop an indicator to identify failure probability of wave energy converter. *Indian Journal of Science and Technology, 10*(31).

Chang, H., & Jung, I. (2010). Spatial and temporal changes in runoff caused by climate change in a complex large river basin in Oregon. *Journal of Hydrology, 388,* 186–207.

Chatterjee, S. N. (1984). *Tripura: A profile* (pp. 5–10). New Delhi: Inter-India Publications.

Coutu, G. W., & Vega, C. (2007). Impacts of land use changes on runoff generation in the east branch of the Brandywine Creek watershed using a GIS-based hydrologic model. *Middle States Geographer, 40,* 142–149.

Das, M. K., Karmakar, B., Paul, R., & Lodh, R. (2014). Assessment of physicochemical characteristics of drinking water sources in Chawmanu RD Block of Dhalai District, Tripura, India.

Derdour, A., Bouanani, A., & Babahamed, K. (2018). Modelling rainfall runoff relations using HEC-HMS in a semi-arid region: Case study in Ain Sefra watershed, Ksour Mountains (SW Algeria). *Journal of Water and Land Development, 36*(1), 45–55.

Dewan, A., & Yamaguchi, Y. (2009). Land use and land cover change in Greater Dhaka, Bangladesh: Using remote sensing to promote sustainable urbanization. *Applied Geography, 29,* 390–401.

Dhalai District Administration. (2020). About district. Retrieved November 1, 2020, from https://dhalai.nic.in/about-district/.

Dietz, M., & Clausen, J.C. (2008). Stormwater runoff and export changes with development in a traditional and low impact subdivision. *Journal of Environmental Management, 87*(4), 560–566.

Ding, X., Jiang, Y., Zhao, H., Guo, D., He, L., Liu, F., et al. (2018). Electrical conductivity of nutrient solution influenced photosynthesis, quality, and antioxidant enzyme activity of pakchoi (*Brassica campestris L.* ssp. Chinensis) in a hydroponic system. *PloS one, 13*(8), e0202090

District Survey Report. (2018). Dhalai. Retrieved November 1, 2020, from https://dste.tripura.gov.in/pdfs/dsr/dhalai.pdf.

Doyle, M., Harbor, J., Rich, C., & Spacie, A. (2000). Examining the effects of urbanization on streams using indicators of geomorphic stability. *Physical Geography, 21,* 155–181.

Enea, M., & Salemi, G. (2001). Fuzzy approach to the environmental impact evaluation. *Ecological Modelling, 136*, 131–147.

EPA, US (2001). United States Environmental Protection Agency. Quality Assurance Guidance Document-Model Quality Assurance Project Plan for the PM Ambient Air 2

Falah, F., Rahmati, O., Rostami, M., Ahmadisharaf, E., Daliakopoulos, I. N., & Pourghasemi, H. R. (2019). Artificial neural networks for flood susceptibility mapping in data-scarce urban areas. In *Spatial modeling in GIS and R for earth and environmental sciences* (pp. 323–336). Elsevier.

Farmer, W. H., & Levin, S. (2018). Characterizing uncertainty in daily streamflow estimates at ungauged locations for the Massachusetts sustainable yield estimator. *JAWRA Journal of the American Water Resources Association, 54*(1), 198–210.
Field, R., Masters, H., & Singer, M. (1982). Porous pavement: Research, development, and demonstration. *Journal of Transportation Engineering, 108*(3), 244–258.
Fohrer, N., Haverkamp, S., Eckhardt, K., & Frede, H.-G. (2001). Hydrologic response to land use changes on the watershed scale. *Physics and Chemistry of the Earth, Part B: Hydrology, Oceans and Atmosphere, 26*(7–8), 577–582.
Fox, D. M., Witz, E., Blanc, V., Soulie, C., Penalver-Navarro, M., & Dervieux, A. (2012). A case study of land cover change (1950–2003) and runoff in a Mediterranean catchment. *Applied Geography, 32,* 810–821.
Gardner, L. L. (2009). Assessing the effect of climate change on mean annual runoff. *Journal of Hydrology, 379,* 351–359.
Gumindoga, W., Rwasoka, D. T., Nhapi, I., & Dube, T. (2017). Ungauged runoff simulation in Upper Manyame Catchment, Zimbabwe: Application of the HEC-HMS model. *Physics and Chemistry of the Earth, Parts A/B/C, 100,* 371–382.
Guzha, A. C., Rufino, M. C., Okoth, S., Jacobs, S., & Nóbrega, R. L. B. (2018). Impacts of land use and land cover change on surface runoff, discharge and low flows: Evidence from East Africa. *Journal of Hydrology: Regional Studies, 15,* 49–67.
Hall, M. J. (1984). *Urban hydrology.* New York: Elsevier Applied Science Publishers.
Halwatura, D., & Najim, M. M. M. (2013). Application of the HEC-HMS model for runoff simulation in a tropical catchment. *Environmental Modelling & Software, 46,* 155–162.
Harbor, J. (1994). A practical method for estimating the impact of land use change on surface runoff, groundwater recharge and wetland hydrology. *Journal of American Planning Association, 60,* 91–104.
Helmi, A. M., Mahrous, A., & Mustafa, A. E. (2019). Urbanization growth effect on hydrological parameters in mega cities. *Advances in sustainable and environmental hydrology, hydrogeology, hydrochemistry and water resources* (pp. 417–419). Cham: Springer.
Jani, M., Baloothiya, K., Mewar, P., & Patel, D. (2018). Flood Potential estimation of poorly gauged Varekhadi Watersheds using HEC-HMS model—A case of Lower Tapi Basin, India. *EGU General Assembly Conference Abstracts, 20,* 7326.
Jiang, Y., Nan, Z., & Yang, S. (2013). Risk assessment of water quality using Monte Carlo simulation and artificial neural network method. *Journal of Environmental Management, 122,* 130–136.
Jimeno-Sáez, P., Senent-Aparicio, J., Pérez-Sánchez, J., & Pulido-Velazquez, D. (2018). A comparison of SWAT and ANN models for daily runoff simulation in different climatic zones of Peninsular Spain. *Water, 10*(2), 192.
Jodar, J., Carpintero, E., Martos-Rosillo, S., Ruiz-Constán, A., Marin-Lechado, C., Cabrera-Arrabal, J. A., & González-Dugo, M. P. (2018). Combination of lumped hydrological and remote-sensing models to evaluate water resources in a semi-arid high altitude ungauged watershed of Sierra Nevada (Southern Spain). *Science of the Total Environment, 625,* 285–300.
Kling, H., Fuchs, M., & Paulin, M. (2012). Runoff conditions in the upper Danube basin under an ensemble of climate change scenarios. *425,* 264–277.
Klingseisen, B., Metternicht, G., & Paulus, G. (2008). Geomorphometric landscape analysis using a semi-automated GIS-approach. *Environmental Modelling and Software.* In Press.
Kontogiannia, A. D., Papageorgioub, E. I., & Tourkoliasa, C. (2012). How do you perceive environmental change? Fuzzy cognitive mapping informing stakeholder analysis for environmental policy making and non-market valuation. *Applied Soft Computing, 12,* 3725–3735.
Kundu, S., Khare, D., & Mondal, A. (2017). Past, present and future land use changes and their impact on water balance. *Journal of Environmental Management, 197,* 582–596.
Lambin, E. F., & Meyfroidt, P. (2011). Global land use change, economic globalization, and the looming land scarcity. *PNAS, 108,* 9.

Lee, S. K., Mogi, G., & Kim, J. W. (2009). Decision support for prioritizing energy technologies against high oil prices: A fuzzy analytic hierarchy process approach. *Journal of Loss Prevention in the Process Industries, 22*, 915–920.

Li, C., Liu, M., Hu, Y., Shi, T., Zong, M., & Walter, M. (2018). Assessing the impact of urbanization on direct runoff using improved composite CN method in a large urban area. *International Journal of Environmental Research and Public Health, 15*(4), 775.

Li, Q., Peng, Y., Wang, G., Wang, H., Xue, B., & Xinqi, Hu. (2019a). A combined method for estimating continuous runoff by parameter transfer and drainage area ratio method in ungauged catchments. *Water, 11*(5), 1104.

Li, F., Chen, J., Liu, Y., Peng, Xu., Sun, H., Engel, B. A., & Wang, S. (2019b). Assessment of the impacts of land use/cover change and rainfall change on surface runoff in China. *Sustainability, 11*(13), 3535.

Mao, D., & Charkauer, K. A. (2009). Impacts of land-use change on hydrologic responses in the Great Lakes region. *Journal of Hydrology, 374*, 71–82.

Meenu, R., Rehana, S., & Mujumdar, P. P. (2013). Assessment of hydrologic impacts of climate change in Tungabhadra river basin, India with HEC-HMS and SDSM. *Hydrological Processes, 27*, 1572–1589.

Meresa, H. (2019). Modelling of river flow in ungauged catchment using remote sensing data: Application of the empirical (SCS-CN), Artificial Neural Network (ANN) and Hydrological Model (HEC-HMS). *Modeling Earth Systems and Environment, 5*(1), 257–273.

Molina-Navarro, E., Andersen, H. E., Nielsen, A., Thodsen, H., & Trolle, D. (2018). Quantifying the combined effects of land use and climate changes on stream flow and nutrient loads: A modelling approach in the Odense Fjord catchment (Denmark). *Science of the Total Environment, 621*, 253–264.

Napoli, M., Massetti, L., & Orlandini, S. (2017). Hydrological response to land use and climate changes in a rural hilly basin in Italy. *CATENA, 157*, 1–11.

Nyaupane, N., Mote, S. R., Bhandari, M., Kalra, A., & Ahmad, S. (2018). Rainfall-runoff simulation using climate change based precipitation prediction in HEC-HMS model for Irwin Creek, Charlotte, North Carolina. In *World Environmental and Water Resources Congress 2018* (pp. 352–363).

Pandey, G., & Ranganathan, T. (2018). Changing land-use pattern in India: has there been an expansion of fallow lands? *Agricultural Economics Research Review, 31*(347-2018-3195), 113–122.

Pampaniya, N. K., & Tiwari, M. K. (2018). Assessment of land use/land cover change in hadaf watershed using different transformation methods employing remote sensing, GIS and HEC-HMS modelling technique. *Journal of Soil and Water Conservation, 17*(3), 226–231.

Patil, V. K., Saraf, V. R., Karad, O. V., Ghodke, S. B., Gore, D., & Dhekale, S. S. (2019). Simulation of rainfall runoff process using HEC-HMS model for upper Godavari Basin Maharashtra, India. *European Journal of Engineering Research and Science, 4*(4), 102–107.

Prieto, C., Vine, N. L., Kavetski, D., García, E., & Medina, R. (2019). Flow prediction in ungauged catchments using probabilistic random forests regionalization and new statistical adequacy tests. *Water Resources Research, 55*(5), 4364–4392.

Ragettli, S., Zhou, J., Wang, H., Liu, C., & Guo, L. (2017). Modeling flash floods in ungauged mountain catchments of China: A decision tree learning approach for parameter regionalization. *Journal of Hydrology, 555*, 330–346.

Ramirez, M. C., Viloria, A., Muñoz, A. P., & Posso, H. (2017). Application of multiple linear regression models in the identification of factors affecting the results of the Chelsea Football Team.

Rangari, V. A., Sridhar, V., Umamahesh, N. V., & Patel, A. K. (2020). Rainfall runoff modelling of urban area using HEC-HMS: A case study of Hyderabad City. *Advances in water resources engineering and management* (pp. 113–125). Singapore: Springer.

Ravansalar, M., Rajaee, T., & Kisi, O. (2017). Wavelet-linear genetic programming: A new approach for modeling monthly streamflow. *Journal of Hydrology, 549*, 461–475.

Reddy, N. A., Seelam, J. K., Rao, S., & Nagaraj, M. K. (2019). Flood estimation at ungauged catchments of western catchments of Karnataka, West coast of India. *ISH Journal of Hydraulic Engineering, 25*(3), 325–335.

Requena, A. I., Chebana, F., & Ouarda, T. B. (2018). A functional framework for flow-duration-curve and daily streamflow estimation at ungauged sites. *Advances in Water Resources.*

Saravanan, P., & Balakrishnan, P. (2013). Design of renewable energy based shunt active filter with multilevel inverter using genetic algorithm. *International Journal of Engineering.*

Schueler, T., & Holland, H. K. (2000). *The practice of watershed protection.* Ellicott City: The Center for Watershed Protection.

Souissi, D., Zouhri, L., Hammami, S., Msaddek, M. H., Zghibi, A., & Dlala, M. (2019). GIS-based MCDM-AHP modeling for flood susceptibility mapping of arid areas, Southeastern Tunisia. *Geocarto International,* 1–25.

Stone, M. C., Hotchkiss, R. H., Hubbard, C. M., Fontaine, T. A., Mearns, L. O., & Arnold, J. G. (2001). Impacts of climate change on Missouri river basin water yield. *Journal of the American Water Resources Association, 37,* 1119–1129.

Tan, Q., Lei, X., Wang, X., Wang, H., Wen, X., Ji, Y., & Kang, A. (2018). An adaptive middle and long-term runoff forecast model using EEMD-ANN hybrid approach. *Journal of Hydrology.*

Tassew, B. G., Belete, M. A., & Miegel, K. (2019). Application of HEC-HMS model for flow simulation in the Lake Tana Basin: The case of Gilgel Abay Catchment, Upper Blue Nile Basin, Ethiopia. *Hydrology, 6*(1), 21.

Tjadera, Y., Maya, J. H., Shanga, J., Vargasa, L. G., & Gaob, N. (2014). Firm-level outsourcing decision making: A balanced scorecard-based analytic network process model. *International Journal of Production Economics, 147*(Part C), 614–623.

Utami, I. W. P. (2016). A model of microteaching lesson study implementation in the prospective history teacher education. *Journal of Education and Practice, 7*(27), 10–14.

Venkata Rao, R. (2007). Vendor selection in a supply chain using analytic hierarchy process and genetic algorithm methods. *International Journal of Services and Operations Management, 3*(3), 355–369.

Wang, Y., Sang, G., Jiao, C., Xu, Y., & Zheng, H. (2018). Flood simulation and parameter calibration of small watershed in hilly area based on HEC-HMS model. In *IOP Conference Series: Earth and Environmental Science* (Vol. 170, no. 3, p. 032093). IOP Publishing.

Wang, J., Zhou, W., Pickett, S. T. A., Yu, W., & Li, W. (2019). A multiscale analysis of urbanization effects on ecosystem services supply in an urban megaregion. *Science of the Total Environment, 662,* 824–833.

Wei, W., Chen, L. D., Fu, B. J., Huang, Z. L., Wu, D. P., & Gui, L. D. (2007). The effect of land uses and rainfall regimes on runoff and soil erosion in the semi-arid loess hilly area, China. *Journal of Hydrology, 335,* 247–258.

Xu, C.-Y. (2000). Modelling the effects of climate change on water resources in Central Sweden. *Water Resources Management, 14,* 177–189.

Xu, W., Hu, Q., Bai, S., Bao, C., Miao, Y., Yuan, Z., et al. (2019). Rational molecular passivation for high-performance perovskite light-emitting diodes. *Nature Photonics, 13*(6), 418–424.

Yao, R., Cao, J., Wang, L., Zhang, W., & Xiaojun, Wu. (2019). Urbanization effects on vegetation cover in major African cities during 2001–2017. *International Journal of Applied Earth Observation and Geoinformation, 75,* 44–53.

Yuan, W., Liu, M., & Wan, F. (2019). Calculation of critical rainfall for small-watershed flash floods based on the HEC-HMS hydrological model. *Water Resources Management,* 1–21.

Zadeh, L. A. (1965). Fuzzy sets. *Information and Control, 8,* 338–353.

Zhao, W., Fu, B., Meng, Q., Zhang, Q., & Zhang, Y. (2004). Effects of land use pattern change on rainfall-runoff and runoff-sediment relations: A case study in Zichang watershed of the Loess Plateau of China. *16*(3), 436–442.

Zhang, X., Zhang, Q., & Zhang, L. (2019). Applying Fuzzy linguistic method to predict river runoff. In *International Conference on Management Science and Engineering Management* (pp. 607–617). Cham: Springer.

Indicator Based Impact Analysis of Urbanization with Respect to Evapo-Transpiration

Ashesh Rudra Paul and Apu K. Saha

Abstract In recent years large-scale conversions of land use have been taking place worldwide to sustain the ever-growing population. As a result, parameters such as vegetation type, air temperature, relative humidity, soil moisture, and movement of the wind have been markedly affected. Many studies have examined the impact of urbanisation on temperature or relative humidity, but few have explored whether or not potential evapotranspiration (PET) is influenced by changes in population density and urbanisation. In this context, in the present study an indicator was proposed to represent impact of urbanization in metro cities with respect to PET. An attempt was made to measure the impact of urbanisation on PET without the application of any subjective models. However, based on the recent literature, a distinct relationship does indeed exist between PET and urbanisation. The most important parameter affected by urbanisation that also influences PET was investigated by the application of multi-criteria decision-making. Finally, a neurogenetic model was applied to reveal the vulnerability of Evapo-transpiration to urbanisation in four cities with differing levels of population density. The results confirmed that the vulnerability is greater in cities with denser populations and greater levels of urbanisation.

Keywords Evapo-transpiration · Vulnerability analysis · Soft computation · Urbanization

1 Introduction

The Total World Population on 2014 is 7,156,392,318 and the rate of change in population density from rural to urban region is 1.98%. According to World Urbanization Prospect 2011 revision, a report published by UN, the maximum change in population density to urban area was observed in Burundi (6.8%) and Moldova (−1.5%).

A. R. Paul (✉)
National Institute of Technology, Agartala, Tripura, India
e-mail: asheshrudrapaul@gmail.com

A. K. Saha
Department of Mathematics, National Institute of Technology, Agartala, Tripura, India
e-mail: apusaha_nita@yahoo.co.in

M. Majumder and G. D. Kale (eds.), *Water and Energy Management in India*,
https://doi.org/10.1007/978-3-030-66683-5_3

In case of China, India and USA, the three countries having the highest number of population in the World, the rate of change to urban population is 2.7, 2.4 and 1.3% respectively. According to population density, Bangladesh (1034 pop/Km^2), Taiwan (646 pop/Km^2) and South Korea (505 pop/Km^2) is the three most populated countries and the rate of change in urban population is respectively 3.5, 0 and 0.6%. All the rate of change in population given above was calculated with respect to the urban population of 2005.

Urban climates are warmer and more polluted than their rural counterparts. These differences are partly due to the urban expansion, which usually removes and replaces crops and natural vegetation with non-evaporating and non-transpiring surfaces such as metal, asphalt and concrete. There is generally low land surface albedo, vegetative cover, and moisture availability in urban areas. These factors, along with the presence of high levels of anthropogenic heating, are associated with the phenomenon known as the urban heat island (UHI), which describes the difference in ambient air temperature between an urban area and its surrounding rural area. UHI often occurs when a large fraction of the natural land cover in an area is replaced by the built surfaces that absorb the incoming solar radiation during the day and then reradiate it at night. UHI has been the most intensively studied climatic feature of cities and has been quantified by calculating air or surface temperature differences between urban and nearby rural areas (Zhan et al. 2013).

The most important anthropogenic influences on climate are the emission of greenhouse gases and changes in land use, such as urbanization and agriculture. But it has been difficult to separate these two influences because both tend to increase the daily mean surface temperature. The impact of urbanization has been estimated by comparing observations in cities with those in surrounding rural areas, but the results differ significantly depending on whether population data or satellite measurements of night light are used to classify urban and rural areas.

For example, changes in air temperature with changes in urbanization were reported by Karl et al. (1988), while Adebayo (1991, 2006) observed that different levels of humidity can be seen in urban and rural areas. Hou et al. (2013) showed that the movement of wind varies with changes in urbanization, and is different among seasons (summer and winter). In a study conducted in Beijing, China, Qiao et al. (2013) examined the contribution of different landscapes to the thermal environment of the city, and found urban zones to be the most important contributor. The vulnerability of rainfall to urbanization was depicted in a study by Shepherd (2013), in which the contribution of aerosol and bifurcation—increased by the change in population density—was attributed to the change in the precipitation regime of the study area. Srinivisan et al. (2013), in their study about the impact of urbanization on water availability, found that population density is an important factor that influences climatic as well as hydrologic parameters. As a result, large-scale urbanization often leads to a shortage of water availability; in their words, "vulnerability to water shortages changes as cities grow". Verbeiren et al. (2013) also stressed the impact of population change on rainfall–runoff relationships using a remote sensing approach. Finally, Pathirana et al. (2014) showed that an increase in the level of urbanizations' likely to increase the severity of extreme events. They simulated four scenarios of urbanization with a coupled climate and land-use model and found that, for three of them, the magnitude of extreme events increased with an increase in urbanization. They concluded that "increasing urbanization causes a significant increase of extreme rainfall values".

All of the above studies related to the impact of urbanization climatic parameters concluded that such parameters are sensitive to urbanization and alter with changes in the density of population. In the present study, we build upon existing knowledge by extending the examination to the parameter of potential Evapo-transpiration (PET). PET refers to the amount of water that could evaporate under conditions of adequate water supply (e.g., precipitation and soil moisture). It is a useful parameter because it can help predict Evapo-transpiration in a given area, and thus assist in the monitoring of potential drought situations. Therefore, we decided to investigate the influence of population-density change on PET and the vulnerability of PET to different levels of urbanization.

In fact, very few studies have discussed PET in this context. However, we know through the literature that factors affecting PET, such as air temperature, relative humidity, movement of wind, soil moisture, and vegetation type, are influenced by urbanization, and so it is logical to expect PET to also be modulated by such changes. For instance, Cui and Shi (2012) investigated the degree of correlation between urbanization and microclimates in Shanghai, China. They found that increases in the amount of paved road, buildings, buses, and population have a decreasing effect on relative humidity and wind speed, and an increasing effect on air temperature and the number of hot days. In an earlier study, Yilmaz et al. (2007) presented similar findings to those of Cui and Shi (2012) by comparing climatic patterns associated with different types of land use. They found that urban land use is usually associated with warmer, drier, and less windy conditions than rural and forested sites. The studies of Tang et al. (2011) and Carlson and Arthur (2000), among others, highlighted that urbanization can increases oil temperature and, as a result, the soil moisture content can decrease. Nikula et al. (2010) examined the impact of urbanization on *Aspen*sp., in which they found that leaf traits and litter decomposition rates are affected by changes in urban population density. Furthermore, according to Buyantuyev and Wu (2012), "urbanization transforms vegetation patterns and ecosystem processes which are controlled primarily by climate in desert regions".

Soukarieh et al. (2018) examined the Impact of Lebanese practices in industry, agriculture and urbanization on soil toxicity. Evaluation of the Polycyclic Aromatic Hydrocarbons (PAHs) levels in soil. Ruoyu wang and Later Kalin (2018) investigated that potential changes in flow, total suspended solid (TSS) and nutrient (nitrogen and phosphorous) loadings under future climate change, land use/cover change. UGu Korkut Pata (2018) examines the dynamic short- and long-term relationship between per capita GDP, per capita energy consumption, financial development, urbanization, industrialization, and per capita carbon dioxide (CO_2) emissions within the framework of the environmental Kuznets curve (EKC) hypothesis. Shujuan et al. (2019) highlighted total water soluble Ca^{2+} content in the soil continuously decreased due to acid rain and the losses of Ca^{2+} from the soil can increase the risk of acidifying soils which may also affect the adjacent regions through changing the chemistry of groundwater and surface water. Teuling et al. (2019) examined how decadal changes in climate (e.g., precipitation, temperature) and land use (e.g., de-/afforestation, urbanization) have impacted the amount and distribution of water resources availability. Khademi et al. (2019) highlighted the Environmental impact assessment of industrial activities on heavy metals distribution in street dust and soil.

Therefore, from the above-mentioned studies, we can see that most of the factors that influence PET are also affected by changes in urbanization. Taking this finding one step further, we were interested in establishing the most important parameter influenced by urbanization that also causes the greatest rate of change in PET. A multicriteria decision-making (MCDM) and artificial neural network (ANN) approach was used to achieve this aim. After first introducing MCDM and ANNs in more detail in Sect. 2, as well as the specific methodology of the study, the results and discussion are presented in Sect. 3. Finally, the conclusion of the study is given in Sect. 4.

2 Methods Used

2.1 The MCDM Approach

The MCDM approach can be applied in various situations whenever and wherever decision-making is required (Kabira et al. 2013). Decision-making generally involves the following five steps:

1. Selection of the criteria/parameters/factors/decider;
2. Selection of the alternatives;
3. Selection of the weighing methods to represent importance;
4. Method of aggregation;
5. Decision-making based on the aggregation results.

All the criteria or factors to be selected must be coherent with the decision objective, independent of each other, measurable, and preferably represented on the same scale. Also, the selection of alternatives must be feasible, practical, available, and not a duplicate of the other alternatives.

Keeping the above conditions in mind, a decision-maker must identify suitable criteria and select ideal alternatives to select the better alternative from the available options. Different methods such as the analytical hierarchy process (AHP) (Alphonce 1997), fuzzy logic decision-making (FLDM) (He et al. 2014), elimination and choice-expressing reality (ELECTRE) (Corrente et al. 2013), and evaluation of mixed data (EVAMIX) (Darji and Rao 2013) are nowadays widely used to solve problems in many fields of science and technology where objective rather than subjective decision-making is required.

In the present investigation, the intuition based FLDM and AHP approaches were utilized to find the most important factor influencing PET that is also affected by the impacts of urbanization. The adoption of FLDM (Bellman and Zadeh 1970) and AHP (Triantaphyllou and Mann 1995) ensures that objective, rather than holistic, decision-making is possible.

The other main objective of the study was to find a method by which the vulnerability of PET can be estimated based on different levels of urbanization. In this regard, a vulnerability index (VI) was prepared and a flexible framework of ANNs was adopted to predict the VI for different urban settlements.

Ferhat et al. (2018) determine the Environmental Risk Assessment of E-waste in Reverse Logistics Systems Using MCDM Methods. Kaya et al. Use the MCDM techniques for energy policy and decision-making problems. Sarfaraj et al. (2018) to this end, a hybrid Multiple Criteria Decision Making (MCDM) model. Mirko et al. (2018) deals with the Application of MCDM Methods in Sustainability Engineering. GULCIN (2019) et al. Analysis of Success Factors in Aviation 4.0 Using Integrated Intuitionist Fuzzy MCDM Methods.

2.2 *Analytical Hierarchy Process*

Strength:

Qualities of AHP: In this AHP is extensively spread in the academic gathering and associated in unmistakable fields like Engineering, Medicine, and different sciences. The characteristics consolidate

I. Its ease of use.
II. It is an easily sensible framework.
III. It unravels a problematic issue by isolating it into little advances.
IV. It doesn't authentic information data sets. The structure of AHP yields a straight forward course for a scholastic individual to deal with complex issues.

Weakness:

AHP uses exact characteristics for decisions. i.e., being used cases, the human feelings are dark and the bosses perhaps not capable fix the cautious numerical characteristics to the assessment decisions. For this circumstance, AHP isn't material. It can settle simply coordinate models. i.e., one whose yield is explicitly relating to its data. It can't unwind non-straight models. i.e., one whose yield isn't direct relating to its data. For e.g.: Weather check.

Quian (2018) Researched on energy conservation and emissions reduction based on AHP-fuzzy synthetic evaluation model: A case study of tobacco enterprises.

Awasthi et al. (2018) Done Multi-tier sustainable global supplier selection using a fuzzy AHP-VIKOR based approach.

Sakhuja et al. (2018) using fuzzy AHP and TOPSIS: a case study in the Indian automotive industry.

Seljuk (2018) performance on evaluation of Turkish airline companies using integrated fuzzy AHP fuzzy topsis model.

2.3 Weighted Sum Method

Strength:

I. This strategy is anything but difficult to utilize, and if the entirety of the loads are sure
II. The WSM is the general model which has been used for different applications such as robotics, handling information, and others.

Weakness:

I. Optimal arrangement could be not worthy, either because of the capacity utilized barring parts of the issue or to a wrong setting of the coefficients;
II. If the arrangement isn't satisfactory, new runs of the streamlining agent are required;
III. A little change in loads may bring about enormous changes in the goal vectors;
IV. Significantly various loads may deliver almost comparative target vectors.

Sorooshian (2018) work with Modified Weighted Sum Method for Decisions with Altered Sources of Information. Weighted Sum-Rate Optimization for Intelligent Reflecting Surface Enhanced Wireless Networks is done by Huayan et al. (2018). A Weighted-Sum PSO Algorithm for HEMS: A New Approach for the Design and Diversified Performance Analysis done by Hussain et al. (2018).

2.4 Weighted Product Method

Strength:

i. The process involved transforming subjective or abstract information into objective or quantitative data that could be effortlessly utilized in settling on choices quicker.
ii. The theory of the weighted strategy being advantageous to the proprietor by giving important venture result was profoundly bolstered.
iii. The methodology could be advantageous in upper administrative situations in the development business or callings where multi-target choices should be made rapidly and effectively absent a lot of hazard since dangers could present genuine outcomes if not very much surveyed and tended to progress of time.

Weakness:

I. The outcomes might be not the same as one individual to the next, which may influence the legitimacy of the weighted item technique.
II. This may represent a situation where it is critical to discover approaches to guarantee the unwavering quality of the strategy in different VE applications in activities, items or administrations.

Weighted Product Method in the Value Engineering Process for Construction Project is done by Ochieng (2018). Comparison of weighted product method and technique for order preference by similarity to ideal solution method: Complexity and accuracy is done by Fitriasari et al. (2018). Simple Additive Weighting and Weighted Product Methods Using Neutrosophic Sets are done by Karajan et al. (2018). Developing decision support systems using the weighted product method for house selection is done by Supriyono (2018).

2.5 *Artificial Neural Networks*

ANNs mimic the human nervous system in solving complex prediction, classification and simulation problems. In recent years, ANN shave been used widely in various types of scientific and technical research. Figure 1 shows the number of ANN-related papers published during the period 2005–2013. The apparent increasing trend demonstrates the growing popularity of this modeling method in solving complex, nonlinear problems.

ANNs involve three phases of development: the identification of network topology; training the network to learn the problem; and testing the network regarding the level of adaptation it is able to achieve in the training process.

In the first phase, the number of hidden layers and nodes are determined with the help of trial and error or cognitive search algorithms such as genetic algorithms, particle swarm etc. The number of hidden layers is responsible for a quick learning of the problem, but also increases the load on the computational infrastructure. Hence, the selection of an optimal number of hidden layers is important for efficient performance in neural network models.

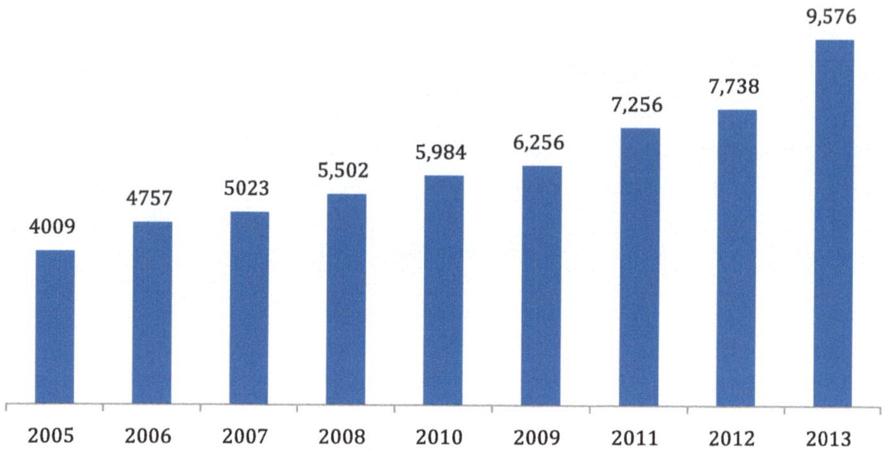

Fig. 1 The number of ANN-related papers published in the period 2005–2013

The second phase involves the equalization of the weighted sum of inputs excited by an activation function and the output by changing the value of the weights based on the difference between the desired and predicted value of the output. There are various algorithms available to vary the weights according to the values of the error. However, broadly, training or learning algorithms can be divided into three classes: propagation, gradient descent, and custom; and quick propagation (QP), conjugate gradient descent (CGD) and Levenberg–Marquadart (LM) are examples of the three types, respectively. Ghaffari et al. (2006) conducted a comparative analysis of the three training algorithm types by examining QP, LM, incremental back propagation, and batch back propagation for the prediction of bimodal drug delivery. From the results the authors concluded that the LM algorithm is better than the QP algorithm in training neural network models. Park and Xu (2013) highlighted the importance of CGD training algorithms in the case of data.

Once the training of the networks is completed, the model is tested with the input–output data-pairs that were not included in the training process. This phase is known as the testing phase, in which the degree of learning is determined by comparing the difference between the predicted and actual output values. It is recommended that testing error is less than training error, which will ensure that the model can estimate the outputs for those inputs that were not included in the learning phase of the model. Generally, 70% of the available dataset of the input–output pair is used for training; while 15% each is utilized for validation and testing purposes.

3 Case Studies

In the present investigation, the vulnerability of evapotranspiration was predicted from the selected attributes that influence PET and are also influenced by urbanization. The weighted sum divided by the sum of the weightage of the attributes was utilized to find the VI of four cities: Agartala (India); Kolkata (India); Brisbane (Australia); and New York (USA) (Table 1).

Table 1 Climatic patterns and population densities of the considered locations

City	Coordinates	Climatic zone	Country	Population density (km^{-2})
Agartala	(23.8333° N, 91.2667° E)	Humid–subtropical (KöppenDfb)	India	10119
Kolkata	(22.5667° N, 88.3667° E)	Tropical–wet/dry (Köppen Aw)	India	23900
Brisbane	(27.4679° S, 153.0278° E)	Humid–subtropical (Köppencfa)	Australia	950
New York City	(40.6700° N, 73.9400° W)	Humid–subtropical (Köppencfa)	United States of America	2050

3.1 Agartala (23.8333° N, 91.2667° E)

Agartala is the capital of the Indian State of Tripura and is the second largest city and municipal body in northeastern India after Guwahati. The city is the seat of the Government of Tripura. Agartala is one of the fastest developing cities of India. There are 8 police stations and 9 legislative assembly constituencies covering various parts of the city.

Agartala is situated on a plain along the Haora River, although the city also extends to the low-lying hills on its northern parts. Agartala has a monsoon influenced humid subtropical climate just short of being hot enough to qualify as a borderline tropical savanna/tropical monsoon climate. Large amounts of rain fall all year except during the dry "winter" or "cool" season. The city experiences long, hot and wet summers, lasting from April to October. Average temperatures are around 28 °C or 82.4 °F, fluctuating with rainfall. There is a short, mild winter from mid- November to early March, with mostly dry conditions and average temperatures around 18 °C (64 °F). The best time to visit is from September to February. Summers are quite long and are extremely hot with a scorching sun and warm day. As mentioned rain is very common in this season and the city sometime remains or can be found flooded. The river Haora flows through the city and remains flooded with water during the time of monsoon.

As of the 2011 Indian census, Agartala city has a total population of 404,004. Total number of literates in the city is 344,711 of which 175,170 are males while 169,541 are females. Average literacy rate of Agartala city is 94.45% of which male and female literacy was 96.16 and 92.75% respectively. The sex ratio of Agartala city is 999 females per 1000 males. Child sex ratio of girls is 950 per 1000 boys (Fig. 2).

3.2 Kolkata (22.5667° N, 88.3667° E)

Kolkata is the capital city of the Indian state of West Bengal. It is the second largest city in India after Mumbai. It is on the east bank of the River Hooghly. When it is called Calcutta, it includes the suburbs. This makes it the third largest city of India. This also makes it the world's 8th largest metropolitan area as defined by the United Nations. Kolkata served as the capital of India during the 1911. Kolkata was once the center of industry and education. However, it has witnessed political violence and economic problems since 1954. Since 2000, Kolkata has grown due to economic growth. Like other metropolitan cities in India, Kolkata struggles with poverty, pollution and traffic congestion.

The total area covered by the city is 71 sq miles or 185 km^2. The metropolitan region of the city covers an area of 728.45 sq miles or 1,886.67 km^2. According to the census conducted in 2011, the population of the city is 4,486,679, making it the 7th most populated city in India. The metropolitan region of the city houses a

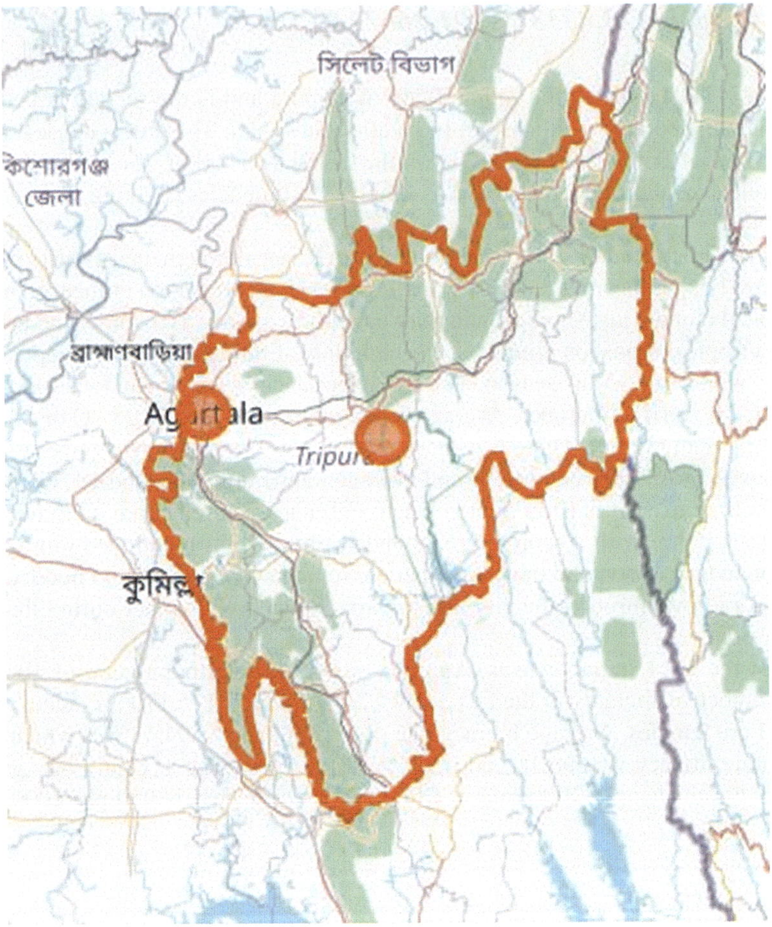

Fig. 2 Figure shows the location of AGARTALA. *Source* https://www.openstreetmap.org/relation/2026458

population of 14,112,536 and the metropolitan region of the city ranks as the third most populated metropolitan region of the country. The density of population of Kolkata is 63,200/sq miles or 24,250/km (Fig. 3).

The city of Kolkata forms a part of Eastern India. The coordinates of the city are 22.82° North and 88.20° east. The mean altitude of the city is 17 feet over sea surface. The shoreline of the Bay of Bengal lies at a distance of 60 km in the south. The Sundarbans National Park is located at a distance of 100 km from the city. Being situated near the sea Kolkata has tropical climate. The summer temperature ranges from 25 to 40 Â °C and in winter it is between 12–25 Â °C. It receives its share of rainfall from monsoon winds.

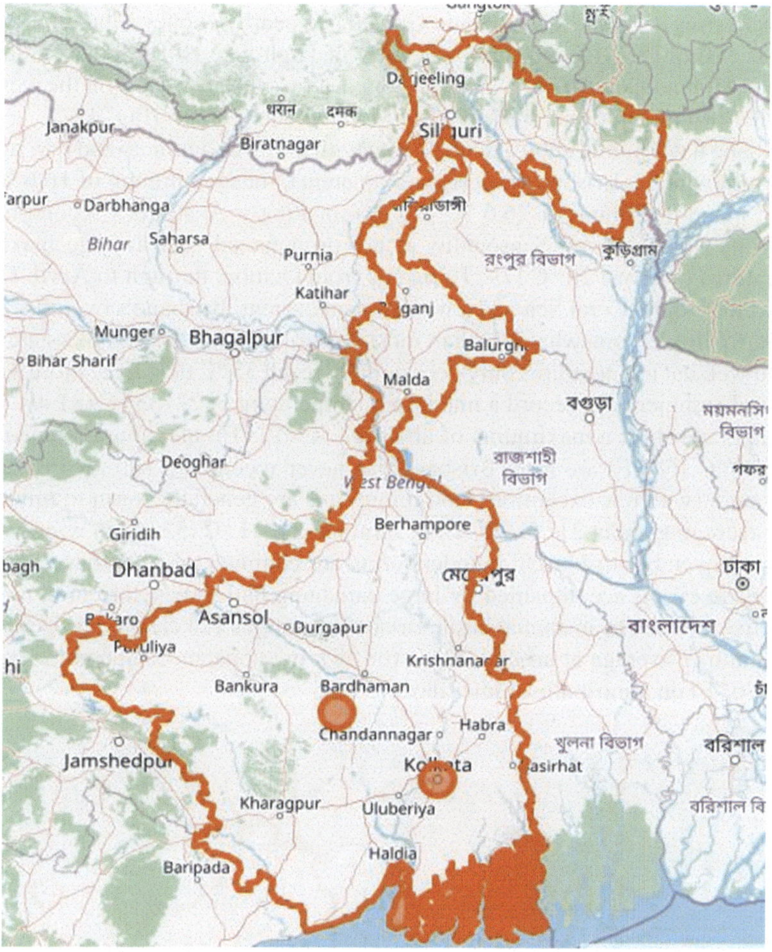

Fig. 3 Figure shows the location of KOLKATA. *Source* https://www.openstreetmap.org/relation/1960177

3.3 *Brisbane (27.4679° S, 153.0278° E)*

Brisbane is the capital of and the most populated city in the Australian state of Queensland, and the third most populous city in Australia. Brisbane's metropolitan area has a population of approximately 2.5 million, and the South East Queensland metropolitan region, centered on Brisbane, encompasses a population of more than 3.6 million.

The city of Brisbane is hilly. The urban area, including the central business district, are partially elevated by spurs of the Herbert Taylor Range, such as the summit of Mount Coot-tha, reaching up to 300 m (980 ft) and the smaller Enoggera Hill. Other prominent rises in Brisbane are Mount Gravatt and nearby Toohey Mountain. Mount Petrie at 170 m (560 ft) and the lower rises of High gate Hill, Mount Ommaney, Stephens Mountain, and Whites Hill are dotted across the city. Also, on the west, are the higher Mount Glorious, (680 m), and Mount Nebo (550 m) (Fig. 4).

Brisbane has a humid subtropical climate with hot, wet summers and dry, moderately warm winters. Brisbane experiences an annual mean minimum of 16.6 °C (62 °F) and mean maximum of 26.6 °C (80 °F), making it Australia's second-hottest capital city after Darwin. Seasonality is not pronounced, and average maximum temperatures of above 26 °C (79 °F) persist from October through to April. Due to its proximity to the Coral Sea and a warm ocean current, Brisbane's overall temperature variability is somewhat less than most Australian capitals. Summers are long, hot, and wet, but temperatures only occasionally reach 35 °C (95 °F) or more. Eighty percent of summer days record a maximum temperature of 27–33 °C (81–91 °F).

Winters are short a maximums of about 22 °C (72 °F); maximum temperatures below 20 °C (68 °F) are rare. Brisbane has never recorded a sub-zero minimum temperature (with one exception), and minimums are generally warm to mild year-round, averaging about 21 °C (70 °F) in summer and 11 °C (52 °F) in winter.

From November to March, thunderstorms are common over Brisbane, with the more severe events accompanied by large damaging hail stones, torrential rain and destructive winds. On an annual basis, Brisbane averages 124 clear days. Dewpoints in the summer average at around 20 °C (68 °F); the apparent temperature exceeds 30 °C (86 °F) on almost all summer days.

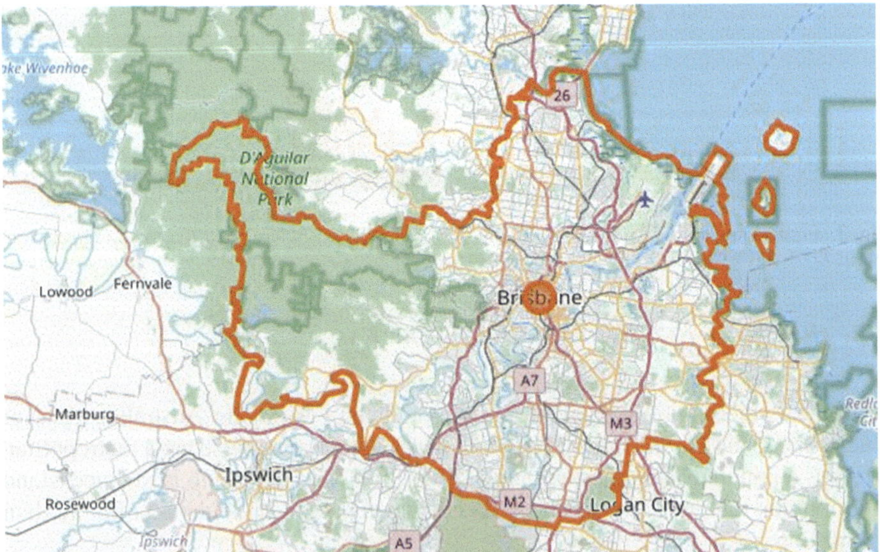

Fig. 4 Figure shows the location of BRISBANE. *Source* https://www.openstreetmap.org/relation/11677792#map=10/-27.4949/153.0739

3.4 *New York City (40.6700° N, 73.9400° W)*

The City of New York, usually called either New York City (NYC) or simply New York (NY), is the most populous city in the United States. With an estimated 2018 population of 8,398,748 distributed over a land area of about 302.6 square miles (784 km^2), New York is also the most dense populated major city in the United States. Located at the southern tip of the state of New York, the city is the center of the New York metropolitan area, the largest metropolitan area in the world by urban landmass and one of the world's most populous megacities, with an estimated 19,979,477 people in its 2018 Metropolitan Statistical Area and 22,679,948 residents in its Combined Statistical Area. A global power city, New York City has been described as the cultural, financial, and media capital of the world, and exerts a significant impact upon commerce, entertainment, research, technology, education, politics, tourism, art, fashion, and sports. The city's fast pace has inspired the term New York minute. Home to the headquarters of the United Nations (Fig. 5).

Under the Koppen climate classification, using the 0 °C (32 °F) isotherm, New York City features a humid subtropical climate, and is thus the northernmost major city on the North American continent with this categorization. The suburbs to the immediate north and west lie in the transitional zone between humid subtropical and humid continental climates. Winters are cold and damp, and prevailing wind patterns that blow sea breezes offshore temper the moderating effects of the Atlantic Ocean; yet the Atlantic and the partial shielding from colder air by the Appalachian Mountains keep the city warmer in the winter than inland North American cities at similar or lesser latitudes such as Pittsburgh, Cincinnati, and Indianapolis. The

Fig. 5 Figure shows the location of NEW YORK. *Source* https://www.openstreetmap.org/relation/175905

daily mean temperature in January, the area's coldest month, is 32.6 °F (0.3 °C); temperatures usually drop to 10 °F (−12 °C) several times per winter, and reach 60 °F (16 °C) several days in the coldest winter month. Spring and autumn are unpredictable and can range from chilly to warm, although they are usually mild with low humidity. Summers are typically warm to hot and humid, with a daily mean temperature of 76.5 °F (24.7 °C) in July. The city receives 49.9 in. (1,270 mm) of precipitation annually, which is relatively evenly spread throughout the year. The record cold daily maximum was 2 °F (−17 °C) on December 30, 1917, while, conversely, the record warm daily minimum was 84 °F (29 °C), last recorded on July 22, 2011 Average winter snowfall between 1981 and 2010 has been 25.8 inches (66 cm); this varies considerably between years.

4 Methodology in Detail

To allocate power resources optimally in an educational institute among various sections of consumers such as residential, hospital, workshop and administrative, the Fuzzy logic decision-making process is adopted. Power supply will be provided to these users based upon their importance which depends on various parameters like demand, consumption pattern and socio-economic aspects. At first, different kinds of consumers are ranked based on their importance in the four respective parameters. After that, a fuzzy scale of importance ratings was used in the four pair wise comparison matrices of the consumers concerning different criteria. As far as deriving the weightage of the importance of the consumers is concerned, The Fuzzy ratings need to be changed into a numerical value. Accordingly, their Crisp Values are computed from a Triangular Membership Function which was suggested by Zadeh and these values replace the fuzzy ratings. Simultaneously, the average of each row was calculated to find the weighted importance of each type of consumer-based upon each of the criteria. The first phase of the work is done. In the second phase, the objective is to optimize the cognitive allocation of power such that the difference between the normal and weighted allocation is minimum. The objective equation is developed and is utilized in the four applied optimization techniques namely, Particle Swarm Optimization, Differential Evolution, Neuro Genetic and GMDH. The competence of these four algorithms is compared and accordingly, the best algorithm for optimization is selected (Fig. 6).

4.1 Study Methodology

The objective of the framework is to estimate the vulnerability of PET in different locations due to change in population density. There are certain parameters which are affected by urbanization and are also responsible for the change in the rate of PET. That means if change in population density is referred as p then to determine

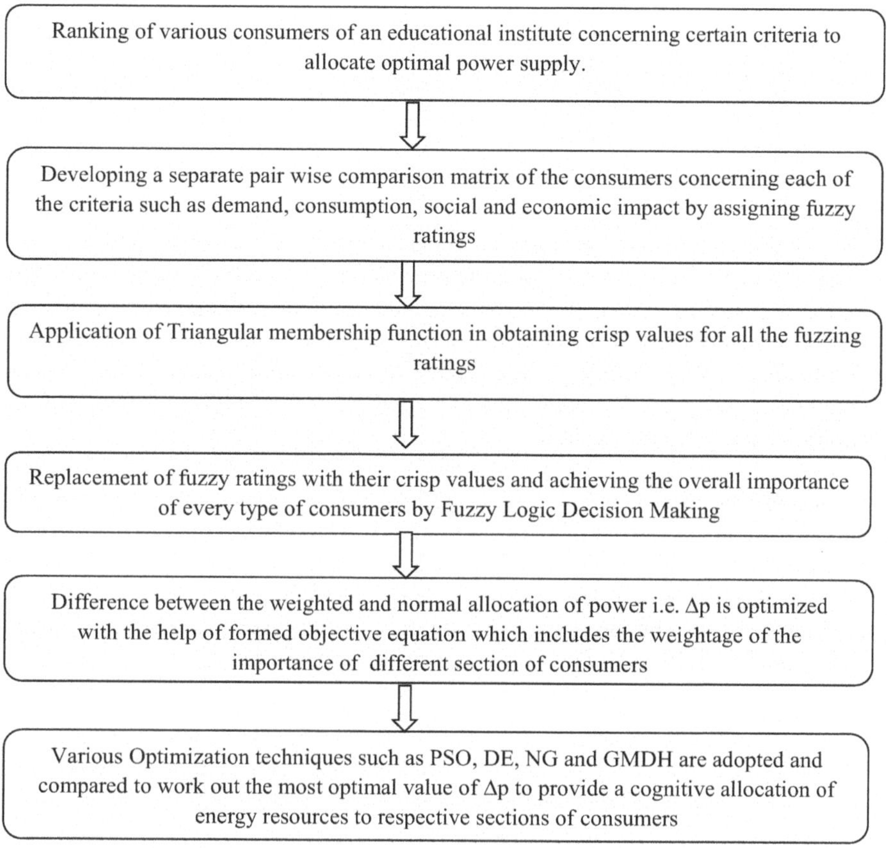

Fig. 6 Detailed methodology of the study

the impact of urbanization on PET the following condition has to be satisfied:

$$PET = F(D)\& \\ D = f(p) \tag{1}$$

where D is various parameters which are affected by urbanization and influences PET of a region. That is why to estimate the impact on PET due to urbanization, D has to be identified first. Again, the influence of all the D parameters on PET and impact of urbanization on the same are not uniform. The objective of the framework can only be achieved if the D parameters and their difference of influence can be identified in an objective manner.

The D parameters can be identified by an extensive literature survey and discussions with the experts in related field. As MCDM methods are widely popular to provide solution to the present type of problems () the present investigation used their strengths to find the difference of influence or importance of each of the *D* parameters in affecting PET.

4.2 Application of the MCDM

In the present study the D parameters can be represented by both qualitative and quantitative manner. Thus, Analytical Hierarchy Process (AHP) and Fuzzy Logic Decision Making (FLDM) method was adopted to find the weightage of importance for each of the parameters which actually separates the D parameters according to their influence on PET and being influenced by urbanization. The steps below provide the methodology adopted to determine the weightage of importance for each of the D parameters by MCDM methods.

Selection of Criteria

In the present study the weightage of importance of the D parameters are required to be estimated. So, all the D parameters are considered as alternatives. To find the weightage of importance some criteria has to be identified with respect to which the alternatives will be compared and the difference in importance can be determined. In this regard the following factors are considered as Criteria:

(i) Expert Survey: A survey was carried out within experts of related fields where participants were asked to suggest about the parameters which can be affected by urbanization and can induce change in PET. The participants were also requested to provide their estimate about the most and least important parameter in this aspect. According to response received from the experts a score was given to the factors according to Eq. 2.

If 'A' is number of experts which mentioned the parameter and 'a' be the number of expert which has referred it as most important parameter and 'a' be the number of expert who have mentioned the parameter as not at all important then,

$$S_E = \left[\frac{A \times \left(\frac{a}{a\prime}\right)}{A_t} \right] \quad (2)$$

where S_E is the score assigned to the parameter and A_t is the total number of experts consulted regarding the present problem.

The score was then normalized and according to the score the top ten parameters were determined. Such parameters will be referred as D_i throughout this manuscript.

(ii) Literatures Considered as a Model Variable: After the expert survey was conducted, the literatures published in reputed journals and chapters in a book or a report in the related subject is extensively analyzed to find the D_i parameters used as an input to estimate PET. If the total number of literatures which mentioned the D_i parameter as an model input be b and the total number of literature surveyed where a model is established or applied to estimate PET be B, then, score S_{Lm} is estimated by Eq. 3.

$$S_{L_m} = \left(\frac{b}{B}\right) \tag{3}$$

The score is then normalized, and the D_i parameters are ranked accordingly.

(iii) Literatures Considered the Variable in Related Studies: The literatures were also surveyed to find the citation of the Di parameters in related studies. If the number of literatures which mentioned the parameter is c and the total number of literatures surveyed be C then the score, S_L, is calculated by Eq. 4.

$$S_L = \left(\frac{c}{C}\right) \tag{4}$$

This score was also normalized, and the D_i parameters are ranked accordingly in a descending manner.

According to the AHP method the criteria are first compared with each other to find the difference of importance between them. Thus a (3 × 3) matrix was formulated and each of the criteria is compared with the other criteria with respect to its importance over the other parameter.

If total number of experts surveyed be E, total number of literatures surveyed for finding the parameter as an input to a model be Lm and the total number of literatures surveyed to find mention of the Di parameter be L then,

$$Sc \text{ For Expert Survey} = \frac{E}{E + L + L_m} \tag{5}$$

$$Sc \text{ For Literatures Considered as Model Variable} = \frac{L}{E + L + L_m} \tag{6}$$

$$Sc \text{ For Literatures Considered as model Variable in related studies} = \frac{L}{E + L + L_m} \tag{7}$$

Where S_c is the score assigned to the criteria.

The score of the criteria are then normalized and rank in a descending manner.

The rank of the criteria is utilized to find the difference of importance between the criteria in the comparison matrix of both AHP and FDM.

Selection of Alternatives

The nine different D_i parameters were selected as the alternatives. All the parameters are measurable, independent of each other, real and is a direct function of the decision objective. Each of the parameter is a direct influence on PET and can be influenced due to the change in the urbanization and thus satisfies the condition expressed in Eq. 1.

Application of Decision-Making Method

In the present investigation AHP and FDM is utilized for identifying the weightage of importance of the parameter due to the availability of both quality and quantitative parameters. A 3 × 3 matrix of criteria is developed to find the weightage of the criteria. The comparative rating was given as per Saaty scale. If C is the Criteria matrix and A is the alternative matrix then,

$$C = \{m, n\} \tag{8}$$

where,

$$m, n = \{E, L, L_m\} \tag{9}$$

The scale proposed by Saaty is used to rate the pair-wise importance of each of the criteria.

Again, the alternatives are compared with each other based on their importance over each other according to each of the criteria and thus,

$$A = \{Di,\ Di\} \tag{10}$$

Where,

$$Di = \{T, RH, WV, SM, Pt, R, St, Lat, \Theta\} \tag{11}$$

and Air Temperature (T), Relative Humidity (RH), Wind Velocity (WV), Soil Moisture (SM), Type of Plant (Pt), Radiation (R), Hours of Solar Radiation (St), Latitude (Lat), Reflection Coefficient (Θ).

The importance was determined by the rank achieved by the alternatives with respect to the criteria. The rating for depicting importance was given according to the Saaty scale.

The geometric sum of each of the rows are calculated and normalized to find the weightage of importance of the alternatives for the criteria according to which the alternatives were compared. Thus for E, L and Lm each of the alternatives will have their weightage of importance. Ultimately a 3 × 9 matrix was drawn where weightage of criteria is multiplied by the weightage of alternatives as estimated when comparing with respect to the same criteria. The geometric mean of all the row of the matrix was also determined and normalized to find the weightage of importance of each of the alternatives.

In case of FDM, the pair-wise rating was performed with the help of littoral fuzzy ratings. The fuzzy rating was then converted to crisp rating by the application of theory of maximization. The weightage of importance of the criteria and alternative in case of FDM is estimated by Eq. 12

$$W = Norm\left[\text{Max}\left\{\frac{r}{R}\right\}\right] \tag{12}$$

Where W is the weightage of importance, r is the score of the littoral rating and R is the maximum score of the row and Max indicates the maximum value of the r/R of a row. The normalization of the maximum values of r/R of each row is taken as weightage of importance of the alternatives or criteria represented by the row.

All the other steps for comparing the criteria (3 × 3), alternative (9 × 9) and resultant (3 × 9) matrix is similar to the AHP method.

4.3 Development of Vulnerability Index

After the weightage of importance is determined an index was developed with the help of the weightage and the magnitude of the Di parameters. The weighted average of all the parameters is proposed as the index for representation of the vulnerability to PET due to urbanization as given in Eq. 13.

$$V = \frac{\sum_{i=1}^{9} w_i \times D_i}{\sum_{i=1}^{9} w_i} \quad (13)$$

Where V is the vulnerability index, w_i is the weightage of importance of the D_i parameters as determined in the previous section.

4.4 Application of Neural Network and Genetic Algorithm

The present study aims to develop an automated framework for representation of urbanization impact on PET. In this regard some algorithms have to be prepared so that V can be automatically calculated once the values of the D_i parameters are given as input. Due to the popularity of ANN, in mapping non-linearity and unknown relationships the said algorithm is applied to estimate the Vulnerability Index (V) once the magnitude of the input parameters are entered in the framework. Another reason for applying ANN is to remove the requirement of repeated application of the MCDM methods once a new alternative is added. In the present study the ANN models were applied to predict the decision for the new alternative based on the existing knowledge that was gained from the available set of data.

4.5 Preparation of the Training Dataset

In development of the ANN model a training dataset is required to be provided. In the present investigation due to the lack of adequate data a random generation of the required values of the Di parameter was performed between the maximum and minimum value of the parameter. The index was calculated with the help of these values and used as an output for the framework's in this case the architecture of mapping the input parameters on the output is important as weightage of the Di parameter is already determined the relationship between the input and output is pre-estimated. The work of the algorithm in this case is to develop the automated flow of calculating V whenever the values of the input parameters are entered.

4.6 Topology Identification

The number of hidden layers and nodes are determined with the help of trial and error or cognitive search algorithms like genetic algorithms, particle swarm etc. The number of hidden layers is responsible for quick learning of the problem but also increase the load on the computational infrastructure. That is why, selection of an optimal number of hidden layers is important for efficient performance of the neural network models and in the present study the said task was performed with the help of genetic algorithms where 50 generations were produced from 40 populations. The cross over rate was fixed at 0.8 whereas the mutation rate was controlled within 0.2.

4.7 Training the Network

The ANN model was performed with three different algorithms, viz, Conjugate Gradient Descent, Quick Propagation and Levenberg Marquadart and Root Mean Square Error (RMSE), Relative Error (RE), Absolute Error (AE), Nash-Sutcliffe Efficiency (NSE) and Covariance (Var) between the predicted and actual output of the model is calculated to find the optimally trained network. The ANN model with selected training algorithm was used to predict V to the values of the D_i parameter provided to the modeling framework.

5 Validation of the Framework

The performance of the framework as well as the validation of the same was conducted by applying sensitivity analysis and finding the Vulnerability Index of various places around the world and comparing their real situation with the software output.

5.1 Sensitivity Analysis

The sensitivity analysis was performed with the help of Multiple Input One output Tornado method developed by SensIt Limited. The range for the input variables were varied between 0 to 1. The impact of each input is then observed on the output and the results were compared with the weights of the variables found from the MCDM analysis.

5.2 Compare with Other Similar Methods

The study results from AHP MCDM were compared with other MCDM techniques and the relative difference between the priorities of the alternative was noted and compared. The better MCDM was selected with the help of the relative difference and comparing the same to identify the method with maximum difference.

5.3 Case Study Analysis

Four cities of different levels of urbanization were selected from all over the world like Kolkata and Agartala (India), Brisbane (Australia) and New York (USA). The population density and type of climate observed and depicts the average annual values of the parameters considered as input to the model. Section 3 describe about the selected case study locations (Figs. 7 and 8, Table 2).

6 Results and Discussions

The score for the criteria according to Eqs. 5–7 is depicted in Table 3 and scores of the alternatives calculated with the help of Eqs. 2–4 is shown in Table 5. The weightage of importance of the criteria and alternatives according to AHP is respectively shown in Table 6 (Table 4).

6.1 MCDM Results

The results from the two different kind of decision-making method clearly depicts that Air Temperature is the most and Reflection Coefficient is the least important parameter in effecting PET and being influenced by change in population density. In similar studies, air temperature has always referred as the commonest parameter in

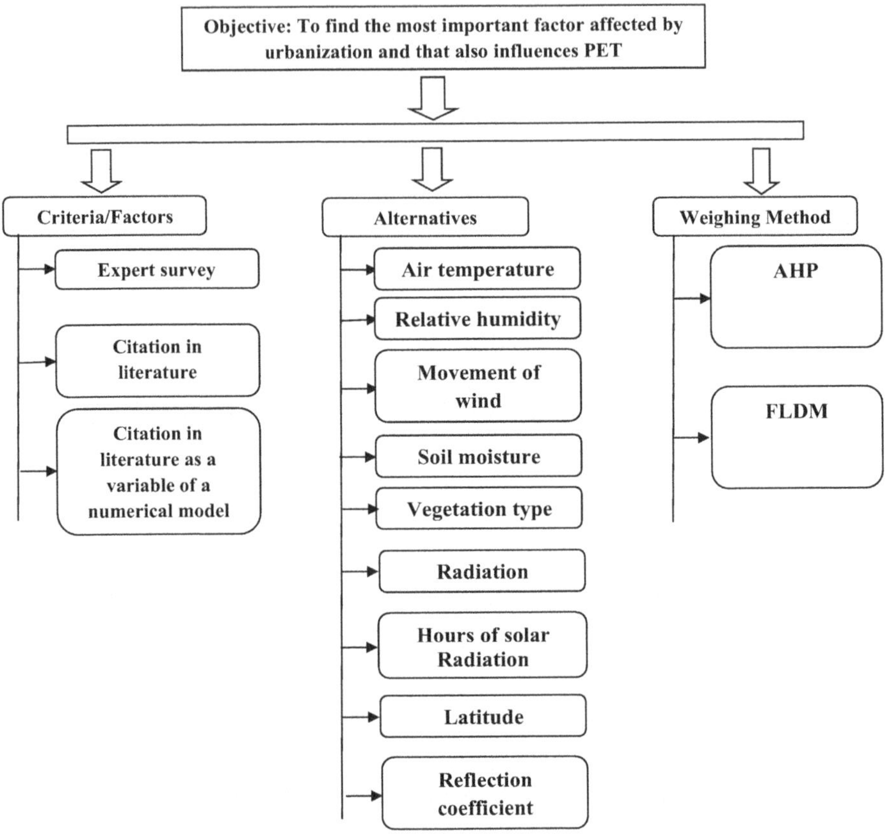

Fig. 7 Flow chart of the development of the MCDM model, designed to find the most important factor affected by urbanization that also influences PET

prediction of PET and in every studies about impact of urbanization on temperature it was clearly observed that change in population density largely effects the annual variation in temperature. Thus, the conclusion that Air Temperature being the most important parameter in estimation of vulnerability of PET due to urbanization seems to be quite certain. Regarding Reflection Coefficient very few studies has actually cited this parameter for estimation of PET or impact of urbanization and most of the related studies the authors has not mentioned this parameter.

In case of criteria selection L seems to be most and E is the least important criteria according to both AHP and FDM method which is obvious as the rank determined from Eqs. 5 to 7 clearly defines the difference of importance between the factors (Table 5).

Although same alternatives were determined as the most and least important by the AHP and FDM method but in case of the former method the parameter Wind Velocity was determined as the most important parameter. The Wind Velocity has noticeable

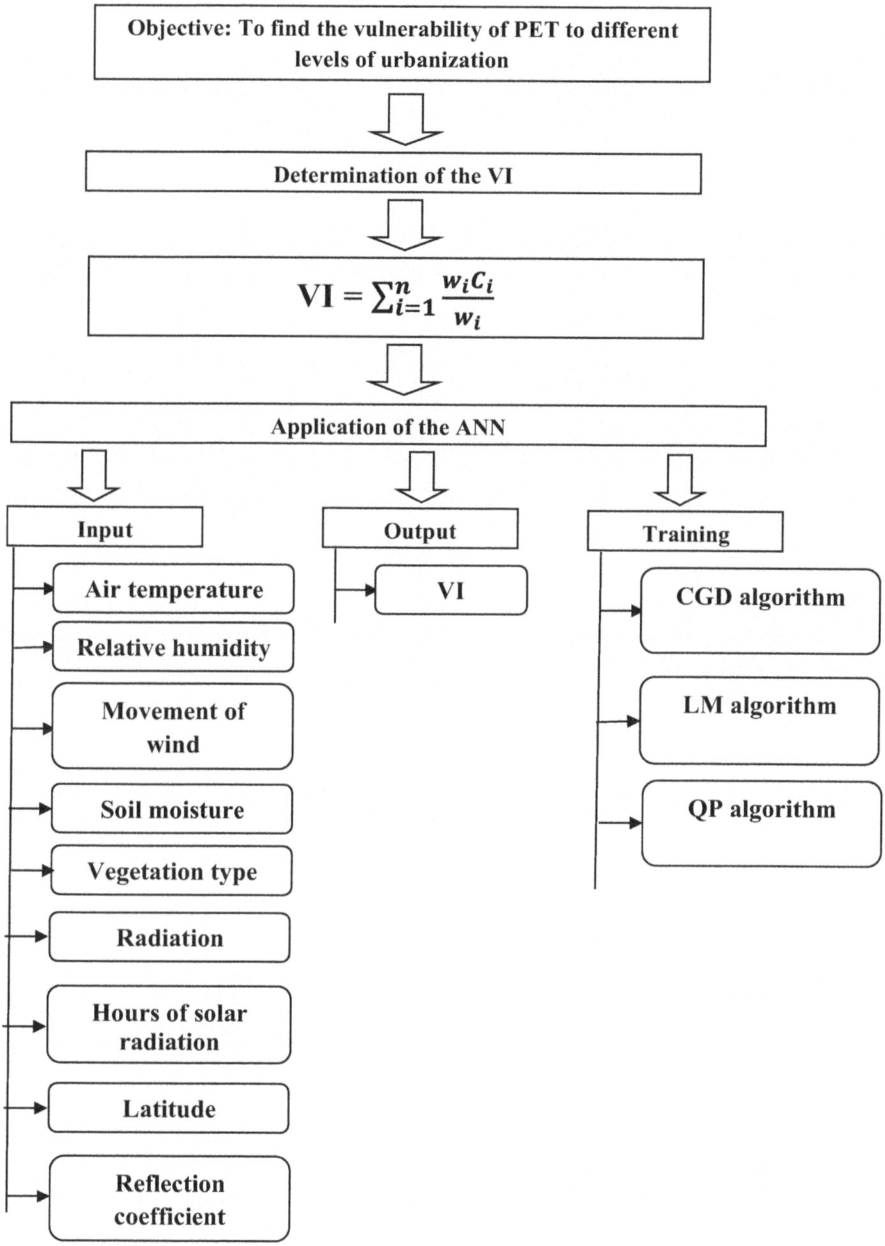

Fig. 8 Flow chart of the development of the neurogenetic model for estimating the VI

Table 2 Table shows the rank and score of the different criteria

Criteria	Score	Rank
Literatures considered the variable in related studies	0.5	1
Literatures considered as a model variable	0.375	2
Expert survey	0.125	3

Table 3 Table shows the rank and priority value of the alternatives according to the different criteria

	L	Rank	Lm variable	Rank	E	Rank
Air temperature	0.186	1	0.161	1	0.135	2
Relative humidity	0.144	3	0.140	2	0.115	4
Wind velocity	0.146	2	0.116	5	0.154	1
Soil moisture	0.095	7	0.090	8	0.096	8
Type of plant	0.095	7	0.125	3	0.135	2
Radiation	0.104	5	0.108	7	0.115	4
Hours of solar radiation	0.104	5	0.116	5	0.115	4
Latitude	0.106	4	0.125	3	0.115	4
Reflection coefficient	0.019	9	0.018	9	0.019	9

Table 4 The weightage of importance by AHP for each of the criteria considered

Criteria	Weightage of importance
E	0.1428571
L_m	0.4081633
L	0.4489796

Table 5 The weightage of importance by AHP for each of the alternatives considered

	Expert survey	Literature considered as a model variable	Literature considered the variable in related Studies
Air temperature	0.205	0.259	0.259
Relative humidity	0.095	0.210	0.208
Wind velocity	0.273	0.167	0.162
Soil moisture	0.037	0.125	0.128
Vegetation type	0.158	0.089	0.093
Radiation	0.090	0.059	0.064
Hours of solar radiation	0.070	0.036	0.040
Latitude	0.050	0.030	0.029
Reflection coefficient	0.023	0.025	0.018

Table 6 The overall weightage of importance by AHP for each of the alternatives considered

Parameter	Weightage	Rank of importance
Air temperature	0.076	1
Wind velocity	0.076	1
Relative humidity	0.048	3
Vegetation type	0.043	4
Radiation	0.026	5
Soil moisture	0.025	6
Hours of solar radiation	0.019	7
Latitude	0.014	8
Reflection coefficient	0.008	9

impact on both PET and it is affected by expansion of urban area or increase in density of high-rises. In this aspect FDM has clearly selected that Air Temperature (weightage of importance: 0.217) is more important than Wind Velocity (weightage of importance: 0.210) and seems to be a more pronounced decision-making method than AHP. The relative difference (Eq. 14) between the importance assigned to the alternatives by FDM and AHP is respectively 0.7 and 0.9% which indicates that decisions from AHP is more prominent than FDM if all the weightage of importance assigned to each of the parameter is taken into account. The relative difference is certainly more extensive measure of prominence than the difference of weightage of the top alternative. Thus, decision from the AHP method was preferred to the same from FDM.

$$Relative\ Difference(\delta) = \sum_{i=1}^{9} \frac{a_i - a_{i-1}}{i} \tag{14}$$

6.2 *Vulnerability Index*

As AHP was found to be better method than FDM the weightage of importance as selected by the former was taken as weightage for estimation of Vulnerability Index (V) by Eq. 13.

The magnitude of the parameters for the region of interest will be required to be entered as D_i to calculate the V for the same region.

6.3 Results from the Neuro Genetic Model

Table 7 depicts the parameters of the neuro-genetic model. The neural network model used genetic algorithm to select the optimal network topology for the current problem. The model has nine input and one output as depicted. The optimal topology for the model was selected as 9–5–5–1 having a network weight of 75. The training of the model was conducted with the help of QP, LM and CGD algorithms. The targeted training error was fixed at 2.22%. In total 50,000,000 iterations were performed to train the model. The training performance of the model by the three algorithms were compared with respect to six performance metrics viz, RMSE, AE, RE, NSE, Covar and Test AE or AE in the testing phase and same in training phase.

According to the literatures, accuracy or efficiency of the model is directly proportional to NSE but inversely proportional to RMSE, AE, RE and Test AE/Train AE. Covariance has to be near to 1 which will show complete variance and 0 will indicate complete invariance.

In the present investigation the LM trained model was found to have satisfactory performance metrics than the other three training algos which is depicted in Table 13. If NSE of the model output from LM, QP and CGD trained model is compared then LM has the highest value of the efficiency criteria, 98.32% whereas for QP and CGD the same is equal to 1.32 and 0.29% respectively. The RMSE values of QP, LM and CGD is respectively 9.39, 1.07 and 9.39% which again indicates that LM trained model has less erroneous output. The covariance also indicated maximum variance in case of LM model (0.80%) but for QP and CGD it is 0.0017 and 0.0010% respectively. The AE for LM (0.39%) is minimum than CGD (72.86%) and QP(0.43%) and so is RE (0.49,144.51 and 3.08%). Test AE and Train AE is same for LM but for CGD former is higher than latter and for QP test AE is less than Train AE which indicates better learning ability of the model for QP. But overall LM has the better performance metrics compared to QP and CGD trained model. The LM trained model was utilized to estimate V for different regions of the World (Fig. 9, Table 7).

7 Validation of the Model

7.1 Sensitivity Analysis

On a spider chart, lines that are nearly horizontal generally indicate an input variable where small percentage changes do not have much effect on the output value. Lines that are more vertical indicate an input variable where small percentage changes have a greater affect on the Output value (Figs. 10 and 11, Table 8).

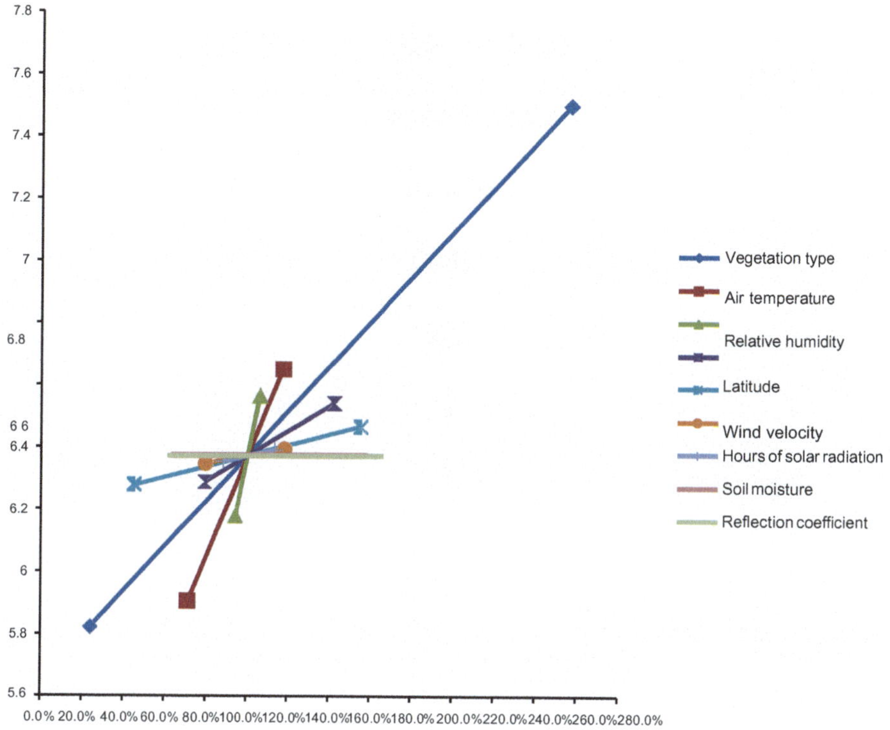

Fig. 9 SensIt spider chart output

Table 7 Table showing the performance characteristics of the neural network model

Parameters	Value for the present investigation					
Input	(T), (RH), (WV), (SM), (P_t), (R), (S_t), (Lat), (Θ)					
Output	V					
Network topology (as determined by Genetic Algorithm)	9–5–5–1					
Number of iterations	50,000,000					
Desired AE	0.022					
Training algorithms	RMSE	AE	RE	NSE	Covar	Test AE/Train AE
CGD	0.0939	0.7286	1.4451	0.0029	0.000017	0.0091/0.0090
LM	0.0107	−0.0039	−0.0049	98.32	0.008034	0.0001/0.0001
QP	0.0939	−0.0043	0.0308	0.0132	0.000010	0.0064/0.0088

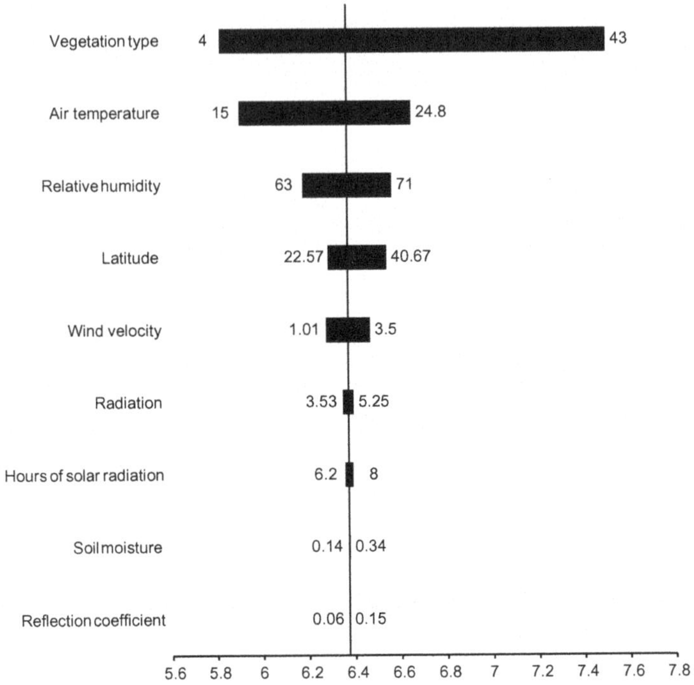

Fig. 10 Tornado chart with vertical lin

8 Compare with Other Similar Methods Show the Results from AHP, WSM and WPM

8.1 Weighted Sum Method (WSM)

See Tables 9, 10, 11, 12 and 13.

8.2 Weighted Product Method (WPM)

See Tables 14, 15 and 16.

Fig. 11 SensIt two-factor tornado sorted by downside risk

Table 8 Table shows the SensIt spider numerical

Input variable	Corresponding input value			Output value			Percent	
	low	Base	High	Low	Base	High	Swing	Swing^2 (%)
Vegetation type	4	16.7875	43	5.8227	6.3725625	7.4997	1.677	77.7
Air temperature	15	21.1375	24.8	5.9061125	6.3725625	6.6509125	0.7448	15.3
Relative humidity	63	67	71	6.1805625	6.3725625	6.5645625	0.384	4.1
Latitude	22.57	28.6325	40.67	6.2876875	6.3725625	6.5410875	0.2534	1.8
Wind velocity	1.01	2.2575	3.5	6.2777525	6.3725625	6.4669925	0.18924	1.0
Radiation	3.53	4.4475	5.25	6.3487075	6.3725625	6.3934275	0.04472	0.1
Hours of solar radiation	6.2	7.05	8	6.3564125	6.3725625	6.3906125	0.0342	0.0
Soil moisture	0.14	0.22	0.34	6.3705625	6.3725625	6.3755625	0.005	0.0
Reflection coefficient	0.06	0.0925	0.15	6.3723025	6.3725625	6.3730225	0.00072	0.0

Table 9 Table shows the rank of the alternatives according to the criteria

	L	Rank	Lm variable	Rank	E	Rank
Air temperature	0.186	1	0.161	1	0.135	2
Relative humidity	0.144	3	0.140	2	0.115	4
Wind velocity	0.146	2	0.116	5	0.154	1
Soil moisture	0.095	7	0.090	8	0.096	8
Type of plant	0.095	7	0.125	3	0.135	2
Radiation	0.104	5	0.108	7	0.115	4
Hours of solar radiation	0.104	5	0.116	5	0.115	4
Latitude	0.106	4	0.125	3	0.115	4
Reflection coefficient	0.019	9	0.018	9	0.019	9

9 Case Study Analysis

With the help of the model, the V for Kolkata, Agartala, Brisbane and New Yorkwere predicted. When all the Visof the cities are compared, it can be seen that PET is most affected by urbanization Kolkata and least affected in Brisbane.

Table 10 Table shows the priority value of the alternatives according to the criteria

	L	Priority value	Lm variable	Priority value	E	Priority value
Air temperature	0.186	10 – 1 = 9	0.161	10 – 1 = 9	0.135	10 – 2 = 8
Relative humidity	0.144	10 – 3 = 7	0.140	10 – 2 = 8	0.115	10 – 4 = 6
Wind velocity	0.146	10 – 2 = 8	0.116	10 – 5 = 5	0.154	10 – 1 = 9
Soil moisture	0.095	10 – 7 = 3	0.090	10 – 8 = 2	0.096	10 – 8 = 2
Type of plant	0.095	10 – 7 = 3	0.125	10 – 3 = 7	0.135	10 – 2 = 8
Radiation	0.104	10 – 5 = 5	0.108	10 – 7 = 3	0.115	10 – 4 = 6
Hours of solar radiation	0.104	10 – 5 = 5	0.116	10 – 5 = 5	0.115	10 – 4 = 6
Latitude	0.106	10 – 4 = 6	0.125	10 – 3 = 7	0.115	10 – 4 = 6
Reflection coefficient	0.019	10 – 9 = 1	0.018	10 – 9 = 1	0.019	10 – 9 = 1

Table 11 The weightage of importance by WSM for each of the criteria considered

Criteria	Weightage of importance
Literature considered the variable in related studies	0.5
Literature considered as a model variable	0.375
Expert Survey	0.125

Table 12 The priority value by WSM for each of the alternatives considered

	Literature considered the variable in related studies	Literature considered as a model variable	Expert survey
Air Temperature	9	9	8
Relative humidity	7	8	6
Wind velocity	8	5	9
Soil moisture	3	2	2
Vegetation type	3	7	8
Radiation	5	3	6
Hours of solar radiation	5	5	6
Latitude	6	7	6
Reflection coefficient	1	1	1

The input to the model is shown in Table 14. The data are then normalized before feeding to the model for estimation of the V. As more the magnitude of V more will be the vulnerability of PET to be affected by urbanization.

The results from the model also seconded the relationship between urbanization and PET. The places with higher population density show greater vulnerability than

Table 13 The overall weightage of importance by WSM for each of the alternatives considered

Parameter	Weightage	Rank of importance
Air temperature	8.87	1
Relative humidity	7.25	2
Wind velocity	7.00	3
Soil moisture	2.50	8
Vegetation type	5.12	5
Radiation	4.37	7
Hours of solar radiation	5.12	5
Latitude	6.37	4
Reflection coefficient	1.00	9

Table 14 The weightage of importance by WPM for each of the alternatives considered

	Literature considered the variable in related studies	Literature considered as a model variable (0.375)	Expert survey (0.125)
Air temperature	9/7 * 0.5	9/8 * 0.375	8/6 * 0.125
Relative humidity	7/8 * 0.5	8/5 * 0.375	6/9 * 0.125
Wind velocity	8/3 * 0.5	5/2 * 0.375	9/2 * 0.125
Soil moisture	3/3 * 0.5	2/7 * 0.375	2/8 * 0.125
Vegetation type	3/5 * 0.5	7/3 * 0.375	8/9 * 0.125
Radiation	5/5 * 0.5	3/5 * 0.375	6/6 * 0.125
Hours of solar radiation	5/6 * 0.5	5/7 * 0.375	6/6 * 0.125
Latitude	6/1 * 0.5	7/1 * 0.375	6/6 * 0.125
Reflection coefficient	1/9 * 0.5	1/9 * 0.375	1/8 * 0.125

Table 15 The overall weightage of importance by WPM for each of the alternatives considered

Parameter	Weightage	Rank of importance
Air temperature	1.23	4
Relative humidity	1.120833	5
Wind velocity	2.833	2
Soil moisture	0.63839	8
Vegetation type	1.28611	3
Radiation	0.850	6
Hours of solar radiation	0.80952	7
Latitude	5.75	1
Reflection coefficient	0.112847	9

Table 16 Comparison studies of different alternatives according to their weightage

Parameter	AHP	WPM	WSM
Air temperature	1	4	1
Relative humidity	3	5	2
Wind velocity	1	2	3
Soil moisture	6	8	8
Vegetation type	4	3	5
Radiation	5	6	7
Hours of solar radiation	7	7	5
Latitude	8	1	4
Reflection coefficient	9	9	9

the regions with lesser population density. Kolkata and Agartala have the maximum population with respect to New York and Brisbane and thus the maximum value of V is observed in this two places and minimum value was predicted for the latter places. These results also indicated that the model is developed satisfactorily and represents the objective of the investigation.

The development of the Vulnerability Index will now enable the scientists and professionals of the related field to estimate the impact of urbanization on PET and can also predict the onset of drought and crop productivity. The cultivators can also conclude the type of crop they have to harvest for maximum yield. The water yield of the watershed will also be possible to expect, and accordingly developmental projects can be sanctioned or planned by the city planners. The model can become a reliable tool to ensure food as well as water security in the coming days of climate change.

Although the model can propose the impact on PET but predicting only one variable may not be sufficient for taking developmental or mitigation activities. That is why it is essential to include the provision to introduce other climatic parameters also so that model can become extensive and applied for a broader prospect and problem solving (Fig. 12, Table 17).

10 Conclusion

In the present investigation, attempted to answer three questions. Is there a relationship between PET and urbanization? If so, which parameter, able to change the pattern of PET, is the most sensitive to urbanization each region? And lastly, how can the impact of urbanization on PET be measured without the application of any of the subjective models currently available and used worldwide to solve similar problems?

A vast amount of literature is available to answer the first question, and an analysis of this literature confirmed that there is indeed a relationship between PET and urbanization, like other climatic parameters such as temperature, humidity, cloud cover etc. The solution to the second question was found by applying comparative MCDM

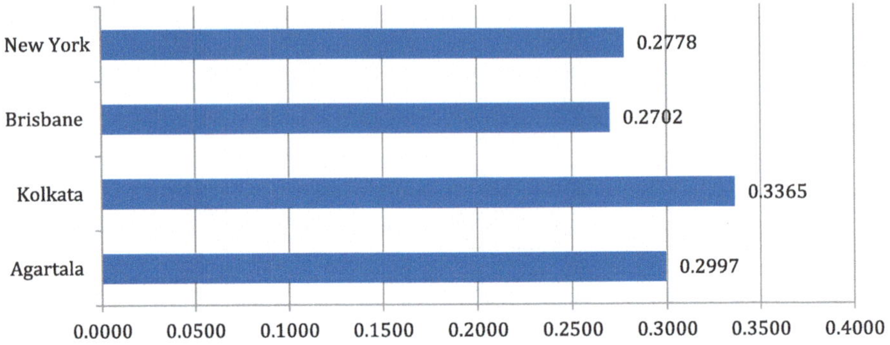

Fig. 12 VI values for the four cities with different population densities, as predicted by the neuro-genetic model

Table 17 Table showing the non-normalized input values fed to the model for prediction of V for the following four cities

	T (in °C)	RH (%)	WV (m/s)	SM (parts per hundred)	P_t (%)	R (Joules)	S_t (hr/day)	Lat	Θ
Agartala	24.50	69.00	1.01	0.25	12.15	4.56	6.20	23.83	0.06
Kolkata	24.80	71.00	1.02	0.15	8.00	4.45	6.50	22.57	0.07
Brisbane	20.25	65.00	3.50	0.34	43.00	5.25	7.50	27.46	0.09
New York	15.00	63.00	5.50	0.14	4.00	3.53	8.00	40.67	0.15

techniques; namely, AHP and FLDM. From the results, we found air temperature to be the most important parameter and the reflection coefficient the least important parameter in analyzing the influence of urbanization on PET. Lastly, the cognitive abilities of neurogenetic algorithms were utilized to find an objective method to predict the influence of urbanization on PET. The model was utilized to predict an index referred to as the vulnerability index (VI), which is directly proportional to the influence of urbanization on the PET parameter. The model's performance was analyzed by determining the VI of four cities having different levels of population density. The results showed PET to be most affected in the city with the highest population density, and vice versa.

Although the purpose of the model can be satisfied with the help of AHP or MCDM, by introducing different cities as alternatives the framework is not very flexible. Every time, a new alternative will be required for the comparison, and so the entire process has to be repeated. However, in the case of neurogenetic models, there is no requirement to repeat the development process. Also, the cognitive framework can work separately without the assistance of any other software or manual interventions. The study can be repeated by estimating the VI of other cities, and perhaps maps can be prepared to help the scientific community in devising different mitigation measures to compensate for the impacts.

References

Alphonce, C. B. (1997). Application of the analytic hierarchy process in agriculture in developing countries. *Agricultural Systems, 53*(1), 97–112, ISSN 0308-521X.

Bellman, R. E., & Zadeh, L. A. (1970). Decision making in a fuzzy environment. *Management Science, 17*, 141–164.

Buyantuyev, A., & Wu, J. (2012). Urbanization diversifies land surface phrenology in arid environments: Interactions among vegetation, climatic variation, and land use pattern in the Phoenix metropolitan region, USA. *Landscape and Urban Planning, 105*(1–2), 149–159.

Carlson, T. N., & Arthur, S. T. (2000). The impact of land use—Land cover changes due to urbanization on surface microclimate and hydrology: A satellite perspective. *Global and Planetary Change, 25*(1–2), 49–65.

Corrente, S., Greco, S., & Słowiński, R. (2013). Multiple criteria hierarchy process with ELECTRE and PROMETHEE. *Omega, 41*(5), 820–846.

Cui, Linli, & Shi, Jun. (2012). Urbanization and its environmental effects in Shanghai. *China, Urban Climate, 2*, 1–15.

Darji, V., & Rao, R. (2013). Application of AHP/EVAMIX method for decision making in the industrial environment. *American Journal of Operations Research, 3*(6), 542–569. https://doi.org/10.4236/ajor.2013.36053.

Ghaffari, A., Abdollahi, H., Khoshayand, M. R., Bozchalooi, I. S., Danger & Rafiee-Tehrani, M. (2006). Performance comparison of neural network training algorithms in modelling of bimodal drug delivery. *International Journal of Pharmaceutics, 327*(1–2), 126–138.

He, Y., Chen, H., Zhou, L., Liu, J., & Tao, Z. (2014). Intuitionistic fuzzy geometric interaction averaging operators and their application to multi-criteria decision making. *Information Sciences, 259*, 142–159. ISSN 0020-0255.

Kabira, G., Sadiqa, R., & Tesfamariama, S. (2013). A review of multi-criteria decision-making methods for infrastructure management. *Structure and Infrastructure Engineering: Maintenance, Management, Life-Cycle Design and Performance*. https://doi.org/10.1080/15732479.2013.795978.

Nikula, S., Vapaavuori, E., & Manninen, S. (2010) Urbanization-related changes in European aspen (Populustremula L.): Leaf traits and litter decomposition. *Environmental Pollution, 158*(6), 2132–2142.

Park, S. K., & Xu, L. (2013). *Data Assimilation for Atmospheric, Oceanic and Hydrologic Applications (Vol II)* (Vol. II, pp. 350–351). Springer.

Pathirana, A., Denekew, H. B., Veerbeek, W., Zevenbergen, C., & Banda, A. T. (2014). Impact of urban growth-driven landuse change on microclimate and extreme precipitation—A sensitivity study. *Atmospheric Research, 138*, 59–72.

Qiao, Z., Tian, G., & Xiao, L. (2013). Diurnal and seasonal impacts of urbanisation on the urban thermal environment: A case study of Beijing using MODIS data. *ISPRS Journal of Photogrammetry and Remote Sensing, 85*, 93–101.

Shepherd, J. M. (2013). Impacts of urbanization on precipitation and storms: Physical insights and vulnerabilities, reference module in earth systems and environmental sciences. *Climate Vulnerability, 5*, 109–125.

Srinivisan, V., Seto, K. C., Emerson, R., & Gorelick, S. M. (2013). the impact of urbanization on water vulnerability: A coupled human–environment system approach for Chennai. *India, Global Environmental Change, 23*(1), 229–239.

Tang, C. S., Shi, B., Gao, L., Daniels, J. L., Jiang, H. T., & Liu, C. (2011). Urbanization effect on soil temperature in Nanjing. *China, Energy and Buildings, 43*(11), 3090–3098.

Triantaphyllou, E., & Mann, S. H. (1995). Using the analytic hierarchy process for decision making in engineering applications: Some challenges. *International Journal of Industrial Engineering: Applications and Practice, 2*(1), 35–44.

Verbeiren, B., Voorde, T. V., Canters, D. F, Binard,. M., Cornet, Y., & Batelaan, O. (2013). Assessing urbanisation effects on rainfall-runoff using a remote sensing supported modelling strategy. *International Journal of Applied Earth Observation and Geoinformation, 21*, 92–102.

Yılmaz, S., Toy, S., Irmak, M. A., & Yilmaz, H. (2007), Determination of climatic differences in three different land uses in the city of Erzurum, Turkey. *Building and Environment, 42*(4), 1604–1612.

Zhan, J., Huang, J., Zhao, T., Geng, X., & Xiong, Y. (2013). Modeling the impacts of urbanization on regional climate change: A case study in the Beijing-Tianjin-Tangshan metropolitan area. *Advances in Meteorology, 2013*. Article ID 849479. http://dx.doi.org/10.1155/2013/849479.

Trend Analyses in Groundwater Levels of the Bikaner District, Rajasthan

Sanju R. Phulpagar, Ganesh D. Kale, Sagar Patel, and Sudhansu Mohanta

Abstract Groundwater (GW) is the essential part of the worldwide freshwater resources. Over-exploitation and insufficiency of GW resources are unfortunately widespread in several parts of the India. As per Central Groundwater Board (2016), groundwater level (GWL) fall of greater than 4 m was found in parts/patches of the Bikaner district, Rajasthan by comparison of depth to water level corresponding to year 2015 (2016 for January) with decennium average over the period 2005–2014 (for January, 2006–2015 is the decadal period) corresponding to January, August, November and May. Thus, in the current study, trend analyses of GWLs of six blocks of the Bikaner district, Rajasthan is carried out for monsoon, post-monsoon kharif, post-monsoon rabi and pre-monsoon temporal scales for the time period (1994–2018). For trend analyses, Mann Kendall (MK)/MK test with correction factor-2, Sen's slope test, innovative trend analysis plot and smoothing curve are used. The results showed presence of statistically significant decreasing trends in GWL time series of the Khajuwal and Dungargarh blocks corresponding to monsoon and pre-monsoon temporal scales, respectively. All statistically significant trends are decreasing, thus it shows improvement in GWLs in given blocks.

Keywords Six blocks · Bikaner · Groundwater · Trend analyses

S. R. Phulpagar (✉) · G. D. Kale · S. Patel · S. Mohanta
Department of Civil Engineering, Sardar Vallabhbhai National Institute of Technology, Surat 395007, Gujarat, India
e-mail: sanjup1213@gmail.com

G. D. Kale
e-mail: gdk@ced.svnit.ac.in

S. Patel
e-mail: sagarpatel2613@gmail.com

S. Mohanta
e-mail: sudhansusekhar94@gmail.com

M. Majumder and G. D. Kale (eds.), *Water and Energy Management in India*,
https://doi.org/10.1007/978-3-030-66683-5_4

Abbreviations

AC	Autocorrelation
ARIMA	Autoregressive integrated moving average
GW	Groundwater
GWL	Groundwater level
ITA	Innovative trend analysis
MK	Mann-Kendall
MK-CF_2	Mann-Kendall test with correction factor 2
POMKH	Post-monsoon Kharif
POMRB	Post-monsoon Rabi
PREMON	Pre-monsoon
SS	Sen's slope
SGI	Standardized groundwater level index
TS	Time series

1 Introduction

Groundwater (GW) is the crucial part of the worldwide freshwater resources. It is the primary source of drinking, industrial and irrigation water requirements in many areas of the world (Panda et al. 2012). Due to the rapid population growth and the accelerated speed of industrialization, the demand for fresh water has risen exponentially in the last few decades. Rapid urbanization, particularly in developing countries such as India, has influenced GW availability and quality due to its over-exploitation and inappropriate disposal of waste, particularly in urban areas (Ramakrishnaiah et al. 2009). In India, more than 90% of rural population and almost 30% of urban population rely on GW to meet their domestic and drinking needs. Over-exploitation and scarcity of GW resources are unfortunately common in many parts of the India (Jha et al. 2010). Thus, study of groundwater levels (GWLs) in India is essential.

The stage of GW development in several blocks is greater than hundred percent in states like Rajasthan, Delhi, Haryana, Punjab, Gujarat, Uttar Pradesh, Tamil Nadu, Andhra Pradesh and Karnataka. In these states, this resulted in rapid decrease in water level (Patle et al. 2015). Maximum GWL fall was found in and around portions of the Rajasthan state by comparison of depth to water level for the year 2016 (2017 for January) with decadal average over the period 2006–2015 (for January decadal period is 2007–2016) corresponding to January, August, November and pre-monsoon (Central Ground Water Board, Year Not Mentioned). Thus, study of GWLs in the Rajasthan state is essential.

Analysis of depth to water level during May 2015 (i.e. pre-monsoon) showed that, water levels of greater than 40 m below ground level were found in 19.20% of stations, which fall mostly in the districts of Jaisalmer, Barmer, Jalore, Jodhpur,

Bikaner, Alwar, Bharatpur, Churu, Hanumangarh, Ganganer, Jaipur, Nagaur, Jhunjhunu presenting western and north-central part of the Rajasthan state (Central Ground Water Board 2016). According to Central Ground Water Board (2016) of the Rajasthan state, GWL fall of more than 4 m was found in patches/parts of the Bikaner district of the Rajasthan state by comparison of depth to water level for year 2015 (2016 for January) with decadal average over the period 2005–2014 (for January decadal period is 2006–2015) corresponding to January, August, November and May. Thus, study of GWLs in the Bikaner district of the Rajasthan state is very important.

Trend detection is a necessary task in hydrological series assessment, as it is not only basis for understanding the long-run alterations of hydrological processes, but also for showing hydrological processes periodicities and other characteristics (Sang et al. 2014). Thus, trend analysis in GWLs of the Bikaner district in the Rajasthan state, India is necessary. Consequently, it is considered as study area in the present study.

The research publications are reviewed to obtain information on the previous research work carried out in the field of trend analysis in GWL time series (TS) outside the India and in India. Trend detection studies performed outside the India are reviewed in the literature review (Abdullahi et al. 2015; Bui et al. 2012; Brocque et al. 2018; Gibrilla et al. 2017; Goode et al. 2013; Hodgkins et al. 2017; Ilia et al. 2018; Kawamura et al. 2011; Lee et al. 2014; Ribeiro et al. 2015; Shamsudduha et al. 2009; Tabari et al. 2012; Vousoughi et al. 2013; Yilmaz et al. 2020; Zafor et al. 2017; Zhou et al. 2016).

Thakur and Thomas (2011) have analyzed GWL trends for the few blocks of the Sagar district, Madhya Pradesh, India. The analysis was performed for the periods of 15–17 years. To identify the trend in GWLs, they have applied the parametric linear regression test and the non-parametric Kendal rank correlation test. Results of the study shown that, at three blocks of the Sagar district (i.e. Bina, Khurai and Sagar) decreasing trends were found.

Panda et al. (2012) have analyzed trends in GWL, maximum temperature and rainfall, to study the response of GW systems to anthropogenic and climatic stresses in the Gujarat state for the period of 1995 to 2005. They have applied non-parametric Kendall slope test for analyzing the trends and variability. The results of the study showed that, GWLs in the majority parts of the study area were declined, while the maximum temperature and rainfall was increased. Noticeable changes were identified in the intensity and frequency of the rainfall and also in percentiles of the temperature for the data sets corresponding to station scale. Therefore, they have concluded that, combined effect of the human induced forcing mechanism of over-withdrawal and climatic extremes may be one amongst the main causes of the decline in observed water table in the study area.

Patle et al. (2015) have analyzed long-term GWLs of pre and post-monsoon seasons to identify trends and predict future GWLs in the Karnal district of the Haryana state, India. They used Sen's Slope (SS) estimator and Mann Kendall (MK) test to examine the trend over the period 1974 to 2010. The non-seasonal Autoregressive Integrated Moving Average (ARIMA) model was used for modelling and forecasting of GWL TS. They found that, GWLs declined significantly during pre and post-monsoon seasons, with average rates of 0.228 m/year and 0.267 m/year, respectively. Results of the study showed that, in 2050, pre and post-monsoon GWLs will decline by 12.97 m and 12 m, respectively over the GWL observed in 2010.

Kumar et al. (2018) have analysed the trends in GWLs of four districts in Lucknow division (i.e. Lucknow, Sitapur, Hardoi, Laxmipur) of the Uttar Pradesh, India, corresponding to the period of 1998 to 2012. They have applied non- parametric tests such as SS estimator test and Modified MK test for analysis of trend in GWL TS. For preparation of spatial maps during post-monsoon and pre-monsoon seasons, they utilized inverse distance weighted interpolation technique and ArcMap 10.4 software. Positive trends were observed at 6 locations and negative trends were observed at 7 locations corresponding to pre-monsoon season. Nonetheless, 4 locations showed negative trend and 9 locations exhibited positive trends in post-monsoon season.

Pathak and Dodamani (2019) have studied GWL trends and estimated regional GW drought characteristics for the drought-prone Ghatprabha river basin in India. They used the MK test to evaluate the seasonal and annual GWL trends. To categorize the wells, in case of long-term monthly GWLs, cluster analysis was performed. For evaluation of GW drought, they utilized standardized GWL index (SGI). More than 61% of wells were found to have significantly decreasing trends with an average decrease of 0.21 m. The SGI indicated a significant number of GW droughts in the study area, nonetheless, the wells of clusters 1 and 2 were found to be suffered by frequent droughts.

The following research gaps are found from the reviewed literature: (a) in some studies linear regression was used to assess the trend magnitude. But the linear regression test is affected by outliers (Sen 1968), (b) majority of studies have not assessed data assumptions, which is required for the selection of appropriate statistical test, (c) previous studies have not used innovative trend analysis (ITA) plot and smoothing curve concurrently to support trend analysis results and (d) trend analysis in GWLs of six blocks in the Bikaner district, Rajasthan was not carried out for the period of more than one decade.

Therefore, in the current study, trend analyses in GWLs of six blocks in the Bikaner district, Rajasthan is carried out by addressing the above research gaps by: (a) using SS test to determine the trend magnitude, as this test is not influenced by outliers as per Sen (1968), (b) assessing the assumption of data independence by using autocorrelation (AC) plot to select suitable statistical test (MK test and MK test with correction factor 2 (MK-CF_2) are utilized for the assessment of trend significance in serially uncorrelated and serially correlated data, respectively), (c) using ITA plot and smoothing curve simultaneously to support results of trend analyses and (d) performing trend analyses in GWLs of six blocks in the Bikaner district, Rajasthan for the period of more than one decade.

2 Study Area

Bikaner district is situated in the north-western part of the Rajasthan state and covers a geographical area of 30247.90 km^2. and have east longitudes 71° 52′ to 74° 15′ and north latitudes 27° 11′ to 29° 03′. As the Bikaner district lies in the desert region, the desert is characterized by cold in winter and extremes of heat in summer. In the Bikaner district temperature varies from 1 °C in winter to 48 °C in summer. Temperatures rises gradually during both day and night time and reach their highest values in June, May and April months. Average annual rainfall (1991 to 2010) of the Bikaner district is 277.55 mm, while normal rainfall (1901 to 1971) of the district is below average rainfall and it is 257.8 mm. The soils of the Bikaner are mainly weak-structured, light textured, sand to sandy loam with the clay content. Soils of the district are usually of desertic type with very low water holding capacity and poor fertility status. In the Bikaner district, soils are usually have good porosity (40%) and good to very good permeability. In the Bikaner district one of the major GW problem is decline in water level (Central Ground Water Board 2013). Location of six blocks of the Bikaner district in the Rajasthan state, India is shown in Fig. 1.

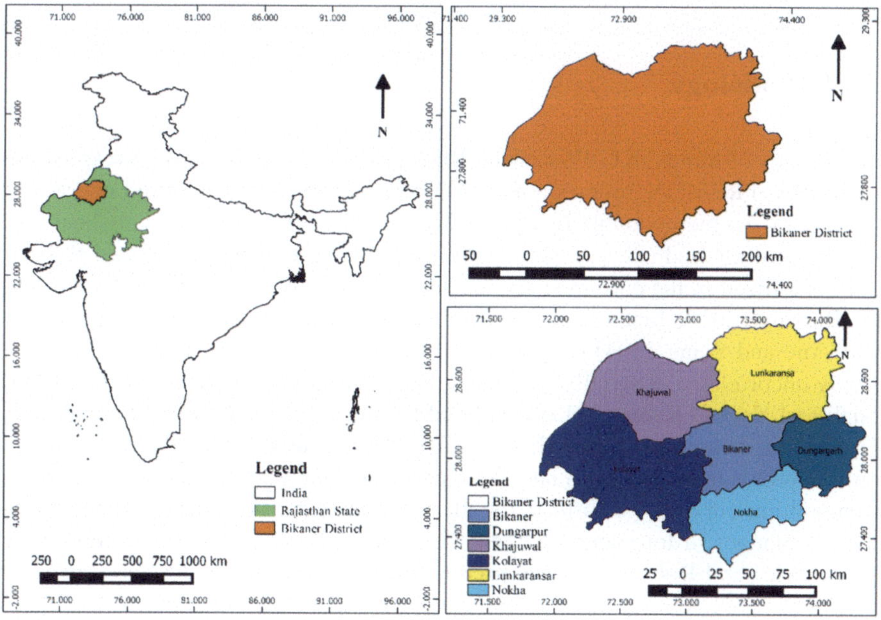

Fig. 1 Location of six blocks of the Bikaner district in Rajasthan state, India

3 Data Collection

In the current study, trend analyses in GWLs of six blocks in the Bikaner district, Rajasthan is carried out for the period of 1994 to 2018. The GWL data is available for four seasons viz. post-monsoon Rabi (POMRB) (January to March), pre-monsoon (PREMON) (April to June), Monsoon (July to September) and post-monsoon Kharif (POMKH) (October to December) (http://59.179.19.250/GWL/GWL.html?UType=R2VuZXJhbA==?UName= last accessed on 10th February 2020"). The GWL data of six blocks in the Bikaner district (1994 to 2018) is downloaded from website 'http://59.179.19.250/GWL/GWL.html?UType=R2VuZXJhbA==?UName=' (last accessed on 10th February 2020). For the preparation of GWL data, average of GWLs of all sites in a block for a given season of the given year is taken as the value for the corresponding season of the given year and likewise GWL data of all seasons for each year of analysis period is prepared. If, there is single site in a block for a given season of the given year, then, it is taken as the value for the corresponding season of the given year. In the data, there were some missing values, which are filled by the average of all available seasonal data of the corresponding season.

4 Methodology

For analyses of trends in GWLs of the Bikaner district (six blocks) corresponding to POMRB, PREMON, monsoon and POMKH seasons over the period of 1994 to 2018, following methodology is applied. For the selection of suitable statistical test, the graphical method (i.e. AC plot) is utilized. AC plot (Kundzewicz and Robson 2000) is utilized in the current study for the evaluation of dependency of data. MK test (Bui et al. 2012; Tabari et al. 2012) or MK-CF_2 test (Sonali and Nagesh Kumar 2013; Yue and Wang 2004) is applied for the assessment of trend significance for serially uncorrelated or serially correlated data, respectively. The SS test (Sen 1968; Sonali and Nagesh Kumar 2013) is applied for the assessment of trend magnitude. ITA (Sen 2012; Sonali and Nagesh Kumar 2013) plot is used to identify monotonic or non-monotonic trend in the TS and to support trend analyses results. When plotting of smoothing curve is performed for the seasonal data, objective is generally to identify changes in long-term (Kundzewicz and Robson 2000). Smoothing curve (Kundzewicz and Robson 2000) is used to support the trend analyses results. The flowchart of methodology adopted in the present study is shown in Fig. 2. Details of statistical tests and graphical method used are given in Sects. 4.1–4.4.

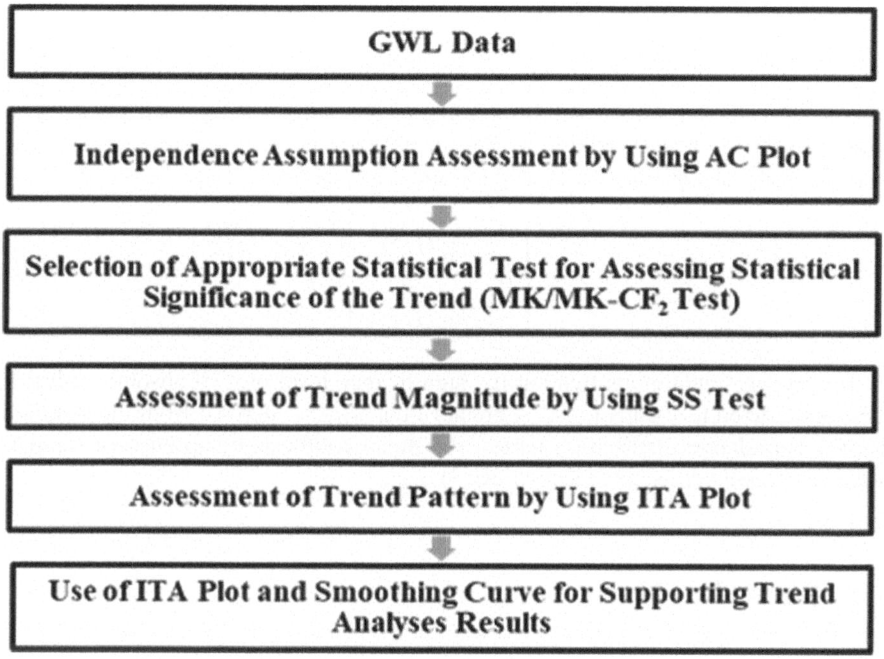

Fig. 2 Flowchart of trend analyses methodology

4.1 Mann Kendall Test

Majority of the studies on trend detection have analyzed trend by applying usually applied technique like MK test (non-parametric test having assumption of independent observations). MK test is rank-based test, which is a non-parametric test. The test statistics ‘S’ is determined by using Eq. 1 given below.

$$S = \sum_{i=2}^{n} \sum_{j=1}^{i-1} \operatorname{sign}(y_i - y_j) \tag{1}$$

where, data series length is ‘n’, y_i and y_j are the consecutive data in the series, and

$$\operatorname{sign}(y_i - y_j) = \begin{cases} -1 \; for \; (y_i - y_j) < 0 \\ 0 \; for \; (y_i - y_j) = 0 \\ 1 \; for \; (y_i - y_j) > 0 \end{cases} \tag{2}$$

$$E(S) = 0 \tag{3}$$

$$\mathrm{Var(S)} = \frac{\mathrm{n(n-1)(2n+5)} - \sum_{\mathrm{p-1}}^{\mathrm{q}} \mathrm{t_p}\left(\mathrm{t_p}-1\right)\left(2\mathrm{t_p}+5\right)}{18} \tag{4}$$

where, 't_p' is the tally of ties corresponding to pth value and 'q' is the tally of tied values. In the variance formula (Eq. 4), second term is for tied censored data. The standardized test statistic i.e. 'Z' is determined by Eq. 5.

$$Z = \begin{cases} \frac{S-1}{\sqrt{Var(S)}} \; if \; S > 0 \\ 0 \qquad if \; S = 0 \\ \frac{S+1}{\sqrt{Var(S)}} \; if \; S < 0 \end{cases} \tag{5}$$

To test for monotonous trend at the significance level of α, if absolute value of the standardized test statistic 'Z' is greater than '$Z_{1-\alpha/2}$' derived from the standard normal cumulative distribution tables, the null hypothesis of absence of trend is rejected (Sonali and Nagesh Kumar 2013).

4.2 Sen's Slope Test

The magnitude of the slope b_{sen} can be derived by using Eq. 6 given below.

$$b_{sen} = Median\left(\frac{X_j - X_i}{j - i}\right) for \; all \; j > i \tag{6}$$

where, X_j = data at time point j and X_i = data at time point i. In the series, if 'n' is the total number of data points, then, there will be n(n−1)/2 slope values and the b_{sen} (i.e. test statistic) is the median of all slope values. If the sign of test statistic is positive (negative) then it indicate that the trend is increasing (decreasing) trend (Sonali and Nagesh Kumar 2013).

4.3 Mann Kendall Test with Correction Factor-2

Existence of negative (positive) serial correlation results in decrease (increase) in the variance of the MK test statistic. Due to this drawback, approach of variance correction was suggested. MK test statistic's modified variance is calculated by using Eq. 7 given below.

$$\mathrm{Var(S)}* = \mathrm{CF}^*\mathrm{Var(S)} \tag{7}$$

where, CF = correction factor. Yue and Wang (2004) given correction factor 2 (CF_2), which is calculated by using Eq. 8.

$$CF_2 = 1 + 2\sum_{k=1}^{n-1}\left(1 - \frac{k}{n}\right)r_k \tag{8}$$

where, r_k is the serial correlation coefficient of the data corresponding to lag-k and 'n' indicate length of the complete series. MK test with CF_2 is known as MK-CF_2 test (Sonali and Nagesh Kumar 2013).

4.4 *Innovative Trend Analysis Plot*

This technique is based on the notion that, if two TS are same, their plot against each other shows a straight line of 45° (i.e. 1:1 line) on the Cartesian coordinates regardless of not holding good corresponding to all the suppositions about distribution, serial correlation and sample length. In the ITA plot, if the data points are on the 45° line (i.e. 1:1 line), it indicate that, trend is not present in the series. If the scatter points are above or below the 45° line (1:1 line), it indicates monotonically increasing trend or monotonically decreasing trend, respectively. If the scatter points lie on both the sides of 45° line, it indicates non-monotonic decreasing or increasing trend hidden at the diverse time scales in the same TS (Sonali and Nagesh Kumar 2013). A monotonic downward (upward) trend means that the variable consistently decreases (increases) throughout the time, however the trend might be or might not be linear (https://vsp.pnnl.gov/help/vsample/Design_Trend_Mann_Kendall.htm).

5 Results and Discussions

In the present study, analyses of trends in GWLs of the Bikaner district (six blocks), Rajasthan is carried out for POMRB, PREMON, monsoon and POMKH seasons corresponding to the period of 1994 to 2018. The results of aforesaid trend analyses are given in Sects. 5.1–5.4. AC plot, ITA plot and smoothing curve are shown only for the statistically significant trend (if present) in a given seasonal TS.

5.1 *Trend Analyses in Monsoon Groundwater Level Time Series (1994–2018) of Six Blocks in the Bikaner District of the Rajasthan*

Statistically significant trend is detected in monsoon GWL TS of one block i.e. Khajuwal block, out of six blocks, by application of MK test as given in Table 1. The magnitude of significant trend found at the Khajuwal block is −0.267 m/year, while

Table 1 Trend analyses results for monsoon GWL TS (1994–2018) of six blocks in the Bikaner District of the Rajasthan

Block name	[a]SS value (m/year)	Pattern of trend	Statistical test	Significant trend
Khajuwal	−0.267	Non-monotonic	MK	Yes
Nokha	0.248	Non-monotonic	MK-CF_2	No
Bikaner	−0.367	Monotonic	MK-CF_2	No
Kolayat	−0.424	Monotonic	MK-CF_2	No
Dungargarh	0.001	Non-monotonic	MK-CF_2	No
Lunkaransa	−0.610	Monotonic	MK-CF_2	No

[a]Negative trend denotes improvement in GWL and positive trend denotes decline in GWL

the pattern of trend found at the Khajuwal block is non-monotonous. Also, statistically significant trends are not detected in monsoon GWL TS of five blocks, namely as Nokha, Bikaner, Kolayat, Dungargarh and Lunkaransa. The negative magnitude of insignificant trends in monsoon GWLs of three blocks, namely as Bikaner, Kolayat and Lunkaransa, indicates an improvement in monsoon GWLs at the given blocks. While magnitude of insignificant trends in monsoon GWLs at the Nokha and Dungargarh blocks is positive, which shows decline in monsoon GWLs at corresponding blocks.

AC plot of monsoon GWL TS at the Khajuwal block for the time period 1994–2018 is shown in the Fig. 3a. The AC plot is utilized for evaluation of independence of data, which is needed for the selection of suitable trend detection test. AC plot displayed in Fig. 3a shows that, the corresponding data is independent and thus, MK test is employed for the given TS. For analysing pattern of the trend in monsoon GWL TS (1994–2018) at the Khajuwal block, ITA plot is used and it is shown in Fig. 3b. Smoothing curve (with window 10) of monsoon GWL TS (1994–2018) at the Khajuwal block is shown in Fig. 3c. The monsoon GWL TS (1994–2018) at the Khajuwal block has shown significant decreasing trend, which is supported by decreasing pattern of data observed in corresponding ITA plot (Fig. 3b) and smoothing curve (Fig. 3c), respectively.

5.2 Trend Analyses in Post-monsoon Kharif Groundwater Level Time Series (1994–2018) of Six Blocks in the Bikaner District of the Rajasthan

Statistically significant trends are not detected in POMKH GWL TS (1994–2018) at all blocks, namely as Khajuwal, Nokha, Bikaner, Kolayat, Dungargarh and Lunkaransa as given in Table 2. Magnitude of trend in POMKH GWL TS at all blocks is negative, which shows an improvement in POMKH GWLs at corresponding blocks.

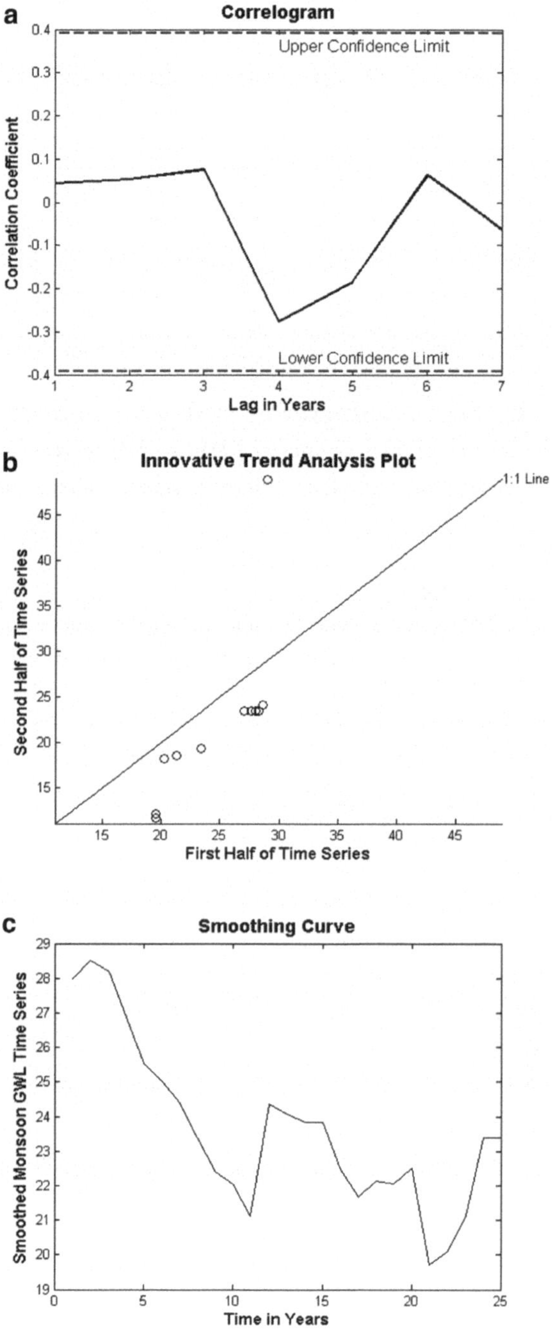

Fig. 3 **a** AC plot of monsoon GWL TS (1994–2018) at the Khajuwal block. **b** ITA plot of monsoon GWL TS (1994–2018) at the Khajuwal block. **c** Smoothing curve of monsoon GWL TS (1994–2018) at the Khajuwal block

Table 2 Trend analyses results for POMKH GWL TS (1994–2018) of six blocks in the Bikaner District of the Rajasthan

Block name	[a]SS value (m/year)	Pattern of trend	Statistical test	Significant trend
Khajuwal	−0.422	Monotonic	MK-CF_2	No
Nokha	−0.093	Non-monotonic	MK-CF_2	No
Bikaner	−0.305	Non-monotonic	MK-CF_2	No
Kolayat	−0.497	Monotonic	MK-CF_2	No
Dungargarh	−0.196	Non-monotonic	MK	No
Lunkaransa	−0.524	Monotonic	MK-CF_2	No

[a]Negative trend denotes improvement in GWL

5.3 *Trend Analyses in Post-monsoon Rabi Groundwater Level Time Series (1994–2018) of Six Blocks in the Bikaner District of the Rajasthan*

Statistically significant trends are not detected in the POMRB GWL TS (1994–2018) at all blocks, namely as Khajuwal, Nokha, Bikaner, Kolayat, Dungargarh and Lunkaransa as shown in Table 3. The negative magnitude of trends in POMRB GWL TS at five blocks, namely as Khajuwal, Bikaner, Kolayat, Dungargarh and Lunkaransa, indicates an improvement in POMRB GWLs at corresponding blocks. Magnitude of trend in POMRB GWL TS at the Nokha block is positive, which shows decline in POMRB GWL at the given block.

Table 3 Trend analyses results for POMRB GWL TS (1994–2018) of six blocks in the Bikaner District of the Rajasthan

Block name	[a]SS value (m/year)	Pattern of trend	Statistical test	Significant trend
Khajuwal	−0.395	Monotonic	MK-CF_2	No
Nokha	0.333	monotonic	MK-CF_2	No
Bikaner	−0.367	Non-monotonic	MK-CF_2	No
Kolayat	−0.447	Non-monotonic	MK-CF_2	No
Dungargarh	−0.122	Non-monotonic	MK	No
Lunkaransa	−0.641	Monotonic	MK-CF_2	No

[a]Negative trend denotes improvement in GWL and positive trend denotes decline in GWL

5.4 *Trend Analyses in Pre-monsoon Groundwater Level Time Series (1994–2018) of Six Blocks in the Bikaner District of the Rajasthan*

Statistically significant trend is detected in pre-monsoon GWL TS (1994–2018) of one block i.e. the Dungargarh block, out of six blocks, by application of MK-CF_2 test, as given in Table 4. The magnitude of significant trend found at the Dungargarh block is −0.185 m/year, while pattern of significant trend found at the Dungargarh block is monotonous. Also, statistically significant trends are not detected in pre-monsoon GWL TS at five blocks, namely as Khajuwal, Nokha, Bikaner, Kolayat and Lunkaransa. The negative magnitude of insignificant trends in pre-monsoon GWL TS at four blocks, namely as Khajuwal, Bikaner, Kolayat and Lunkaransa indicates an improvement in pre-monsoon GWLs at corresponding blocks. While magnitude of the trend in pre-monsoon GWL TS at the Nokha block is positive, which shows decline in pre-monsoon GWL at the given block.

AC plot of pre-monsoon GWL TS at the Dungargarh block for the time period 1994–2018 is shown in the Fig. 4a. The AC plot shown in Fig. 4a corresponding to pre-monsoon GWL TS (1994–2018) at the Dungargarh block shows that, the corresponding data is dependent and thus, the MK-CF_2 test is employed for the given TS. For analyzing pattern of the trend in pre-monsoon GWL TS (1994–2018) at the Dungargarh block, ITA plot is used and it is shown in the Fig. 4b. Smoothing curve (with window 10) of the pre-monsoon GWL TS (1994–2018) at the Dungargarh block is shown in the Fig. 4c. The pre-monsoon GWL TS (1994–2018) at the Dungargarh block has showed a significant decreasing trend, which is supported by decreasing pattern of data observed in corresponding ITA plot (Fig. 4b) and smoothing curve (Fig. 4c).

Table 4 Trend analysis results for pre-monsoon GWL TS (1994–2018) of six blocks in the Bikaner District of the Rajasthan

Block name	[a]SS value (m/year)	Pattern of trend	Statistical test	Significant trend
Khajuwal	−0.448	Monotonic	MK-CF_2	No
Nokha	0.131	Non-monotonic	MK	No
Bikaner	−0.440	Monotonic	MK-CF_2	No
Kolayat	−0.163	Non-monotonic	MK-CF_2	No
Dungargarh	−0.185	Monotonic	MK-CF_2	Yes
Lunkaransa	−0.795	Monotonic	MK-CF_2	No

[a]Negative trend denotes improvement in GWL and positive trend denotes decline in GWL

Fig. 4 **a** AC plot of pre-monsoon GWL TS (1994–2018) at the Dungargarh block. **b** ITA plot of pre-monsoon GWL TS (1994–2018) at the Dungargarh block. **c** Smoothing curve of pre-monsoon GWL TS (1994–2018) at the Dungargarh block

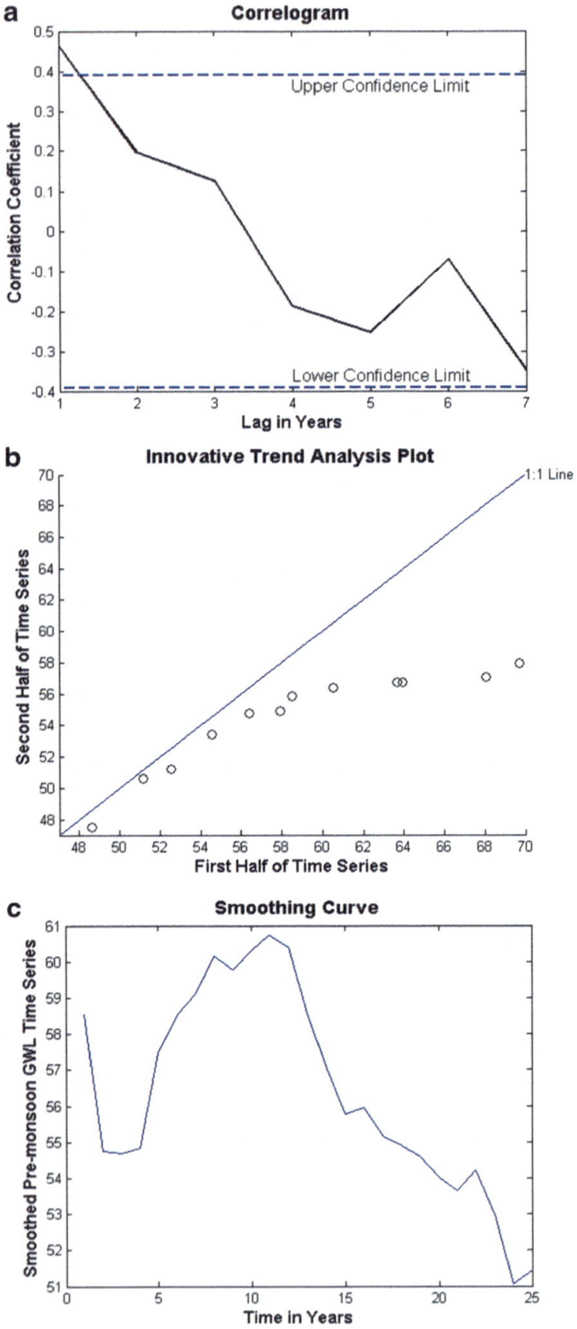

6 Conclusions

Statistically significant decreasing trend is found in GWL TS (1994–2018) at the Khajuwal block corresponding to monsoon temporal scale. Statistically significant decreasing trend is also found in pre-monsoon GWL TS (1994–2018) at the Dungargarh block. As given above, all statistically significant trends are decreasing, thus showing an improvement in GWLs at given blocks in corresponding seasons.

Increasing insignificant trends are found in GWL TS (1994–2018) at the Nokha block corresponding to temporal scales of monsoon, POMRB and pre-monsoon. Also increasing insignificant trend is found in GWL TS (1994–2018) at the Dungargarh block corresponding to monsoon temporal scale. These aforesaid increasing insignificant trends indicate decline in GWLs at given blocks in corresponding seasons. Insignificant decreasing trends found in GWL TS (1994–2018) at various blocks corresponding to temporal scales of monsoon, POMKH, POMRB and pre-monsoon, shows an improvement in GWLs at the given blocks in corresponding seasons. Attribution of statistically significant decreasing trends found in GWL TS at the Khajuwal and the Dungargarh blocks corresponding to monsoon and pre-monsoon temporal scales, respectively, can be performed in future.

In the present study, the effect of climate change on GWLs of six blocks in the Bikaner district of the Rajasthan is studied through trend analyses. Thus, this book chapter is related to this book titled "Water and Energy Management in India Artificial Neural Networks and Multi-Criteria Decision Making Approaches."

Acknowledgements Authors are thankful to the Central Ground Water Board, India for making necessary data available, which is used in the present study.

References

Abdullahi, M. G., Toriman, M. E., Gasim, M. B., & Garba, I. (2015). Trends analysis of groundwater: using non-parametric methods in Terengganu Malaysia. *Earth Science and Climatic Change, 6*(1), 1–3. Available at https://doi.org/10.4172/2157-7617.1000251.

Brocque, L. A. F., Kath, J., & Reardon, S. K. (2018). Chronic groundwater decline: A multi-decadal analysis of groundwater trends under extreme climate cycles. *Journal of Hydrology, 561,* 976–986.

Bui, D. D., Kawamura, A., Tong, T. N., Amaguchi, H., & Nakagawa, N. (2012). Spatio-temporal analysis of recent groundwater-level trends in the Red River Delta Vietnam. *Hydrogeology Journal, 20,* 1635–1650.

Central Ground Water Board. (2013). Ground water information Bikaner district, Rajasthan. Western region Jaipur, district ground water brochure. Ministry of water resources, Government of India.

Central Ground Water Board. (2016). Ground water year book 2015–2016, Rajasthan State. Regional office data centre, Western region Jaipur, ministry of water resources, water development and Ganga rejuvenation, Government of India.

Central Ground Water Board (Year Not Mentioned) Ground water year book-India 2016–2017. Ministry of Water Resources, Water Development and Ganga Rejuvenation, Government of India, Faridabad. Available at doi:http://cgwb.gov.in/Ground-Water/Groundwater%20Year%20Book%202016-17.pdf.

Gibrilla, A., Geophrey, A., & Dickson, A. (2017). Trend analysis and ARIMA modelling of recent groundwater levels in the White Volt a River basin of Ghana. *Groundwater for Sustainable Development*. https://doi.org/10.1016/j.gsd.2017.12.006.

Goode, D. J., Senior, L. A., Subah, A., & Jaber, A. (2013). *Groundwater-level trends and forecasts, and salinity trends, in the Azraq, Dead Sea, Hammad, Jordan Side Valleys, Yarmouk and Zarqa groundwater basins, Jordan*. Open-File Report 2013-1061, U.S. Department of the Interior and U.S. Geological Survey.

Hodgkins, G. A., Dudley, R. W., Nielsen, M. G., Renard, B., & Qi, S. L. (2017). Groundwater-level trends in the US glacial aquifer system, 1964–2013. *Journal of Hydrology, 553,* 289–303.

Ilia, I., Loupasakis, C., & Tsangaratos, P. (2018). Land subsidence phenomena investigated by spatiotemporal analysis of groundwater resources, remote sensing techniques, and random forest method: the case of Western Thessaly, Greece. *Environmental Monitoring and Assessment, 190,* 623.

Jha, M. K., Chowdary, V. M., & Chowdhury, A. (2010). Groundwater assessment in Salboni Block, West Bengal (India) using remote sensing, geographical information system and multi-criteria decision analysis techniques. *Hydrogeology Journal, 18*(7), 1713–1728.

Kawamura, A., Bui, D. D., Tong, T. N., Amaguchi, H., Nakagawa, N. (2011). Trend detection in groundwater levels of Holocene unconfined aquifer in Hanoi, Vietnam by Non-Parametric Approaches. In *World environmental and water resources congress, Palm Springs, California, United States* (pp 914–923). Available at https://doi.org/10.1061/41173(414)94.

Kumar, P., Chandniha, S. K., Lohani, A. K., Krishan, G., & Nema, A. K. (2018). Trend analysis of groundwater level using non-parametric tests in alluvial aquifers of Uttar Pradesh, India. *Current World Environment, 13*(1), 44.

Kundzewicz, Z. W., & Robson, A. J. (2000). *Detecting trend and other changes in hydrological data*. World Climate Programme Data and Monitoring, WMO/TD-No. 1013, Geneva.

Lee, B., Hamm, S. Y., Jang, S., Cheong, J. Y., & Kim, G. B. (2014). Relationship between groundwater and climate change in South Korea. *Geosciences Journal, 18*(2), 209–218.

Panda, D. K., Mishra, A., & Kumar, A. (2012). Quantification of trends in groundwater levels of Gujarat in western India. *Hydrological Sciences Journal, 57*(7), 1325–1336.

Pathak, A. A., & Dodamani, B. M. (2019). Trend analysis of groundwater levels andassessment of regional groundwater drought: Ghataprabha river basin, India. *Natural Resources Research, 28*(3), 631–643.

Patle, G. T., Singh, D. K., Sarangi, A., Rai, A., Khanna, M., & Sahoo, R. N. (2015). Time series analysis of groundwater levels and projection of future trend. *Journal of Geological Society of India, 85,* 232–242.

Ramakrishnaiah, C. R., Sadashivaiah, C., & Ranganna, G. (2009). Assessment of water quality index for the groundwater in Tumkur Taluk, Karnataka State, India. *Journal of Chemistry, 6*(2), 523–530.

Ribeiro, L., Kretschmer, N., Nascimento, J., Buxo, A., Rotting, T., Soto, G., et al. (2015). Evaluating piezometric trends using the Mann-Kendall test on the alluvial aquifers of the Elqui River basin, Chile. *Hydrological Sciences Journal, 60*(10), 1840–1852.

Sang, Y., Wang, Z., & Liu, C. (2014). Comparison of the MK test and EMD method for trend identification in hydrological time series. *Journal of Hydrology, 510,* 293–298.

Sen, P. K. (1968). Estimates of the regression coefficient based on Kendall's tau. *Journal of the American Statistical Association, 63*(324), 1379–1389.

Sen, Z. (2012). Innovative trend analysis methodology. *Journal of Hydrologic Engineering, 17*(9), 1042–1046.

Shamsudduha, M., Chandler, R. E., Taylor, R. G., & Ahmed, K. M. (2009). Recent trends in groundwater levels in a highly seasonal hydrological system: The Ganges-Brahmaputra-Meghna Delta. *Hydrology and Earth System Sciences, 13,* 2373–2385.

Sonali, P., & Nagesh Kumar, D. (2013). Review of trend detection methods and their application to detect temperature changes in India. *Journal of Hydrology, 476,* 212–227.

Tabari, H., Nikbakht, J., & Some'e, B. S. (2012). Investigation of groundwater level fluctuations in the north of Iran. *Environmental Earth Sciences, 66,* 231–243.

Thakur, G. S., & Thomas, T. (2011). Analysis of groundwater levels for detection of trend in Sagar district, Madhya Pradesh. *Journal of Geological Society of India, 77,* 303–308.

Vousoughi, F. D., Dinpashoh, Y., Aalami, M. T., & Jhajharia, D. (2013). Trend analysis of groundwater using non-parametric methods (case study: Ardabil plain). *Stochastic Environmental Research and Risk Assessment, 27,* 547–559.

Yilmaz, A. G., Shanableh, A., Al-Ruzouq, R. I., & Kayemah, N. (2020). Spatio-temporal trend analysis of groundwater levels in Sharjah, UAE. *International Journal of Environmental Science and Development, 11*(1).

Yue, S., & Wang, C. (2004). The Mann-Kendall test modified by effective sample size to detect trend in serially correlated hydrological series. *Water Resources Management, 18,* 201–218.

Zafor, M., Alam, M., Bin, J., Rahman, M., & Amin, M. N. (2017). The analysis of groundwater table variations in Sylhet region, Bangladesh. *Environmental Engineering Research, 22*(4), 369–376.

Zhou, Y., Xiao, W., Wang, J., Zhao, Y., Huang, Y., Tian, J., & Chen, Y. (2016). Evaluating spatiotemporal variation of groundwater level in Beijing plain, a groundwater-fed area from 2001 to 2010. *Advances in Meteorology, 11,* Article ID 8714209.

http://59.179.19.250/GWL/GWL.html?UType=R2VuZXJhbA==?UName=.

https://vsp.pnnl.gov/help/vsample/Design_Trend_Mann_Kendall.htm.

Climate Change Impact on Virtual Water Availability: A Categorized Polynomial Neural Network Approach

Mrinmoy Majumder, Apu K. Saha, Prachi D. Khobragade, Deepjyoti Deb, and Khakachang Tripura

Abstract The uncontrolled extraction followed by technological advancements and increasing population has induced an increase in Green House Gas concentration in the atmosphere and the outcome was global warming followed by the change in climatic pattern in different places of the world has reduced per capita availability of water. Due to climate change, water accessibility for manufacturing also gets affected. The volume of water which is also avowed as virtual water is significantly impacting the industrial output. Few studies tried to estimate the consequences of variation in climatic conditions change on virtual water utilization but at present, no study has considered the same impact on the availability of virtual water. That is why climate change impact was estimated on the availability of virtual water per consumer of the industrial output of the region in the current study. The estimation was depicted in cardinal rather than ordinal output. Also, the Polynomial Neural Network architecture-based Group Method of Data Handling was utilized as the optimal classifier of climate change impact on virtual water availability per consumer. Four metro cities of the Indian Sub-continent were considered as the study area. According to the results New Delhi is worst and Kolkata least affected for the Intergovernmental Panel on Change of Climate proposed B2 scenario. In the case of the A2 scenario, here also New Delhi was found to be the worst affected and Mumbai least affected. The classifier was found to be more efficient than Stepwise

M. Majumder (✉) · P. D. Khobragade · D. Deb · K. Tripura
Hydro-Informatics Engineering (under Department of Civil Engineering), National Institute of Technology Agartala, Jirania, Agartala, India
e-mail: mmajumder15@gmail.com

P. D. Khobragade
e-mail: prachikhobragade111@gmail.com

D. Deb
e-mail: ddeb210@gmail.com

K. Tripura
e-mail: khaka.tp@gmail.com

A. K. Saha
Department of Mathematics, National Institute of Technology Agartala, Agartala, Tripura, India
e-mail: apu_saha@gmail.com

M. Majumder and G. D. Kale (eds.), *Water and Energy Management in India*,
https://doi.org/10.1007/978-3-030-66683-5_5

Forward Regression, Decision Forest, and Logistic Regression techniques utilized in the present investigation for the same purpose. But due to the equal sign, that was considered for all the selected input variables the Percentage Correct Classification Rate was found to not above 85% which can be improved by incorporating the importance of the inputs and their effect on the output in the model algorithm.

1 Introduction

The notion of virtual water aid to realize what quantity of water is required to produce different goods and services. New concepts have been developed regarding water affairs and issues. One such form is the 'virtual water' concept and this was considered as a considerable water saving methodology in goods production. Virtual water trade directs towards the idea that when the goods and services are exchanged, so the virtual water. For example, if a country imports two tons of wheat instead of producing it domestically, it saves nearly 2,600 cubic meters of real local water and if the scarcity of water is a problem in that country then that water can be used for some other purpose. If the exporting country face the problem of water scarcity as it has exported 2,600 cubic meters of virtual water, therefore, the absolute quantity of water needs to cultivate the wheat will not be available anymore for other purposes (Phillips et al. 2008).

Hoekstra and Chapagain have delineated the virtual-water accomplished of a product as the freshwater quantity required to develop the product which was evaluated at the manufacturing site. The water is supposed to be virtual since once the wheat is matured, the real water which is utilized while cultivating it is not contained in the wheat. In the barren and semi-arid regions where the problem of water scarcity arises, knowing the value of virtual water as a good or service can be beneficial. Virtual water is described as the amount of water required to manufacture a product from the start to end and which is generally neglected as well as a hidden component of manufacturing goods (Sonesson et al. 2010). The total quantity of water essential for the management of a specific quantity of output product depends entirely on the succeeding manufacturing conditions, as well as the place and time of production, water use efficiency (Hoekstra and Chapagain 2003). The calculation considers water used directly in production from all sources, and including the water that is indirectly used concerning the virtual water quantity of other production objects. Thus, the water content from the surface reserves, precipitation, and groundwater resources are evaluated (Frontier Economics Pty Ltd. 2019). The global scenario of virtual water is described in detail in Sect. 1.1.

1.1 Global Scenario of Virtual Water Availability

Nowadays food trading is acknowledged as the virtual water transfer (Ercin et al. 2009; Aldaya et al. 2010) and newly as implied virtual land export (Fader et al. 2011). The term 'virtual water' was established by Allan (1993) nearly two decades ago, but it took another decade to invade in global water resource estimation (Horlemann and Neubert 2007). Virtual water is classified into three types: blue, green, and grey. Water that flows in surface water bodies is termed as blue water and this water is treated by the water-related companies and then supplied to public and commercial demand, thus this kind of water is applied in the food processing industry. Green water is the type of virtual water refined by water-based enterprises to meet social and commercial requirements which also plays a pivotal role in the food processing industry. Water pouring in the form of rainfall and infiltrating in the soils is termed as green water and well yielded agricultural activities depend on this green water. The concept of greywater is a bit more baffling and its meaning changes with the viewpoint: actually, grey water refers to wastewater that comes from domestic or commercial sources, which is contaminated lightly with detergents and other pollutants, but that can still be released to the public sewer, e.g. bathwater and dishwashing water. The problem of water scarcity in countries like Israel put off the freight of oranges having heavy water guzzlers to block the export of an enormous volume of water to various chunks of the world.

Global water availability patterns are also influenced by annual and seasonal average rainfall and evaporation. For an already highly variable rainfall region, if the projected seasonal mean precipitation increases, it will eventually create a pattern of seasonally fickle regions becoming more variable (Konapala et al. 2020).

Water use and its productivity are evaluated for cotton, maize, soybeans, and wheat crops in Brazil as it is a huge exporter of virtual water using the EPIC crop model. The output results based on the study of the four selected crops show green water as the primary asset for the yielding of biomass. Moreover, an increase in the agricultural yield will accordingly show growth in green water productivity also (Flach et al. 2016).

The notion of virtual water business has acquired weight both in the scientific as well as in the political discussion in recent years. The concept of the topic is debatable. The definition alters among a systematic, eloquent concept and economical persuaded approach. In the systematic concept, virtual water commerce means an instrument that concedes a recognition and appraisal of policy choice in both scientific and political analysis. Hoekstra and Hung (2003) have done extensive work on quantifying virtual water flows between nations through the trade of crops.

About 700–1100 km^3 of virtual water motions linking nations and through this trade, about 455 km^3 of water is saved per year (Oki et al. 2003). The developers of the virtual water trade conception support the economic argument by referring to the international trade theory of comparative advantage; that nations that are relatively well endowed with water will be exporting water-intensive products and those who are not will be importing. In reality, the situations where water-abundant nations are virtual water importers and water-scarce nations are net virtual exporters. For example, Norway and Switzerland are net virtual water importers while they have no water scarcity issues and on the other hand, water-scarce nations such as Afghanistan and Malawi exports virtual water (Ansink 2010). In these cases, the theory which is comparative advantage does not stand strong (Reimer 2012).

1.2 The Criticality of Virtual Water Concept

The idea of Virtual Water relies on an assumption that all sources of water, whether in the form of rainfall or provided through an irrigation system, are of equal value. Implicitly suspect that the water that would be liberated by decreasing a high-water use activity that could necessarily be available for use in less water quantity activity. The implicit hypothesis says that water used in rangeland beef production that is available to be used to generate another option, less water quantity activity. Practically this may not be the case, nor might the substitute be economic. This will not cause any harm to the environment nor does it give any criteria of whether water resources are being used inside sustainable extraction limitations. The utilization of virtual water estimates therefore does not provide any guidance for the policymakers pursuing to make sure that environmental objectives are fulfilled.

Nezamoleslami and Hosseinian (2020) in their study have utilized the concept of water footprint which incorporates a life cycle appraisal framework in a steel plant placed in the central part of Iran. Blue Water footprint was primarily prioritized to invoke to groundwater or surface water used for steel manufacturing. To measure the footprint of personnel's foods, concepts from eco-friendly footprint were utilized. The authors found that the water footprint value increases significantly when the water footprint of the personnel's food was taken into consideration in the steel plant.

This study evaluated the future virtual water flows from a period of 2030–2050 associated with barley and imports of meat to the United Kingdom by taking the impact of projected climate, land use, and population changes into consideration. The results depict that the future virtual water inflows linked with barley imports to balance sedentary inadequacy are larger than the total quantity of water used in domestic barley yield in the United Kingdom. The outcome in the present study can bring a positive change in the UK's commercial sector and business to persuade food security while reducing environmental influence in exporting countries (Yawson et al. 2020).

There is a need to understand the interrelationship between the population dynamics and drought especially in the areas where the population is very high and this leads to the drought as the climate keeps on changing. Thus, we have inspected the relation between population and drought using hydrology anthropology exemplary that has helped to measure the decrease in population of Ancient Maya of

Central America. The conclusion of the study shows that economic, social, and technological development plays a vital role to reduce the climate change impact on the population of human (Kuil et al. 2019).

In this study, the authors scrutinize the comprehensive climate change impact and patterns of land use on green water resources in the watershed from 1995 to 2015. In the evaluation phase, The ENS (Nash–Sutcliffe model efficiency coefficient) and R_2 were 0.94 and 0.89 respectively which was changed to 0.89 and 0.88 in the verification stage. The results indicate towards an eminent level of simulation precision, alteration in the discharge of green water and the green water storage due to climate fluctuation report for an increment of 2.07 and 1.28 mm/a (Lyu et al. 2019).

The article proposes a bottom-up proposal that attempts to associate the Water Footprint idea with climate variability adaptation and capacity development. This approach was then expanded and verified by taking assistance from a school in Austria to provide an initial point for WF analysis and formulate an upgrade the results. The pupils who worked on this have decreased their WF by 9% and came with the change agents. This approach helps young people to establish self-assurance by inventing the connection between their activities at the local level and feature climate changes at a global level. This will help to enhance this perception and put up to the adaptation of various impacts of climate variability and minimize vulnerability (Haida et al. 2019).

In China, Jiangsu province is a developed one is withstanding both energy and water stress. The impact of the generation of energy on water resource remains is observed incorporating climate change. A comprehensive Energy Alternatives Planning System model in combination with plant-level data to examine the effect of energy scheme on water resources management. The conclusion of this study depicts that evolution in energy adaptability and optimization in industry structure results in 40 and 33% water removal savings in the future which depends on the level of optimization. Overall water utilization can conform to the 2030 targets by enforcing coordinated measures, which was determined by the Jiangsu government. To attain water savings benefits and fulfilling water use control targets, the shifts in the cooling technology should be done significantly (Zhou et al. 2019).

1.3 Motivation

Climate change has an impact on virtual water availability but the research on the effect of climate change over virtual water availability was seldom performed. As can be seen from the definition of Virtual Water by Hoekstra and Chapagain, in the estimation of virtual water the entire freshwater that helped produce goods is considered as Virtual Water.

However, climate change will impact water availability not the consumption of water. The consumption of water by the industries will depend on the population dynamics and demand for the product produced. So, to analyze climate change consequences, the ease of access to water that can be consumed by the industries to yield the product has to be defined as virtual water. Any change in the accessibility of the water allocated for industrial consumption will impact the amount of virtual water and the trade based on this concept. But in previous years all the studies involving climate change impact virtual water was conducted considering virtual water as that quantity of water which is consumed for the production of the intended product.

In the case of earlier works on virtual water estimation, as proposed by the proposer and his successors, the unitary equation was used to approximate the amount of virtual water of a manufactured product. Although Artificial Neural Networks were often used to estimate virtual waters, no studies have considered the application of polynomial neural networks to estimate the amount of virtual water available for allocation.

Again, the quantity of virtual water in the estimation is assumed in a range of values which reduce the authenticity of the output for further calculation but is used for cognitive discussions only. Crisp interpretation is difficult from the estimation of virtual water. As a result, cardinal output models will perform better compared to those other models. That is why the present investigation has two objectives which are described in Sect. 1.4.

1.4 Objective and Novelty

The objective of the present investigation is to estimate the virtual water availability concerning climate change impact. The study also wants to analyze the potential of Polynomial Neural Network (PNN) based classification algorithms (Classifiers) in an approximation of the amount of virtual water availability.

The novelty of the current study is its pioneering analysis of the climate change impact on virtual water availability with the help of the PNN based Group Method of Data Handling algorithm. The model was also the first model which uses dummy outputs to represent the situation of the virtual water availability in different scenarios of climate change as proposed by the Intergovernmental Panel on Climate Change (IPCC).

In this regard, the present investigation is also the first approach where virtual water availability per capita of consumers was analyzed. In previous studies, virtual water is treated as the quantity of freshwater utilized for industrial production. However, in the present investigation virtual water availability per consumer was approximated instead of virtual water utilized for production. The benefit of predicting availability instead of quantity consumed is it depicts the preparedness of the region in face of the industrial demand and degradation in the supply of freshwater due to changes in climate change. The detailed methodology of the paper is described in Sect. 2.

2 Methodology

The present investigation aims to estimate the climate change impact on virtual water availability. In this regard, the procedure followed to achieve the objective of this present work can be divided into two phases. The first phase involved the development of the predictive model with the help of polynomial neural network-based algorithms and the second phase involved the analysis of climate change impacts on virtual water availability. The procedure adopted to develop the predictive model for determination of Virtual Water Availability Prediction (VWAP) was described in Sect. 2.1 and the methodology followed for climate change analysis was explained in Sect. 2.2.

2.1 Development of the Virtual Water Availability Prediction Model

The objective of the model was to estimate virtual water availability. Figure 1 describes the entire procedure adopted to achieve the study objective. The input and output variable of the model was selected to be Total water availability (A1), total freshwater availability (A1f), total freshwater allocated to agriculture (A1fa), domestic (A1fd) and another type of users (A1fo) whereas the output is selected to be the total quantity of virtual water or amount of water allocated to industrial consumers. As the main objective was to estimate the total quantity of basic water which is the amount of freshwater allotted to industries as raw material the input variables were so selected to represent the impact of both quantity and number of users. That is why the volume of water feasible and utilized by a different type of consumer was divided by the total population of the area (See Fig. 1). That means all the input and output variable is described in cubic meters per capita per month.

Figure 2 describes the various classes of the output variable (virtual water quantity) such as Extremely High, Very High, High, Medium, Low, Very Low, and Extremely Low based on its availability which is represented with the help of a schematic diagram. If the amount of virtual water is more than eighty-five percent, class of the virtual water comes under Extremely High (EH) whereas if the percentage of the quantity of virtual water is less than eleven, it indicates extremely low availability of Virtual Water.

During the model development process, the dummy output instead of crisp numerical was used as it is known that the exact prediction of virtual availability will be difficult. And error-prone as a result dummy values of the output were used at the time of model development.

In the present study four input variables were used and one output variables. The model development procedure followed a polynomial neural network architecture and the algorithms used to develop the model is Group Method of Data Handling (Ivakhnenko 1995; Mehri et al. 2019), Stepwise Forward Regression (SFR) (McCarthy et al. 2019), and Decision forest (DF) (Pham et al. 2019) The logistic

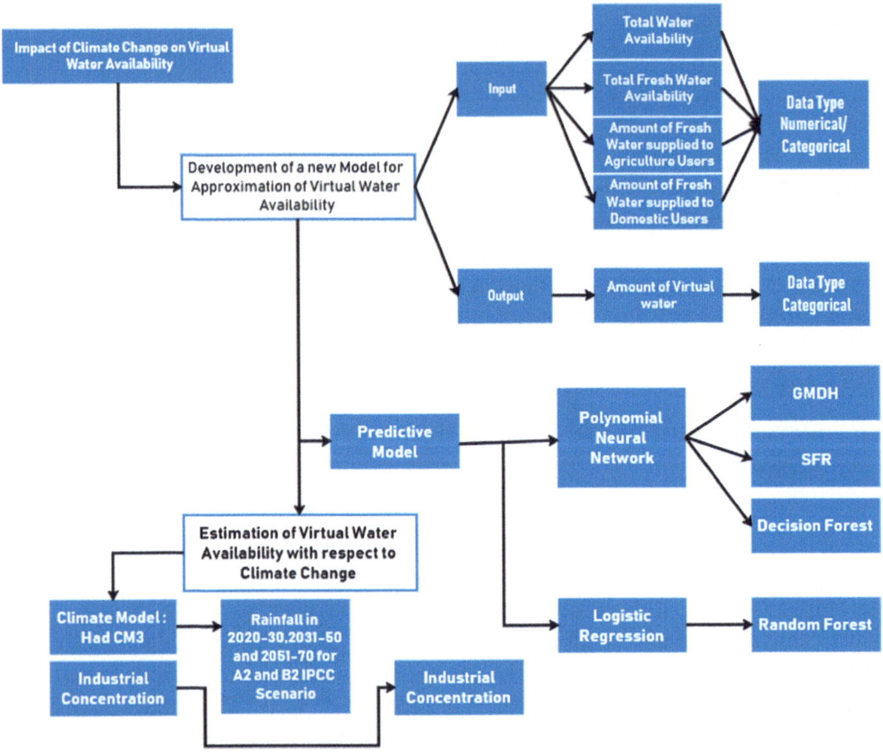

Fig. 1 Schematic diagram of the procedure followed to estimate the consequences of climate change on virtual water availability

regression (Chen et al. 2019) was also utilized for comparison and identification of the best output.

In total 1000 data rows were utilized for training the PNN based algorithms where 80% was assigned for training and the remaining for testing the developed model. The best algorithm out of the three algorithms was required to be identified. In this aspect the performance metrics as described in Sect. 2.1.1. Adding on to the above algorithm and in this regard, the logistic regression algorithm was also used.

Before the training process by the algorithms, two types of preprocessing on the same data was conducted. In the first type, the outputs from the models were converted to seven different groups and in the case of the second type, the input, as well as output parameters, were converted into a dummy. In total seven different clusters were proposed and the entire data of output and in case of both variables, then the data of both the parameters would be classified.

The output from the PNN models as well as linear models was compared using the help of performance metrics to identify the best model and the climate change impact analysis was conducted by the chosen model. Section 2.1.1 describes the methods used in the study.

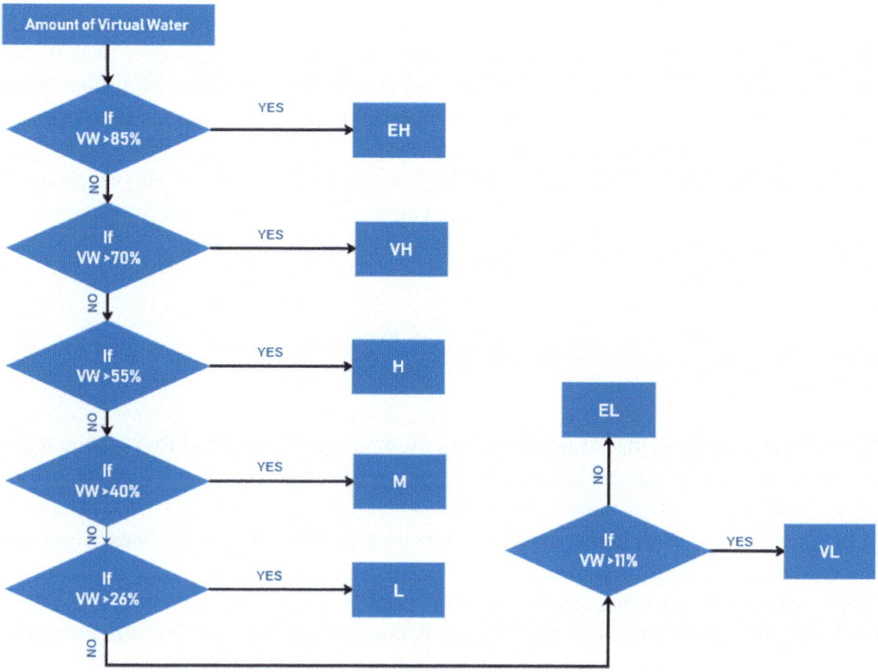

Fig. 2 Decision tree to classify virtual water availability

2.1.1 Performance Metrics

The accuracy of the predicted classification results was compared with the help of the Correct Classification Rate, KAPPA Coefficient of agreement, and F-score which are described in Sections "Correct Classification Rate", "KAPPA Coefficient of Agreement" and "Weighted F-Measure" respectively.

Correct Classification Rate

The Correct Classification Rate (CCR) is the ratio of the number of times the category of the data pattern has been correctly classified to the total number of times classes were estimated (Hisao et al. 1999).

If the total number of classes that are predicted correctly is C_t and the total dataset is C, then the CCR is depicted by

$$\%CCR = \left(\frac{C_t}{c}\right) \times 100 \quad (1)$$

KAPPA Coefficient of Agreement

Kappa is used for measuring the proportion of input values in the principal diagonal of the table for accommodating these values for compliance that can be expected due to desolate chances. The value of Kappa is designated as the numerator that exhibits the disparity between the detected probability of profit and the success probability under the hypothesis of an exceedingly atrocious case. Independence implies that a couple of raters concede about recurrently as the two pairs of persons who prepare their valuation by adequately tossing a coin.

Cohen's kappa is used to evaluate the affinity between two raters who segregate N elements into C mutually exclusive classes. Galton (1892) used KAPPA like statistics for the first time (Smeeton 1985).

The equation for κ is:

$$\kappa = \frac{\Pr(a) - \Pr(e)}{1 - \Pr(e)} \tag{2}$$

where $\Pr(a)$ is the contingent noticed agreement among raters, and $\Pr(e)$ is the theoretical probability of casual agreement, utilizing the observed data for probabilities calculation of respective viewer randomly of each category. $\kappa = 1$, if the raters are all in agreement. $\kappa = 0$, If the raters are not in agreement except the ones which are expected coincidentally (Carletta 1996).

The value of kappa is maximum when the perceived level of agreement is 1, therefore it makes the numerator as large as the denominator. The numerator deteriorates when the observed probability of agreement drops.

Weighted F-Measure

F-Measure or F1 Score brings balance among the precision and the recall. Precision is the number of True Positives divided by the number of True Positives and False Positives. A recall is the number of True Positives divided by the number of True Positives and the number of False Negatives. The multinomial classification problems involve approximation of multiple classes and the constitution of as many decision matrices. Finally, the average of true and false positives and true and false negatives are calculated to estimate the equivalent precision and recall value. Then (Eq. 3) is used to predict the Weighted F-Measure, the average of F-measure for each class. The Weighted F-Measure is directly proportional to the accuracy of the classification by the classifier algorithm.

$$f = \frac{2 \times (p \times r)}{(p + r)} \tag{3}$$

where p and r indicate the value of precision and recall of the classification results and f depicts the F-measure (Sokolova et al. 2006).

2.2 *Impact of Climate Change on Virtual Water Availability*

Once the model for VWAP is successfully developed the second phase was initiated for analysis of the climate change impact on virtual water availability. In this aspect, the output from the climate prediction model, HadCM3 was used for two different IPCC SRES scenarios, A2 and B2 (Pachauri and Reisinger 2007) during the time slabs 2020–30, 2031–50, and 2051–70. The industrial concentration of in the same time slabs under the same scenario was approximated from the description of the scenarios which was explained in the IPCC Climate Change Assessment Report 2007. The population density of the selected study areas for the same time slabs and scenarios were collected from the Word Bank Population Reports. The study area was selected from four different metro cities located in four separate climate zones of the Indian Sub-Continent.

The predicted rainfall by the Had CM3 model for the two scenarios in the selected time slabs was used to estimate the total water available and from total water availability, the amount of freshwater was calculated based on the water quality report of the four selected cities (Maiti and Agrawal 2005). The number of agriculture and domestic consumers was approximated from the scenario description given in the IPCC Assessment Report 2007. Table 1 exhibits the magnitude of the input parameters for the four metro cities in the future time slabs under the A2 and B2 scenarios.

New Delhi, Kolkata, Mumbai, and Chennai was selected as the case study area. The climatic and industrial characteristics can be retrieved from Khosla and Bhardwaj (2019) and Bose (2019) respectively.

3 Results and Discussions

Table 2 depicts the performance metrics of the five classifiers which were utilized to predict the class of virtual water availability as per Fig. 2 both with ordinal and cardinal input data. Table 3 depicts the predicted virtual water availability of the selected four study areas as predicted by the best classifier.

According to results from Table 2, it was observed that the metrics of the GMDH classifier with numerical input data is better compared to all the other classifiers with numerical or categorical input data. The GMDH classifier has the highest %CCR, F, and KAPPA value compared to all the other models. If %CCR of GMDH classifier and SFR algorithm is compared, then it was found that %CCR of former is 1.051 times higher than that of the latter model when the same metric was contrasted with the best classifier among the model trained with cardinal data, i.e., SFR, it was observed that GMDH has %CCR 1.094 times more than the SFR model.

F-score and KAPPA value of GMDH model has the highest value compared to all other models. As a result, the GMDH classifier was selected for the prediction of

Table 1 Magnitude of the input variables in the four different case study areas

Case study area	Time slab	A1	A1f	A1fa	A1fd	A1fo
New Delhi	B2_2051–70	0.228	0.183	0.132	0.085	0.066
	B2_2031–50	0.236	0.189	0.183	0.142	0.099
	B2_2021–30	0.193	0.155	0.003	0.000	0.000
	Baseline	0.019	0.015	0.001	0.000	0.000
	A2_2021–30	0.072	0.058	0.053	0.049	0.025
	A2_2031–50	0.140	0.112	0.098	0.031	0.002
	A2_2051–70	0.040	0.032	0.018	0.013	0.010
Case study area	Time slab	A1	A1f	A1fa	A1fd	A1fo
Kolkata	A2_2051–70	0.445	0.356	0.053	0.028	0.010
	A2_2031–50	0.557	0.445	0.428	0.144	0.141
	A2_2021–30	0.668	0.535	0.505	0.093	0.033
	Base line	0.770	0.616	0.167	0.037	0.036
	B2_2021–30	0.755	0.604	0.085	0.042	0.038
	B2_2031–50	0.907	0.726	0.476	0.037	0.027
	B2_2051–70	0.978	0.783	0.060	0.003	0.002
Case study area	Time slab	A1	A1f	A1fa	A1fd	A1fo
Chennai	A2_2051–70	0.146	0.117	0.050	0.018	0.000
	A2_2031–50	0.365	0.292	0.225	0.082	0.036
	A2_2021–30	0.397	0.318	0.043	0.033	0.025
	Baseline	0.298	0.238	0.180	0.028	0.023
	B2_2021–30	0.424	0.339	0.099	0.056	0.030
	B2_2031–50	0.308	0.246	0.238	0.086	0.083
	B2_2051–30	0.499	0.399	0.017	0.006	0.005
Case study area	Time slab	A1	A1f	A1fa	A1fd	A1fo
Mumbai	A2_2051–70	0.576	0.461	0.146	0.088	0.079
	A2_2031–50	0.603	0.483	0.063	0.053	0.028
	A2_2021–30	0.555	0.444	0.386	0.026	0.003
	Baseline	0.505	0.404	0.332	0.094	0.037
	B2_2021–30	0.693	0.554	0.025	0.019	0.013
	B2_2031–50	0.513	0.411	0.029	0.019	0.003
	B2_2051–70	0.907	0.726	0.106	0.002	0.000

Note Average normalized value of the input variables for the time slabs of present or baseline (2010–18) and climate change scenario are depicted

Table 2 Performance metrics of the selected classifiers

Name of the classifier	Description of training data	%CCR	F	KAPPA
GMDH	Dummy O/P	88.700	0.874	0.763
SFR	Dummy O/P	84.400	0.830	0.683
DF	Dummy O/P	82.600	0.799	0.621
Logistic regression	Dummy O/P	76.830	0.654	0.234
GMDH	Dummy I/P and O/P	80.300	0.756	0.577
SFR	Dummy I/P and O/P	81.100	0.797	0.631
DF	Dummy I/P and O/P	65.700	0.521	0.000
Logistic regression	Dummy O/P	56.230	0.254	0.134

I/P: Input variables, O/P: Output variables, Dimension of the Dummy as per Fig. 2

virtual water availability for the selected four metro cities and to analyze the climate change impact.

According to the results depicted in Table 3, for most of the metro cities, virtual water is found to be available in extremely low quantity except Kolkata where the amount of virtual water is in the second-lowest category, i.e., "L" if the present situation is considered. However, if the prediction of the B2 scenarios were compared which is conceptualized based on the assumption of strict environmental policies then the worst category was at New Delhi where for all the three-time slabs the class representing the lowest availability of virtual water was predicted by the selected model, i.e., GMDH. The best condition of virtual water availability was found to be at Kolkata in the 2051–70 time slab when the average virtual water availability will be under the third-highest category, i.e., "H".

If the A2 scenarios were compared it can be found that the worst situation was found to be in New Delhi for the time slabs 2031–70 when the worst category of availability" was observed. The best condition was predicted in Mumbai in the time slab 2051–70 when the third-worst class of virtual water availability was predicted.

Based on the prediction the availability of virtual water is not very high in any of the metro cities even in the present scenario. In the future, also due to climate change the availability will not improve. That is why proper conservation measures are required to be adopted on an urgent basis.

4 Conclusion

In the present investigation, two different objectives were attempted. In the first objective, a classifier algorithm was developed to predict Virtual Water availability and in the next objective, the consequences of climate change on the virtual water availability were analyzed using the help of the same model developed for the accomplishment of the first objective. In this aspect, the polynomial neural networks-based

Table 3 Effect of climate change on virtual water for the selected IPCC scenarios (A2 and B2) during the three-time slabs (2020–30, 2031–50 and 2051–70) and also for the present or baseline scenario

Case study area	Time slab	Predicted virtual water availability (A1fvw_P)
New Delhi	B2_2051–70	EL
	B2_2031–50	EL
	B2_2021–30	VL
	Baseline	*EL*
	A2_2021–30	EL
	A2_2031–50	EL
	A2_2051–70	EL
Case study area	Time slab	A1fvw_P
Kolkata	A2_2051–70	L
	A2_2031–50	EL
	A2_2021–30	EL
	Baseline	*L*
	B2_2021–30	M
	B2_2031–50	EL
	B2_2051–70	H
Case study area	Time slab	A1fvw_P
Chennai	A2_2051–70	EL
	A2_2031–50	EL
	A2_2021–30	VL
	Baseline	*EL*
	B2_2021–30	VL
	B2_2031–50	EL
	B2_2051–30	L
Case study area	Time slab	A1fvw_P
Mumbai	A2_2051–70	VL
	A2_2031–50	L
	A2_2021–30	EL
	Baseline	*EL*
	B2_2021–30	M
	B2_2031–50	L
	B2_2051–70	M

Note Average normalized value of the input variables for the time slabs of present or baseline (2010–18) and climate change scenarios are depicted. Category of Output variable A1fvw_P as per Fig. 2

classification algorithms along with a logistic regression model were applied to approximate the cardinal output representing the virtual water availability. A decision tree was used for representing the situation of virtual water availability and the cardinality was predicted by the best model among the five models utilized for prediction of the output variable.

According to the result, the GMDH classifier was found to be the better algorithm compared to the other four models based on the three different metrics like %CCR, F-score, and KAPPA. The climate change impact was predicted with the help of a selected classifier for four metro cities of the Indian sub-continent. This was the first attempt to estimate the virtual water availability of the four metro cities in the different climate change scenarios as proposed by the IPCC SRES. According to the results, it was found that during the ecologically sensitive B2 scenario the better availability was proposed by the selected model in the case of Kolkata and the worst condition was estimated for New Delhi.

For the industrially active A2 scenario Mumbai was found to have a better condition compared to the worst affected situation in New Delhi. The present investigation was found to be the first such attempt to estimate the virtual water availability in cardinal output. The advantage of cardinality for climate change prediction studies lies in the fact that instead of point estimation such analysis requires range predictions which help to create an idea of the situations during the future time slabs. In the present work, virtual water is predicted as a resource available for utilization not already utilized. Hence, the results from the study can be used by city managers to intelligently and cognitively distribute the available freshwater such that the industrial output of the region is not compromised.

However, it can be noted that not all the input variable considered in the present model is equally important and have an impact on the availability of virtual water. But the application of all the input variables and related data is required for a reliable estimation from the model. This might be the reason for the %CCR of the selected GMDH algorithm not being higher than 84%. The researchers interested to work in this aspect can try to mitigate these lacunae by utilizing different Multi-Criteria Decision Making methods for estimation of priority of the selected input variable and as a result to predict virtual water availability only the most significant variables can be used.

References

Aldaya, M. M., Allan, J. A., & Hoekstra, A. Y. (2010). Strategic importance of green water in international crop trade. *Ecological Economics, 69*, 887–894. http://dx.doi.org/10.1016/j.ecolecon.2009.11.001.

Ansink, E. (2010). Refuting two claims about virtual water trade. *Ecological Economics, 69*(10), 2027–2032.

Bose, S. (2019). A history of the Indian economy in Asian and global contexts, 1810s–2010s. *Emerging states and economies* (pp. 139–151). Singapore: Springer.

Carletta, J. (1996). Assessing agreement on classification tasks: The kappa statistic. *Computational Linguistics, 22*(2), 249–254.

Chen, W., Yan, X., Zhao, Z., et al. (2019). Spatial prediction of landslide susceptibility using data mining-based kernel logistic regression, naive Bayes, and RBFNetwork models for the Long County area (China). *Bulletin of Engineering Geology and the Environment, 78*(1), 247–266. https://doi.org/10.1007/s10064-018-1256-z.

Ercin, A. E., Aldaya, M. M., & Hoekstra, A. Y. (2009). A pilot in corporate water footprint accounting and impact assessment: the water footprint of a sugarcontaining carbonated beverage. UNESCO-IHE, Delft, the Netherlands.

Fader, M., Gerten, D., Thammer, M., Heinke, J., Lotze-Campen, H., Lucht, W., & Cramer, W. P. (2011). Internal and external green-blue agricultural water footprints of nations, and related water and land savings through trade. *Hydrology and Earth System Sciences, 15*, 1641–1660. http://dx.doi.org/10.5194/hess-15-1641-2011.

Flach, R., Ran, Y., Godar, J., Karlberg, L., & Suavet, C. (2016). Towards more spatially explicit assessments of virtual water flows: Linking local water use and scarcity to global demand of Brazilian farming commodities. *Environmental Research Letters, 11*, 075003.

Frontier Economics Pty Ltd. (2019). The concept of 'virtual water'—A critical review. Retrieved from https://agriculture.vic.gov.au/agriculture/farm-management/soil-and-water/water/virtual-water.

Galton, F. (1892). Finger Prints Macmillan, London.

Haida, C., Chapagain, A. K., Rauch, W., et al. (2019). From water footprint to climate change adaptation: Capacity development with teenagers to save water. *Land Use Policy, 80*, 456–463. https://doi.org/10.1016/j.landusepol.2018.02.043.

Hoekstra, A. Y., & Chapagain, A. K. (2003). Virtual water trade: A quantification of virtual water flows between nations in relation to international trade of livestock and livestock products. In: this volume.

Hoekstra, A. Y., & Hung, P. Q. (2003). Virtual water trade: A quantification of virtual water flows between nations in relation to international crop trade. In: this volume.

Horlemann, L. & Neubert, S. (2007). Virtual water trade: A realistic concept for resolving the water crisis? Bonn, German Development Institute.

Ivakhnenko, A. G., & Ivakhnenko, G. A. (1995). The review of problems solvable by algorithms of the group method of data handling (GMDH). *Pattern Recognition and Image, 5*(4), 527–535.

Khosla, R., & Bhardwaj, A. (2019). Urbanization in the time of climate change: Examining the response of Indian cities. *Wiley Interdisciplinary Reviews: Climate Change, 10*(1), e560. https://doi.org/10.1002/wcc.560.

Kuil, L., Carr, G., Prskawetz, A., et al. (2019). Learning from the Ancient Maya: Exploring the impact of drought on population dynamics. *Ecological Economics, 157*, 1–16.

Konapala, G., Mishra, A. K., Wada, Y., et al. (2020). Climate change will affect global water availability through compounding changes in seasonal precipitation and evaporation. *Nature Communications, 11*, 3044.

Lyu, L., Wang, X., Sun, C., et al. (2019). Quantifying the effect of land use change and climate variability on green water resources in the Xihe River Basin, Northeast China. *Sustainability, 11*(2), 338.

Maiti, S., & Agrawal, P. K. (2005). Environmental degradation in the context of growing urbanization: A focus on the metropolitan cities of India. *Journal of Human Ecology, 17*(4), 277–287. https://doi.org/10.1080/09709274.2005.11905793.

Mehri, Y., Soltani, J., & Khashehchi, M. (2019). Predicting the coefficient of discharge for piano key side weirs using GMDH and DGMDH techniques. *Flow Measurement and Instrumentation, 65*(2019), 1–6. https://doi.org/10.1016/j.flowmeasinst.2018.11.002.

McCarthy, R. V., McCarthy, M. M., Ceccucci, W., et al. (2019). Predictive models using regression. *Applying Predictive Analytics*. Cham: Springer. https://doi.org/10.1007/978-3-030-14038-0_4.

Nezamoleslami, R., & Hosseinian, S. M. (2020). An improved water footprint model of steel production concerning virtual water of personnel: The case of Iran. *Journal of Environmental Management, 260,* 110065. https://doi.org/10.1016/j.jenvman.2020.110065.

Oki, T., Sato, M., Kawamura, A., Miyake, M., Kanae, S., & Musiake, K. (2003). Virtual water trade to Japan and in the world. In: *Virtual Water Trade: Proceedings of the International Expert Meeting on Virtual Water Trade, Value of Water Research Report Series*, Hoekstra, AY, (Vol. 12).

Pachauri, R. K., & Reisinger, A. (2007). *IPCC fourth assessment report*. IPCC, Geneva.

Pham, B. T., Bui, D. T., & Prakash, I. (2019). Landslide susceptibility modeling using different advanced decision trees methods. *Civil Engineering and Environmental Systems, 35*(1–4), 139–157. https://doi.org/10.1080/10286608.2019.1568418.

Phillips, D. J. H., Allan, J. A., Claassen, M., et al. (2008). *The transcend-TB3 project: A methodology for the trans-boundary waters opportunity analysis (The TWO Analysis).* Prepared for the Ministry of Foreign Affairs, Sweden.

Reimer, J. J. (2012). On the economics of virtual water trade. *Ecological Economics, 75*, 135–139.

Sokolova, M., Japkowicz, N., & Szpakowicz, S. (2006). Beyond accuracy, F-score, and ROC: A family of discriminant measures for performance evaluation. In: A. Sattar & B. Kang (Eds.), *AI 2006: Advances in artificial intelligence, AI 2006*. Lecture Notes in Computer Science (Vol. 4304). Berlin, Heidelberg: Springer. https://doi.org/10.1007/11941439_114.

Sonesson, U., Berlin, J., & Ziegler, F. (Eds.). (2010). *Environmental assessment and management in the food industry: Life Cycle Assessment and related approaches*. Elsevier.

Yawson, D. O., Mohan, S., Armah, F. A., et al. (2020). Virtual water flows under projected climate, land use, and population change: The case of UK feed barley and meat. *Heliyon, 6*(1), e03127. https://doi.org/10.1016/j.heliyon.2019.e03127.

Zhou, Y., Ma, M., Gao, P., et al. (2019). Managing water resources from the energy-water nexus perspective under a changing climate: A case study of Jiangsu province, China. *Energy Policy, 126*, 380–390.

Development of ANN Model for Simulation of the Runoff as Affected by Climatic Factors on the Jamuna River, Assam, India

Mridusmita Debnath, Arup Kumar Sarma, and Chandan Mahanta

Abstract Quantifying rainfall-runoff relationships in an orographic area is challenging. Meanwhile, estimation of runoff in a river is urgent as the river supports a human's livelihood through various ways, including agriculture. Complex physiographic nature of watersheds limits the development as well use of physical based models to predict runoff in a river. Jamuna river is a sub-tributary of Brahmaputra river fed through snow and glacier melt. A barrage is constructed in the river for providing irrigation water to agricultural land. Artificial Nueral Network (ANN) model network has been developed for identification of the impacts of climate change in the runoff process of the Jamuna river using six predictors (surface upward latent heat flux, specific humidity, precipitation, wind speed, maximum air temperature and minimum air temperature) as input variables in the models. The results showed that the network was optimized with a network structure of 5–1–1. The RMSE and R^2 value was found to be 0.352 and 0.88 respectively. Further, the results suggested that the study provides practical significance to the water resources managers under changing climate and long-term prediction of hydrological processes.

Keywords ANN · Agriculture · Climate change · Runoff

M. Debnath (✉) · A. K. Sarma · C. Mahanta
Civil Engineering Department, Indian Institute of Technology Guwahati, Guwahati 781039, Assam, India
e-mail: mridu.ars590@gmail.com

M. Debnath
Scientist, Land and Water Management Division, ICAR-Research Complex for Eastern Region Patna, Patna 800014, India

M. Majumder and G. D. Kale (eds.), *Water and Energy Management in India*,
https://doi.org/10.1007/978-3-030-66683-5_6

1 Introduction

Globally hydrological processes are affected significantly by changing climate like shift in the pattern of precipitation, evapotranspiration and soil moisture storage (Minville et al. 2008; Srikanthan and McMahon 2001; Xu and Singh 2004; Ghosh and Majumdar 2007). The runoff that is water flow per unit of time, in a river from a watershed is dependent on climatic and physiographic factors of the drainage basin. In some cases manmade structures like reservoir, dams present in the river play a role in regulating the rate of water flow. However precipitation and evapotranspiration are the major climatic factors responsible for amount of runoff in the river along with other climatic factors.

The term "artificial intelligence" is used to define informally the machine language that imitates "cognitive" functions of humans associating with other human minds, for example "learning" and "problem-solving". One of the AI that is Analytical AI uses learning based on past experience to predict future decisions. Artificial Intelligence (AI), in the last two decades, has shown tremendous progress in predicting and modeling nonlinear hydrological applications as well as in solving data complexity (Yaseen et al. 2015). The most widely used AI approaches are artificial neural networks (ANNs) (Haykin 1994), Support Vector Machine (SVM) (Vapnik 1995), fuzzy logic approach (Zadeh 1965; Haykin 1994). Among all, artificial neural network (ANN) model, is the most widely used approach that is applied recently in the forecasting and modeling of various hydrological processes like stream flows (Kis 2007), evaporation (Kisi 2008), water quality (Zhao 2016), rainfall (Dahamsheh and Aksoy 2014), groundwater level changes (Govindaraju 2000; Garcia and Shigidi 2006; Yoon et al. 2011; Chang et al. 2015a). Furthermore, ANN yields significantly robust result for hydrological analysis, which involves non-linear systems of data and primarily when the physical laws governing the processes are not known precisely (Genxu et al. 2009). ANN hardly considers the detailed knowledge of the watershed's physical and climatic characteristics, which is in contrast to physically-based numerical models. ANNs can be used very conveniently in case of imprecise and complex data. Besides, hybrid ANN models like SWAT-ANN, ANN with various other models are developed to predict discharge of a river more precisely (Fathian et al. 2019; Kassem et al. 2020).

Rivers are lifeline of any area supporting land, agricultural activity and aquatic life. Fluctuations in river flows have vast consequences for the hydrologic cycle, crops as well as the environment. Besides various other uses like domestic, power generation, etc., most agricultural production is through irrigation from groundwater and river by the construction of barrage or dam. However, changing climatic conditions affect the river flow and that leads to change in agricultural system that the river sustains. ANN model has been used for various Indian rivers for predicting runoff (Khan et al. 2016; Shetty et al. 2018). Alok et al. (2013) applied ANN with Elman through cascade characteristics for simulating the runoff for the River Brahmani. In addition to this, ANN was used for stage–runoff relation modelling (Sudheer and Jain 2003; Bhattacharya et al. 2005). Further, ANN models can help understand the influence of environmental changes on the hydrological processes in hilly regions. However, many of these research studies are more concentrated for predicting rainfall-runoff relationships (Xu et al. 2014; Sattari et al. 2012; Yilmaz et al. 2011; Abudu et al. 2011; Chokmani et al. 2008; Nilsson et al. 2006). Few studies use ANN model to simulate runoff without considering directly climatic factors like precipitation or temperature rather than the driving force that moulds these climatic factors. Furthermore, due to the orographic variation in study region, getting the accurate rainfall and temperature data is relatively difficult (Chang et al. 2015b). Therefore, in this study, an ANN model is used to (1) develop a runoff ANN model and simulate runoff (2) examine the performance of the model to forecast runoff variations of hilly regions considering various other climatic factors besides precipitation and temperature, on the Jamuna river of Assam.

2 Materials and Methods

2.1 Study Area

The Jamuna command area and watershed feeding the command area lies between 25° 45′ N 92° 48′ E and 26° 4′ N 93° 45′ E. The command area lies on the left bank of Jamuna river, a sub tributary of Brahmaputra river of Assam. The command area covers two districts of Assam; Hojai and Karbianglong. The map of Jamuna watershed up to diversion weir was prepared by the use of ArcGIS software 9.1 and Jamuna command area map was collected from Jamuna Command Area Development Division (JCADD) which was digitized in ArcGIS software. The index map of the study area is shown in Fig. 1.

A diversion weir constructed at Bakuliaghat is used to tap the water of Jamuna river to irrigate a command area of 27,705 ha through main canal, distributaries, minors, and sub minors running in east to west direction. The watershed area of Jamuna river up to Bakuliaghat is 1,90,800 ha.

The Jamuna command area is uniformly graded and has a slope less than 2%. The general elevation of the command area ranges from RL 97.5 m to RL 62 m. It is covered with laterite soil. The command area is located in the rainshadow area of the Khasi Jayantia hills and the Barail ranges and receives a relatively less average annual rainfall of 1,179 mm. About 80% of rainfall is experienced during June to September that is during the monsoon months with peak rainfall during July and August while November to February is generally dry. The annual normal temperature in the command area ranges from 27 to 33 °C during summer and 16–24 °C during winter. The monthly average pan evaporation data is found to be 63 mm (JCADD Project Report).

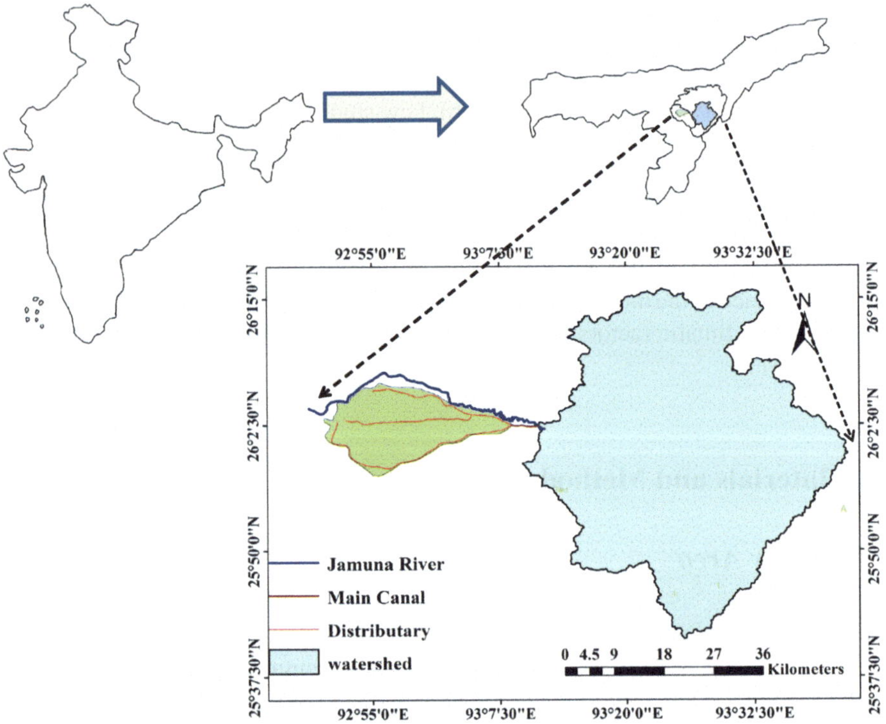

Fig. 1 Location map of study area

2.2 Data Collection

In the study region, daily runoff data were collected from the hydrological station at the weir site (Table 1).

The selection of predictors depends on the predictand of the climate change model and the area investigated. However, depending on various relationship coefficients between predictors and predictand, selection of some quantity of predictors can be made (Fistikoglu and Okkan 2010). Then Best Regression approach is used for determining the best regression models based on some performance criteria like R^2, Adj R^2 (Neter et al. 1996).

The GCM selected for the study area is *Goddard Institute for Space Studies* (GISS-E2-H), having resolution 2.5° × 1°, as it has sufficient historical and future data sets of RCPs. Monthly data for climate change have been procured from the

Table 1 Particulars of Jamuna river gauge station

Sl. No.	Station name	Location	Period of data	Time step
1.	Jamuna river	2.4 km upstream of weir site	1/1/2001–31/12/2016	Daily

site http://pcmdi9.llnl.gov/esgf-web-fe/. For the historical data HistoricalMisc for the year 1990–2015 has been downloaded.

How well a model fits the observed data usually is determined by pair wise comparison of model simulated or predicted values with the observed values. Both graphical and statistical methods were used to compare the simulated and observed runoff values. There are many statistical indicators among which the statistical indicators described below were used in this study for model training, validation and testing.

1. Co-efficient of Determination (R^2)

$$R^2 = 1 - \frac{\sum_{i=1}^{N}(O_i - P_i)^2}{\sum_{i=1}^{N}\left(O_i - O_{avg}\right)^2} \tag{1}$$

   Here, N = total number of observations; O_i = observed runoff at *i*th day; P_i = computed runoff at *i*th day and O_{avg} is the mean of O_i
   The co-efficient of determination value may vary from 0 to 1, with 1 indicating a perfect fit.
2. Root Mean Square Error (RMSE):

$$RMSE = \sqrt{\frac{\sum_{i=1}^{N}(O_i - P_i)^2}{N}} \tag{2}$$

   Here N = total number of observations; O_i = observed runoff at the *i*th day; P_i = computed runoff at the *i*th day
   RMSE can be used to measure the differences between observed and predicted values by a model. Its value starts from 0.0 and lower values indicating better agreement between the model and observations.

2.3 Data Analysis

The downloaded files are in NETCDF format which are then converted to tabular format using ArcGIS 9.1. The Jamuna watershed falls under two grids that are 93.75° E, 25.00° N and 93.75° E, 27.00° N as shown in Fig. 2. The area covering two grids is digitized and respective area was calculated in ArcGIS 9.1 (Hwang et al. 2013). The weighted average of the required data is calculated and then converted to the required units. Finally the processed outputs for 2001–2015 (baseline) can be used to generate climate change scenarios for the future. All the selected predictor data based on selection criteria (Table 2) of historical climate change model used to predict the predictand (runoff) consist of 15 years; 2001–2015. Analyses were performed through ANN using MATLAB Neural Network Toolbox. ANN model has partitioned the data according to the partitioning ratio of 70:15:15 for training, validation and test data respectively.

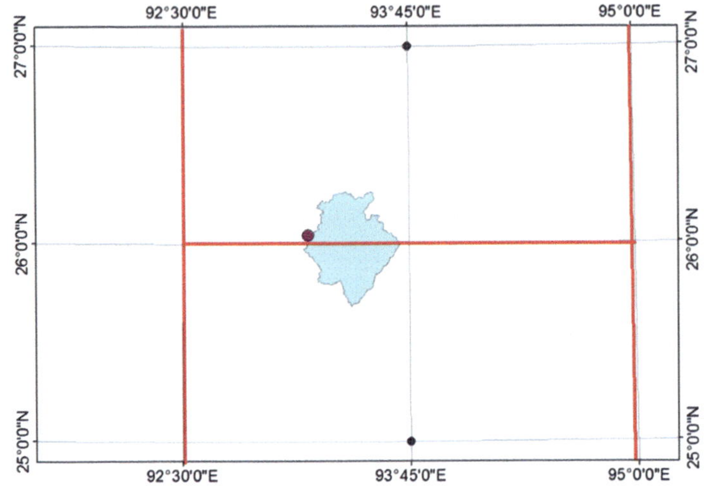

Fig. 2 Jamuna Watershed area with red coloured gridbox and grid point of GISS model

2.4 Development of an ANN-Based Runoff Forecasting Model

The ANN downscaling models can be considered as complex and nonlinear regression models structured between large scale atmospheric inputs and station-based parameters or predictand such as precipitation, temperature. The ANN structure designed here in MATLAB consists of three types of layers; the first one is an input layer consisting of predictors. The second layer is the hidden layer that may have several neurons. Here 5 neurons have been considered (Fig. 3), and all hidden neurons transform the inputs nonlinearly into another dimension employing weight and a bias term shown in Eq. 3. The last layer is the output layer that contains the station based parameter like runoff, which is the case in this study

$$y_k = f_2\left(\sum_{j=1}^{k}\left\{f_1\left(\sum_{i=1}^{n} x_i w_{ij} + w_j\right)\right\} w_{jk} + w_k\right) \tag{3}$$

where x_i = inputs; j = no. of neurons at the hidden layer; w_{ij} = weight of input x_i; f_1 = activation function at the hidden layer; f_2 = activation function at the output layer; k = no. of neurons at the output layer; w_{jk} = weight of neurons at the output layer and w_k = bias term.

The n inputs are transformed to the k outputs using the weights, the bias terms and activation functions (f). The activation functions are usually continuous, non-decreasing and bounded. Various activation function selections are possible (e.g., Haykin 1994; Hsu et al. 1995; Govindaraju and Rao 2000).

Table 2 List of parameters and their regression analysis

	Parameters	No. of parameters in the best regression models								
		1	2	3	4	5	6	7	8	9
	Surface upward latent heat flux (hfls)			•	•	•	•	•	•	•
	Surface upward sensible heat flux (hfss)				•	•		•	•	•
	Surface air pressure (ps)					•		•		•
	Relative humidity (hurs)	•						•	•	•
	Specific humidity (huss)						•	•	•	•
	Precipitation (pr)						•		•	•
	Wind speed (sfc)						•		•	•
	Maximum air temperature (tasmax)		•	•	•	•	•	•	•	•
	Minimum air temperature (tasmin)		•	•	•	•	•	•	•	•
R^2		0.572	0.626	0.659	0.659	0.664	0.665	0.666	0.667	0.667
Adj R^2		0.570	0.622	0.653	0.651	0.654	0.654	0.653	0.651	0.649

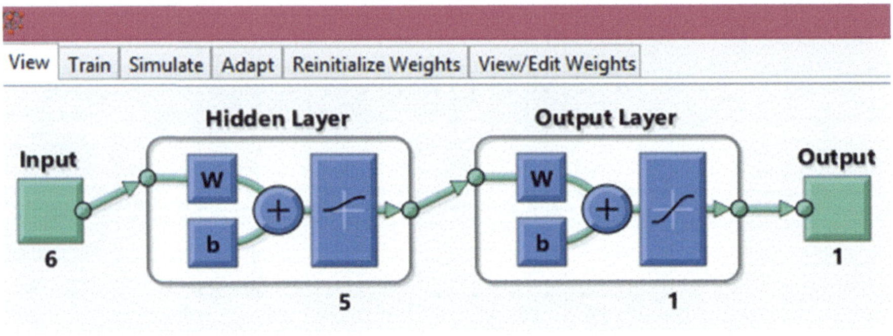

Fig. 3 Neural Network model for the Jamuna watershed in MATLAB

In this study, three widely used transfer functions, namely tangent sigmoid, linear, and log-sigmoid are evaluated in ANN construction trials. The best results are achieved by using the log- sigmoid function as given below:

$$f(z) \approx \frac{1}{1 + e^{-z}} \tag{4}$$

Searching the optimal value of the weights and biases is called the learning or the ANN training. Training is executed using a set of known inputs and output. In each training step, a set of inputs are passed forward through the network to result in the trial outputs which are then compared to the observed outputs or target data. If the difference residual is higher than the desired value, the error is passed backward through the network. The training algorithm uses the error to adjust the connection weights. This method is called the back-propagation algorithm, which is also used in this study. Once the comparison error is reduced to an acceptable level for the whole training set, the training phase ends (Haykin 1994; Demuth and Beale 1998).

In this study, the Levenberg-Marquardt feed forward-backpropagation algorithm is used for training of the network. The Levenberg-Marquardt feed forward-back-propagation algorithm is a second-order nonlinear optimization technique that is usually faster and more reliable than any other back-propagation process (Hagan and Menhaj 1994).

3 Results

The first set of 70% data is used for training and the remaining 30% of data is used equally for validation and testing (Fig. 4). From the correlation matrix it can be seen that 5 and 6 numbers of parameters have relatively good values of R^2 and Adj R^2 (Table 2). ANN model was then run using both the sets of parameters. However the set having the 6 numbers of parameters gave the best results. Recent monthly streamflow

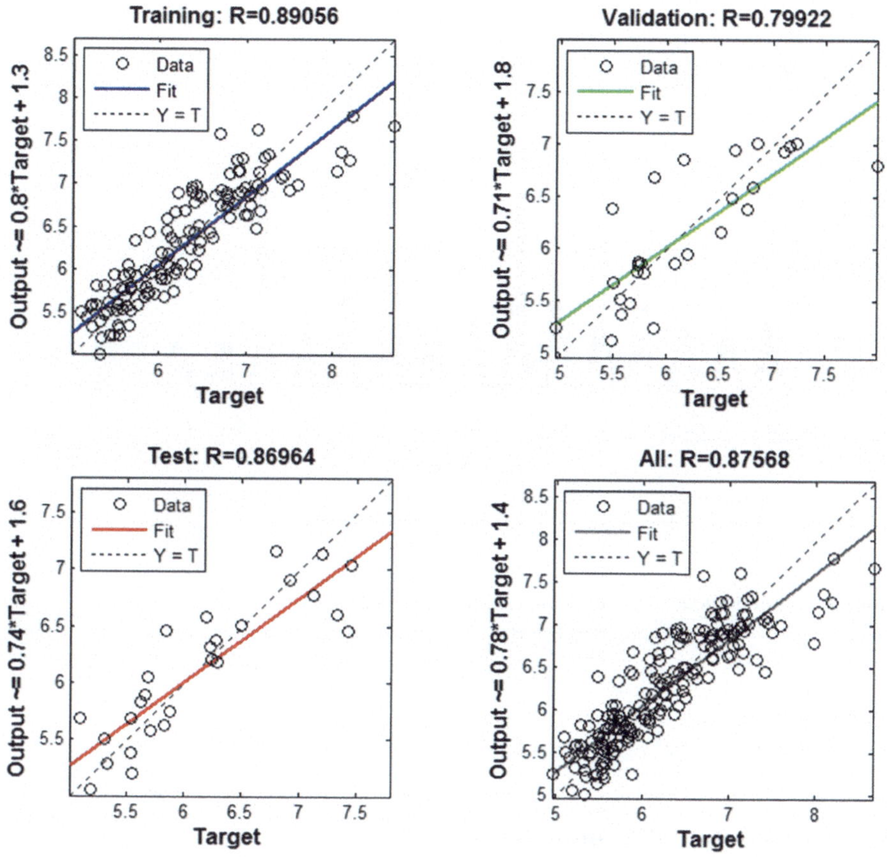

Fig. 4 View of Training, Validation, testing and overall data result for the Artificial Neural Network model in MATLAB

data from 2001–2015 was used for developing the ANN model (Fig. 4) using input data: surface upward latent heat flux, specific humidity, precipitation, wind speed, maximum air temperature and minimum air temperature. The runoff of Jamuna catchment increased with surface upward latent heat flux, specific humidity, precipitation, maximum and minimum temperature and decreased with wind speed. The overall correlation coefficient (R) between the target or log of observed streamflow data and output data was 0.88.

The results obtained after running the ANN model was graphically represented, as shown in Fig. 5. The pattern of the simulated and the observed value matched well except in some cases of peak runoff.

Table 3 shows the statistical analysis between observed and downscaled monthly streamflow data for the Jamuna watershed. The results overall suggested a good correlation of observed and downscaled streamflow values for the watershed.

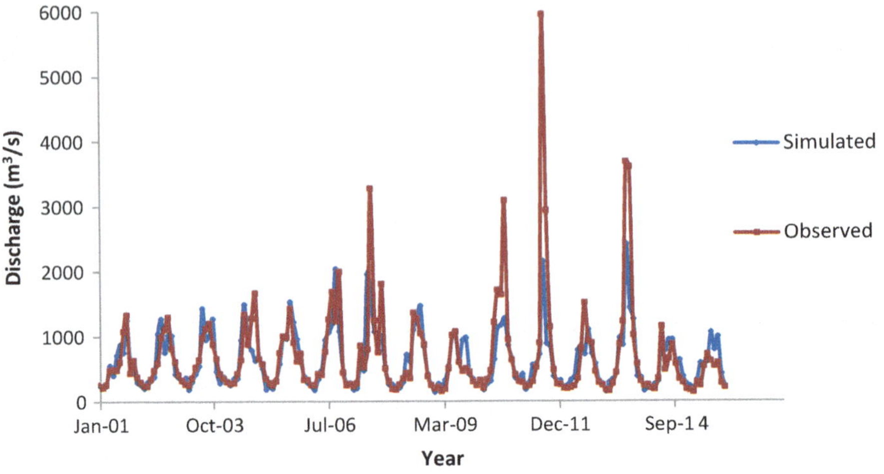

Fig. 5 ANN simulated and observed runoff values for overall data set

Table 3 Result of model evaluation for Jamuna watershed

	Training (2001–2015)				
	Mean	Std. Dev.	N	R^2	RMSE
Observed	6.25	0.73	180	0.77	0.352
Downscaled	6.25	0.65			

4 Discussion and Conclusions

In this study, an ANN-based runoff model was used to study the runoff response of Jamuna river to climate change model. The selection of input predictors is very important for predictand determination using ANN models as the unlinked variables could affect the predicting capacity of ANN models. Six predictors have worked efficiently to simulate the runoff in the Jamuna river at the weir site. Studies showed that runoff increased with an increase in temperature and precipitation as the Brahmaputra, of which Jamuna river is a sub-tributary, is fed through melting of snow and glaciers (Lin et al. 2019). Also runoff decreased with an increase in wind speed as an increase in near-surface wind speed at 2 m height increased the pan evaporation rate, thereby reducing runoff (Liu et al. 2011, 2014). Besides, the runoff is seen to increase with specific humidity and surface upward latent heat flux for a significant relationship of simulated runoff with the observed runoff.

Further, ANN models need to be appreciated for deriving robust simulation of runoff according to observed runoff inspite of the complex nature of data. This is true, especially for water managers, where more importance is given to the response of water resources to climate change rather than the process involved. Results suggested that ANN-based hydrological models have a better grasp over hydrological processes due to climate change than traditional hydrological models (Dawson and Wilby 1998; Imrie et al. 2000; Rajurkar et al. 2004; Solomatine and Dulal 2010; Mishra et al. 2014). Jamuna river, a typical river in orographic region, was used as a case to study the impact of climate changes on water resources where water resources are a vital factor for sustaining the aquatic life and agricultural productivity of the region. The developed ANN model could be expected for robust planning and management of future water resources after predicting future runoff under various climate change scenarios.

The main conclusions are summarized as follows:

1. With the variables of surface upward latent heat flux, specific humidity, precipitation, wind speed, maximum air temperature and minimum air temperature, the developed ANN runoff model could simulate runoff process well.
2. The runoff of Jamuna catchment increased with increase in surface upward latent heat flux, specific humidity, precipitation, maximum and minimum temperature, and decreased wind speed.
3. A network structure of 5–1–1 was developed which simulated runoff of the Jamuna river using climate change model data.
4. The coefficient of determination during the training period of streamflow data for the Jamuna watershed was 0.77.

References

Abudu, S., King, J. P., & Bawazir, A. S. (2011). Forecasting monthly streamflow of spring-summer runoff season in Rio Grande headwaters basin using stochastic hybrid modeling approach. *Journal of Hydrologic Engineering, 16*(4), 384–390.

Alok, A., Patra, K. C., & Das, S. K. (2013). Prediction of runoff with Elman and cascade neural networks. *Research Journal of Recent Sciences, 2,* 279–284.

Bhattacharya, B., Price, R., & Solomatine, D. (2005). Data-driven modelling in the context of sediment transport. *Physics and Chemistry of the Earth, 30,* 297–302.

Chang, J., Wang, G., Li, C., & et al. (2015a). Seasonal dynamics of suprapermafrost groundwater and its response to the freeing thawing processes of soil in the permafrost region of Qinghai Tibet Plateau. *Science China Earth Sciences, 58*(5), 727–738.

Chang, J., Wang, G., & Mao, T. (2015b). Simulation and prediction of suprapermafrost groundwater level variation in response to climate change using a neural network model. *Journal of Hydrology, 529,* 1211–1220.

Chokmani, K., Ouarda, T. B. M. J., Hamilton, S., et al. (2008). Comparison of ice-affected streamflow estimates computed using artifcial neural networks and multiple regression techniques. *Journal of Hydrology, 349*(3–4), 383–396.

Dahamsheh, A., & Aksoy, H. (2014). Markov chain-incorporated artifcial neural network models for forecasting monthly precipitation in arid regions. *Arabian Journal for Science and Engineering, 39*(4), 2513–2524.

Dawson, C. W., & Wilby, R. (1998). An artificial neural network approach to rainfall-runoff modeling. *Hydrological Sciences, 43,* 47–66.

Demuth, H., & Beale, M. (1998). *Neural network toolbox: For use with MATLAB user's guide.* Natick, MA: The Math Works Inc.

Fathian, F., Mehdizadeh, S., Sales, A. K., & Safari, M. J. S. (2019). Hybrid models to improve the monthly river flow prediction: Integrating artificial intelligence and non-linear time series models. *Journal of Hydrology, 575,* 1200–1213.

Fistikoglu, O., & Okkan, U. (2010). Statistical downscaling of monthly precipitation using NCEP/NCAR reanalysis data for Tahtali River Basin in Turkey. *Journal of Hydrologic Engineering, 16*(2), 157–164.

Garcia, L. A., & Shigidi, A. (2006). Using neural networks for parameter estimation in ground water. *Journal of Hydrology, 318*(1–4), 215–231.

Genxu, W., Shengnan, L., Hongchang, H., et al. (2009). Water regime shifts in the active soil layer of the Qinghai-Tibet Plateau permafrost region, under different levels of vegetation. *Geoderma, 149*(3–4), 280–289.

Ghosh, S., & Majumdar, P. P. (2007). Nonparametric methods for modeling GCM and scenario uncertainty in drought assessment. *Water Resources Research, 43.*

Govindaraju, R. S. (2000). Artifcial neural networks in hydrology. In: preliminary concepts. *Journal of Hydrologic Engineering, 5*(2), 115–123.

Hagan, M. T., & Menhaj, M. B. (1994). Training feed forward network with the Marquardt algorithm. *IEEE Transactions on Neural Networks, 5*(6), 989–993.

Haykin, S. (1994). *Neural networks: A comprehensive foundation.* New York, NY, USA: Macmillan College Publishing Company.

Hsu, K. L., Gupta, H. V., & Sorooshian, S. (1995). Artificial neural network modeling of the rainfall-rainoff process. *Water Resources Research, 3110,* 2517–2530.

Hwang, S., Wendy, D. G., Adams, A., et al. (2013). Assessment of the utility of dynamically-downscaled regional reanalysis data to predict streamflow in west central Florida using an integrated hydrologic model. *Regional Environmental Change, 13,* 69–80.

Imrie, C. E., Durucan, S., & Korre, A. (2000). River flow prediction using artificial neural networks: generalisation beyond the calibration range. *Journal of Hydrology, 233,* 138–153.

Jamuna Command Area Development Division (JCADD) (1961) Project Report.

Kassem, A. A., Raheem, A. M., Khidir, K. M., & Alkattan, M. (2020). Predicting of daily Khazir basin flow using SWAT and hybrid SWAT-ANN models. *Ain Shams Engineering Journal, 11*(2), 435–443.

Khan, M. Y. A., Hasan, F., Panwar, S., et al. (2016). Neural network model for runoff and water-level prediction for Ramganga River catchment of Ganga Basin, India. *Hydrological Science Journal, 61*(11), 2084–2095.

Kis, O. (2007). Streamflow forecasting using different artifIcial neural network algorithms. *Journal of Hydrologic Engineering, 12*(5), 532–539.

Kisi, O. (2008). Te potential of different ANN techniques in evapotranspiration modelling. *Hydrological processes, 22*(14), 2449–2460.

Lin, Y., Wen, H., & Liu, S. (2019). Surface runoff response to climate change based on artificial neural network (ANN) models: a case study with Zagunao catchment in Upper Minjiang River, Southwest China. *Journal of Water and Climate Change, 10*(1), 158–166.

Liu, X., Zhang, X. J., Tang, Q., & Zhang, X. Z. (2014). Effect of surface wind speed decline on modeled hydrological conditions in China. *Hydrology and Earth System Sciences, 18,* 2803–2813.

Liu, X. M., Luo, Y. Z., Zhang, D., Zhang, M. H., & Liu, C. M. (2011). Recent changes in pan-evaporation dynamics in China. *Geophysical Reseach Letters, 38,* L13404.

Minville, M., Brissette, F., & Leconte, R. (2008). Uncertainty of the impact of climate change on the hydrology of a nordic watershed. *Journal of Hydrology, 358,* 70–83.

Mishra, S., Gupta, P., Pandey, S. K., et al. (2014). An efficient approach of artificial neural network in runoff forecasting. *International Journal of Computer Applications, 92,* 9–15.

Neter, J., Kutner, M., Nachtsheim, C., & Wasserman, W. (1996). *Applied linear statistical models.* New York: McGraw Hill.

Nilsson, P., Uvo, C. V., & Berndtsson, R. (2006). Monthly runoff simulation: Comparing and combining conceptual and neural network models. *Journal of Hydrology, 321*(1–4), 344–363.

Rajurkar, M. P., Kothyari, U. C., & Chaube, U. C. (2004). Modeling of the daily rainfall-runoff relationship with artificial neural network. *Journal of Hydrology, 285,* 96–113.
Sattari, M. T., Apaydin, H., & Ozturk, F. (2012). Flow estimations for the Sohu Stream using artifcial neural networks. *Environmental Earth Sciences, 66*(7), 2031–2045.
Shetty, S. T., Samanta, C. S., Timane, S. N., et al. (2018). Assessment and estimation of flood runoff using ANN. *International Journal for Scientific Research & Development, 6*(1), 1648–1653.
Solomatine, D. P., & Dulal, K. N. (2010). Model trees as an alternative to neural networks in rainfall–runoff modelling. *Hydrological Sciences, 48,* 399–411.
Srikanthan, R., & McMahon, T. A. (2001). Stochastic generation of annual, monthly and daily climate data: a review. *Hydrology and Earth System Sciences, 5*(4), 653–670.
Sudheer, K. P., & Jain, S. K. (2003). Radial basis function neural network for modelling rating curves. *Journal of Hydrologic Engineering, 8*(3), 161–164.
Vapnik, V. N. (1995). *The nature of statistical learning theory.* Springer.
Xu, C. Y., & Singh, V. (2004). Review on regional water resources assessment models under stationary and changing climate. *Water Resources Management, 18,* 591–612.
Xu, J., Chen, Y., Li, W., et al. (2014). Integrating wavelet analysis and BPANN to simulate the annual runoff with regional climate change: a case study of Yarkand River, Northwest China. *Water Resources Manage, 28*(9), 2523–2537.
Yaseen, Z. M., El-shafe, O., Jaafar, H. A., et al. (2015). Artifcial intelligence based models for stream-flow forecasting: 2000–2015. *Journal of Hydrologic Engineering, 530,* 829–844.
Yilmaz, A. G., Imteaz, M. A., & Jenkins, G. (2011). Catchment flow estimation using artifcial neural networks in the mountainous Euphrates Basin. *Journal of Hydrologic Engineering, 410*(1–2), 134–140.
Yoon, H., Jun, S., Hyun, Y., et al. (2011). A comparative study of artifcial neural networks and support vector machines for predicting groundwater levels in a coastal aquifer. *Journal of Hydrology, 396*(1–2), 128–138.
Zadeh, L. A. (1965). *Fuzzy sets.* Retrieved June 12, 2019, from http://dx.doi.org/10.1109/2.53.
Zhao, Y., Guo, L., Liang, J., & Zhang, M. (2016). Seasonal artificial neural network model for water quality prediction via a clustering analysis method in a wastewater treatment plant of China. *Desalination and Water Treatment, 57*(8), 3452–3465.

Modelling of Reference Evapotranspiration for Semi-arid Climates Using Artificial Neural Network

Adeeba Ayaz, Sharath Chandra, Pratik Mandlecha, and Rehana Shaik

Abstract Reference Evapotranspiration (ET_0) is one of the prominent hydrologic variables affecting water and energy balances and critical factors for crop water requirements and irrigation scheduling. Evapotranspiration is a complex hydrological variable defined by various climatic variables. Various empirical formulations have been developed to estimate ET_0 depending upon the availability of meteorological variables. Such empirical formulations are region-specific and are for particular climatic conditions. In this context, mathematical models have emerged as simple and readily implementable for the estimation of ET_0 with measured meteorological parameters as independent variables. Such data-driven models can be valuable to predict ET_0 when climate data is insufficient. The present study compared various empirical models and data-driven algorithms to predict ET_0 using various climate variables. Artificial neural networks (ANN) were adopted to estimate reference ET_0. Four empirical methods Penman-Monteith, Hargreaves, Turc, and Priestley-Taylor were used to estimate ET_0 at a daily time scale. Dataset consists of daily meteorological data over a period of 51 years (1965–2015) for Hyderabad, the largest city of the Indian state, Telangana, with semi-arid climate. The input variables for the ANN model consist of maximum and minimum air temperatures, relative humidity, solar radiation, and wind speed. The Penman-Monteith method was considered as the standard method to compare the ANN and various empirical models of ET_0. ANN model was trained and tested with climate variables as input variables and various empirical models as reference models. The most influencing climate variables on ET_0 were found in the order of temperature, solar radiation, wind speed, and relative humidity based on correlation coefficients. These variables have formed as the basis to choose different datasets to train over ANN model. Validation has been carried out using the coefficient of determination (R^2) which is obtained for the training (1965–2000) and testing period (2001–2015) period as 0.97 and 0.96 respectively. Temperature and radiation-based models of Turc and Priestley-Taylor methods can be used to estimate ET_0 when all other climate variables are not available as they

A. Ayaz · S. Chandra · P. Mandlecha · R. Shaik (✉)
Lab for Spatial Informatics, International Institute of Information Technology, Hyderabad 500032, Telangana, India
e-mail: rehana.s@iiit.ac.in

M. Majumder and G. D. Kale (eds.), *Water and Energy Management in India*,
https://doi.org/10.1007/978-3-030-66683-5_7

also correlate well with the Penman-Monteith method. Advancement towards artificial intelligence techniques in water resources engineering has motivated to simulate reference ET_0 using limited meteorological variables to produce accurate results. Such data-driven algorithms developed based on standard empirical models can be implemented for prediction with limited climate data.

Keywords Artificial neural network · Penman-Monteith · Evapotranspiration · Turc · Priestley-Taylor · Hargreaves

1 Introduction

Evapotranspiration (ET) is one of the most critical components of the hydrological water cycle affecting terrestrial water-energy balances. Actual evapotranspiration, potential evapotranspiration, and reference evapotranspiration are significant types of evapotranspiration. Actual Evapotranspiration (AET) is a significant component of the water balance and utilized generally in fields such as agronomy, hydrology, climatology, meteorology, ecology, and environmental sciences (Chiew and McMahon 2002; Liu et al. 2018; Peng et al. 2019; Tasumi 2019). Two more closely related types of evapotranspiration are potential evapotranspiration (PET) and reference evapotranspiration (ET_0). Although both PET and ET_0 provide estimates of atmospheric evaporative demand, they are based on different ideas, concepts, application fields and have different equations that can help to differentiate the terms. However, many researchers have treated PET and ET_0 as identical concepts and used similar equations for their estimation (Allen & Food and Agriculture Organization of the United Nations 1998; Irmak and Haman 2003a, b; Yates 1997; Zhang et al. 2017) The first idea of PET was proposed by Thornthwaite (1948) and that core idea with improvements are being used now. The PET equations were classified as mass-transfer, temperature, and radiation-based, while the reference ET_0 equations were classified as temperature, radiation, and pan-evaporation based (Chiew and McMahon 2002). PET has been applied mostly in hydrology, meteorology, and climatology. Whereas, the ET_0 has been applied mostly in agronomy, agriculture, irrigation, and ecology. The accurate estimation of ET_0 is essential in irrigation planning, scheduling, hydrological balance studies, and watershed hydrology (Feng et al. 2018; Yao et al. 2018). It has a broader significance in numerous fields of research including crop yield simulation, optimization of water lost, management and irrigation system design, water usage improvement in agriculture, and hydrologic water balance.

The ET_0 can be estimated based on energy balance and water vapour mass flux transfer methodologies (Rehana et al. 2020). Various methods to estimate reference ET_0 have been developed and are being utilized, depending upon the availability of meteorological variables. Empirical models for ET_0 estimation, i.e., statistical functions of approximation between meteorological variables and values, can overcome the difficulties associated with data availability for ET_0 estimation (Magliulo et al. 2003; Naoum and Tsanis 2003a, b). Among these Priestley-Taylor (Hargreaves and

Samani 1985; Penman 1948; Priestley and Taylor 1972) are well-established models. These empirical models vary in terms of solar radiation, temperature considering the physical processes of radiation and transport characteristics of natural surfaces.

The modified Penman-Monteith 56 equation has been recommended for the calculation of ET_0 and calibration of other equations by various international organizations such as the United Nations Food and Agriculture Organization (FAO) and World Meteorological Organization (Allen & Food and Agriculture Organization of the United Nations 1998; Walter et al. 2001). The Penman-Monteith equation has two critical advantages. First, it can be used in a wide variety of environments and climate scenarios without the need for any local calibrations because of its physical basis. Second, it is a well-documented method that has been validated using lysimeters under a wide range of climate conditions (Landeras et al. 2008). The main drawback of this equation is that it requires data on a large number of climate variables that are unavailable in many regions. An empirical model such as The Priestley-Taylor equation (Priestley and Taylor 1972) can estimate regional monthly ET_0 provided that the adjustment factor is adapted to different site conditions (Castellvi et al. 2001).

Nevertheless, the superiority of the Penman-Monteith method over the Priestley Taylor equation has recently been demonstrated (Alexandris et al. 2006) carried out surface polynomial regression analysis using hourly solar radiation, air temperature, and relative humidity (RH) to estimate ET_0. A much simpler alternative is the Thornthwaite scheme (Thornthwaite 1948) as it requires the only temperature as input data. However, this approach has been found to underestimate ET_0 under arid conditions and overestimate in a humid climate (Pereira and Pruitt 2004). The Hargreaves and Samani equation are an empirical approximation of the ET_0 calculation based on temperature and extraterrestrial radiation data (Gafurov et al. 2018).

Empirical models can be the best choice for estimating the ET_0 given the availability of meteorological variables. However, for ungauged basins where meteorological data is insufficient data-driven algorithms have proven to be valuable tools. Such data-driven models work with various climate factors as input variables and ET_0 estimates as reference models. In the past decades, there has been a widespread interest in the application of data-driven modelling and machine learning techniques in the field of water resources and hydrology (Kumar et al. 2020; Rehana 2019). In this context, the Artificial Neural Network (ANN) has been a widely applied machine learning algorithm in water resources engineering, including evapotranspiration (Kumar et al. 2020). ANNs are mathematical models whose architecture is inspired by biological neural networks and are highly appropriate for the modelling of nonlinear processes and are being used to predict and forecast water variables in the last decades and have been successfully used in hydrological processes, water resources management, water quality prediction and reservoir operation (Antonopoulos and Antonopoulos 2017). In recent years, ANN algorithms have been applied in the field of ET_0 estimation. Kumar et al. (2002) developed ANN models for the estimation of ET_0 and found that the ANNs could predict ET_0 better than the conventional empirical methods. More recently, Kisi (2007) investigated the modelling of ET_0 using ANNs with the Levenberg-Marquardt training algorithm and inferred that ANNs could be employed successfully in modelling ET_0 from available climate data. Jain et al. (2015) interpreted the physical meanings of ANNs for ET_0 estimation. Some of them utilized the comparable climatic data required for the application of the FAO Penman-Monteith method (Kumar et al. 2002; Odhiambo et al. 2001a, b; Trajkovic 2005). These researchers reported that the ANN can anticipate ET_0 ever better than the FAO Penman-Monteith conventional method. Sudheer et al. (2003) and Zanetti et al. (2007) simplified the input variables, and ET_0 was evaluated as a function of air temperature, extraterrestrial solar radiation, and daylight hours. Chauhan and Shrivastava (2009) compared the performance of four climate-based methods and Artificial Neural Networks (ANNs) for estimation of ET_0 when input climatic parameters are insufficient to apply the FAO Penman-Monteith method. They concluded that ANN models performed better than climatic methods. Suryavanshi et al. (2014) examined the trend in temperature and potential evapotranspiration over the Betwa basin, India. Sonali and Nagesh Kumar (2016) analysed the trend of maximum and minimum temperature of annual, monthly, winter, pre-monsoon, monsoon, and post-monsoon. The studies were carried out for three-time slots 1901–2003, 1948–2003, and 1970–2003, for India as a whole and seven homogeneous regions of India. Authors considered the effect of serial correlation, trend detection analysis while applying the Mann-Kendall test, Sen's slope estimator, and other non-parametric methods. Bandyopadhyay et al. (2020) have carried out the trend analysis of ET_0 using the Mann-Kendall trend test for India. The authors indicated that the leading cause of the rising trends of ET is due to an increase in relative humidity and a decrease in wind speed for the study duration. In another study, Rahimikhoob (2010) applied the ANN technique to estimate ET_0 based on air temperature data under humid subtropical conditions on the southern coast of the Caspian Sea situated in the north of Iran. The study showed that ANN successfully estimated the daily ET_0 better than the Hargreaves classical equation. Adamala (2018) made a comparison of developed models with the artificial neural network models and also with the linear and wavelet regression and conventional methods to estimate evapotranspiration using temperature-based generalized wavelet-neural network models. Estimation of the ET_0 of Punjab was done based machine learning models and was compared in predicting daily ET_0 with the performance of the Deep Learning model and was compared to Penman-Monteith model. The Generalized Linear Model (GLM), Random Forest (RF), and Gradient-Boosting Machine (GBM) models were also used in the study as various machine learning algorithms and concluded that the deep learning model performed better than the considered models for training, validation and testing sets. Pal and Deswal (2009) and Saggi and Jain (2019) investigated the different data-driven based regression approaches to model daily ET_0 using four inputs, including solar radiation, average air temperature, average relative humidity, and average wind speed. Results from their study suggested that the different data-driven and machine learning models could successfully be employed in modelling the ET_0.

The present study made efforts to implement the ANN model for the estimation and prediction of ET_0 in a semi-arid climate of India. The objectives of this study are to (1) develop ANN models with available climate factors for ET_0 estimation using long-term meteorological data; (2) to assess the applicability and validity of different ET_0

methods such as Penman-Monteith, Priestley-Taylor, Hargreaves, and Turc methods. Since the maximum and minimum air temperature and relative humidity records are more readily available around the globe, these records with extraterrestrial radiation are being used as input in the above models for the estimation of ET_0. Extraterrestrial radiation reflects the seasonality of ET_0 and can theoretically be calculated as a function of the local latitude and Julian data, according to the equations presented by Allen & Food and Agriculture Organization of the United Nations (1998). Therefore, for the models suggested in this study, only temperature and relative humidity are the parameters that require monitoring. Here, the FAO Penman-Monteith method was used as a substitute for measured ET_0 data, as this is the standard procedure used when no measured lysimeter data is available (Irmak and Haman 2003a, b). The study has been implemented on the semi-arid climate conditions of Hyderabad, Telangana, India.

2 Data and Case Study

The area under study was Hyderabad, the largest city of the Indian state of Telangana which lies between latitude 17.3850° N and longitude 78.4867° E located on the Deccan Plateau in the northern part of South India and covers an area of 650 km^2 at an elevation of 542 m. Based on the Koppen climate classification, the climate is tropical wet and dry bordering on a hot semi-arid, with an average annual precipitation of about 171 mm (Fig 1).

Daily meteorological data were obtained from January 1965 through December 2015 (51 years) (612 months) from weather station situated in Professor Jayashankar Telangana State Agricultural University, Rajendranagar Mandal, Hyderabad, Telangana. The annual average weather data of the meteorological station is presented in Table 1. Five monthly meteorological variables were recorded including: (1) mean maximum air temperature (T_x °C); (2) mean minimum air temperature (T_n °C); (3) mean relative humidity (RH %); (4) mean wind speed (U_2 m s^{-1}); (5) solar radiation (R_s, MJ m^{-2} d^{-1}) and (6) Evapotranspiration (ET_0 mm/day). Measurements were made at the height of 2 m (air temperature and relative humidity) and 10 m (wind speed) above the soil surface. Wind speed data at 2 m (U_2) were obtained from those taken at 10 m using the log-wind profile equation.

3 Materials and Methods

This study mainly implemented the ANN model for estimation and prediction of ET_0. Evapotranspiration is calculated using the following methods with the limited meteorological parameters by considering Penman-Monteith method as standard method as it requires radiation, wind speed, relative humidity and temperature.

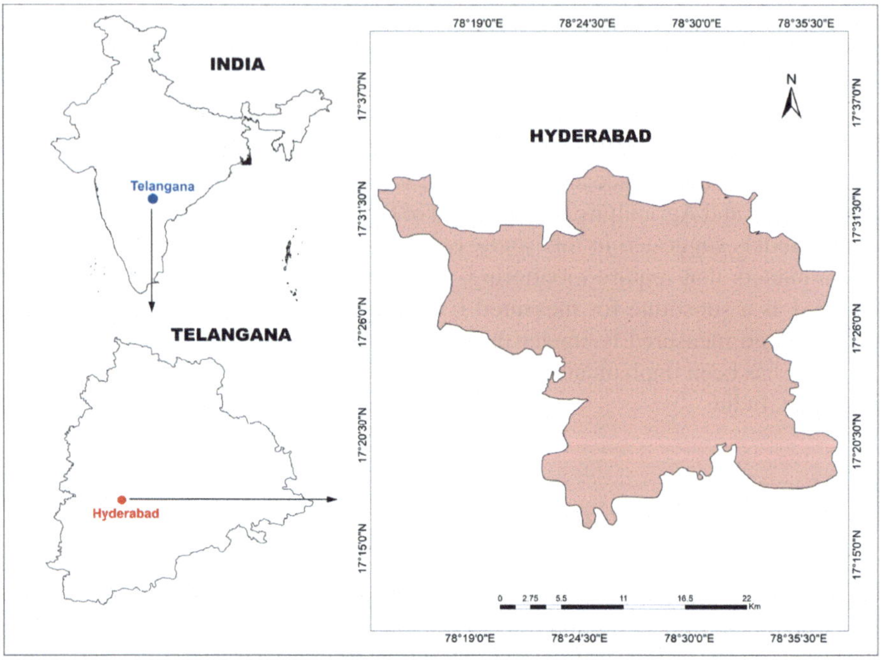

Fig. 1 Case study: Hyderabad, Telangana, India

Table 1 Statistical parameters of available meteorological variables and ET_0 at Hyderabad

Parameters	T_x	T_n	RH_{mean}	U_2	R_s	ET_0
Maximum	45.5	33.0	139	36.0	14.45	13.16
Minimum	17.6	5.0	10	0	4.00	0.48
Mean	32.37	19.88	60.70	4.69	9.32	3.76
Standard deviation	4.1	4.79	14.93	4.62	2.44	1.72

3.1 *FAO-56 Penman-Monteith Method*

The Penman-Monteith (Penman 1948) method was recommended by the FAO. It is calculated on a daily time scale. The formulation can be expressed as follows:

$$ET_0 = \frac{0.408\Delta(Rn - G) + g\left(\frac{900}{T+273}\right)U_2(e_s - e_a)}{\Delta + g(1 + 0.34U_2)} \tag{1}$$

where, Rn is net radiation (MJ m^{-2} d^{-1}), G is soil heat flux (MJ m^{-2} d^{-1}), T is average temperature at 2 m height (°C), U_2 is wind speed measured at 2 m height (m s^{-1}), ($e_s - e_a$) is pressure deficit for measurement at 2 m height (k Pa), Δ is slope vapor pressure curve (k pa °C^{-1}), g is psychrometric constant (k pa °C^{-1}), 900 is coefficient for the reference crop (l J^{-1} kg K d^{-1}), 0.34 is wind coefficient for the reference crop (s m^{-1}).

3.2 *Turc Method*

Turc (1961) method estimates ET_0 based on mean temperature and solar radiation on daily time scale. The formulation can be expressed as follows

$$ET_0 = 0.013\frac{Tm}{Tm + 15}(23.88Rs + 50) \quad (2)$$

where Tm is mean temperature (°C), solar radiation (Rs) is [0.25 + 0.5 (n/N)] Ra, Ra is extraterrestrial radiation (mm day^{-1}), n is actual hours of bright sunshine (h), N is maximum possible hours of sunshine (h).

3.3 *Priestly and Taylor Method*

Priestley and Taylor (1972) method is calculated using net radiation and latent heat of vaporization on a daily time scale. The formulation can be expressed as follows

$$ET_0 = A\left(\frac{D}{D + g}\right)\left(\frac{Rn - G}{L}\right) \quad (3)$$

$$D = \frac{4098\left[0.6108exp\left(\frac{17.27*Tm}{Tm+237.3}\right)\right]}{(Tm + 237.3)^2} \quad (4)$$

where D is slope vapour pressure curve (k pa °C^{-1}), g is psychrometric constant (k pa °C^{-1}), Rn is the net radiation at crop surface (MJ m^{-2} d^{-1}), A is a calibration constant 1.26, L is the latent heat of vaporization and can be considered as 2.45 (MJ/kg) which is constant.

3.4 *Hargreaves Method*

Hargreaves (1972) method which was modified in 1985 (Hargreaves 1983) estimates ET_0 based on temperature and radiation is calculated on a daily time scale. The formulation can be expressed as follows:

$$ET_0 = 0.0023Ra\left(\frac{Tm}{2} + 17.8\right)(Td^{0.5}) \quad (5)$$

where, Td is difference between maximum temperature and min temperature (°C), Tm is mean temperature (°C), Ra is extra-terrestrial radiation (mm day^{-1})

3.5 Artificial Neural Networks (ANN)

Artificial Neural Networks has gained much attention in hydrology for the prediction of various conceptual processes such as rainfall-runoff, streamflows, water quality and ground water modelling, etc. (Kurian et al. 2020). ANN is a computational model inspired by networks of biological neurons, wherein the neurons compute output values from inputs (Heddam and Kisi 2018). It learns from its past experience and errors in a nonlinear parallel processing manner (Gupta and Singh 2011). ANNs are fully connected neural nets that consist of an input layer, hidden layers (multiple or single), output layer. Each node can be considered as a neuron. The neuron is the basic calculating entity that computes from a number of inputs and delivers one output compared with a threshold value and turned on (fired). The computational processing is done by internal structural arrangement consisting of hidden layers that utilize the backpropagation and feed-forward mechanism to deliver output close to accuracy. Fully connected neural nets are those where each node in a layer is connected to every other node in the next layer (right). Each node takes the weighted sum of its inputs which then passes through a nonlinear activation function (like RELU, sigmoid, tanh, etc.), which then becomes the input of other nodes in the next layer (Rumelhart et al. 1986). In Eq. 6 the function, f, represents the activation function and w is the weight matrix, X is the set of input vectors (Fig. 2).

$$\mathrm{Z} = \mathrm{f}\,(\mathrm{x.\ w}) = \mathrm{f}\left(\sum_{i=1}^{n} x_i w_i\right) \quad x \in d_{1\times n}, w \in d_{n\times 1}, z \in d_{1\times 1}, \tag{6}$$

The present study used a feed-forward backpropagation neural network. The weights are initially randomly assigned. The train: test spilt on the dataset is 7:3. A forward pass is performed for every training data using the current weights, and the output is calculated for each node. At the last node, the final output is acquired, and the error is calculated with a loss function. Now, a backward pass is performed to calculate the contribution of each node in error calculated. The error is propagated to every single node using backpropagation. Once, the contribution of each node has been calculated the weights are adjusted accordingly using gradient descent. The present study used gradient descent with momentum and adaptive linear regression. The procedure is repeated until the loss function gives an error which is less than the threshold value and the weights and bias of the required network are thus obtained. Thus, the model converges, and a definite result can be obtained for any type of testing dataset.

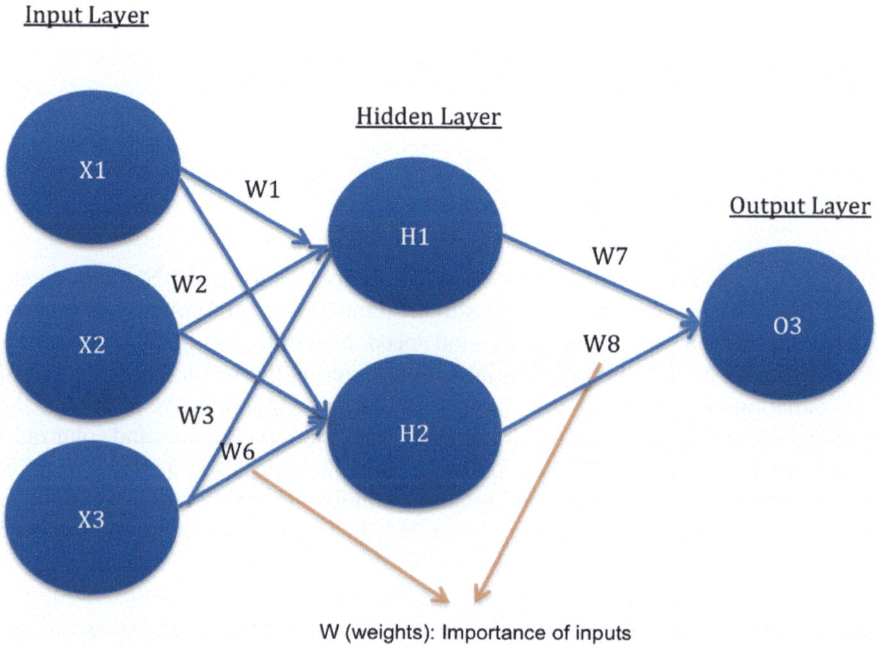

Fig. 2 Structure of ANN used for training a model with hidden layer and weights and the output layer showing a feed-forward pass

3.6 Performance Metrics

The model's performance criteria were validated using different standard statistical methods. In this study coefficient of Determination (R^2) (Krause et al. 2005) Root Mean Square Error (RMSE) (Legates and McCabe 1999) the Mean Absolute Error (MAE) were used as validation criteria. The equations for these methods are as follows:

$$\mathbf{R^2} = \left(\frac{\sum_{i=1}^{n} (O_i - O_{av_9})(S_i - S_{av_9})}{\sqrt{\sum_{i=1}^{n} (O_i - O_{av_9})^2} \sqrt{\sum_{i=1}^{n} (S_i - S_{av_9})^2}} \right)^2 \quad (7)$$

$$\mathbf{RMSE} = \sqrt{\sum_{i=1}^{n} (O_i - S_i)^2} \quad (8)$$

$$\mathbf{MAE} = \frac{1}{N} \sum_{i=1}^{n} (O_i - S_i) \quad (9)$$

where O is the observed values (the reference evapotranspiration), S is the simulated values by the other methods, and O_{avg} and S_{avg} are the mean observed and computed values, respectively.

4 Results and Discussions

The climate variables considered for estimating daily ET_0 using ANN and Penman-Monteith methods were the daily maximum temperature, minimum temperature, relative humidity, solar radiation and wind speed. Similarly, for the Turc and corresponding ANN model, the input variable considered is the mean temperature and solar radiation. Whereas, for the Hargreaves and corresponding ANN model, the input variables considered are maximum and minimum temperatures and solar radiation. Furthermore, for the Priestly Taylor method, the input variables used in ANN are temperature, solar radiation, and relative humidity. The ANN model used in the present study is Multi-Layered Perceptron (MLP) imported from the sci-kit-learn library in python. The study used three hidden layers with several neurons same as the number of features or parameters, i.e. 6 (maximum air temperature, minimum temperature, relative humidity, solar radiation), and ran the model for 500 iterations. Convergence was obtained for the datasets of all the four empirical methods. The prediction values have been calculated by fitting the test data on the trained model. As the number of meteorological variables for each empirical method is different, therefore, for each empirical model, an ANN model was trained, and results were tested. Input vector has the features considered in each method (Penman, Hargreaves, Turc, and Priestley-Taylor) 3 hidden layers have been used for each method and the output vector is the expected reference evapotranspiration value calculated from each method. The optimal node number in the hidden layer of the network was determined using a trial and error method by considering the MAE, RMSE, and R^2 values from a test sample. In this study, ANNs were trained for 500 epochs with one to 6 nodes in the hidden layer and mentioned before, statistical parameters were calculated using only the whole test data set after each training run. The training period considered is from 1965 to 2000 and the testing period considered from 2001 to 2015. The validity and efficiency of the model can be seen when the training dataset is fit on the trained model, and high accuracy and minimal values of RMSE were obtained. The performance of each empirical model corresponding with the ANN model in terms of R^2, RMSE, and MAE was listed in Table 2. Figures 3, 4, 5, 6 and 7 shows the comparison between daily ET_0 values form empirical models of Penman-Monteith, Priestley-Taylor, Hargreaves, Turc, and ANN methodologies for training and testing datasets.

Figure 3a shows the comparison of ET_0 daily values predicted by the ANN model versus the ET_0 values of the Penman-Monteith method for both testing and training periods. A good correlation was observed with R^2 values as higher than 0.95, RMSE as 0.03, and MAE as 0.009 between the Penman-Monteith method and ANN for the testing period. The trained and tested ANN model performs very well when compared

Table 2 Statistical summary of testing and training period for ANN

Empirical methods	Artificial neural network (training)			Artificial neural network (testing)		
	R^2	RMSE	MAE	R^2	RMSE	MAE
Penman-Monteith	0.97	0.02	0.008	0.96	0.03	0.009
Turc	0.96	0.03	0.007	0.95	0.04	0.012
Hargreaves	0.94	0.05	0.015	0.94	0.06	0.017
Priestley Taylor	0.91	0.10	0.022	0.92	0.12	0.025

with Penman-Monteith estimates. The comparison shows that neither overestimation nor underestimation was produced in the range of the values studied. This verifies that the ANN models can be used to estimate ET_0 values. Thus, compared to all other empirical models, the Penman-Monteith has been predicted well with the data-driven algorithm of ANN. It can be noted that, as the Penman-Monteith method accounts for all climate variables into modelling, such accuracies were expected to be comparable to other empirical models.

Furthermore, the present study tried to understand the sensitivity and dependency of each meteorological variable on the modelled ET_0 using the Penman-Monteith model. The study plotted the scatter plots between each climate variable and ET_0 modelled based on the Penman-Monteith method, as shown in Fig. 8.

As shown in Fig. 8, the temperature and solar radiation followed by relative humidity have the most substantial influence on ET_0 estimations based on the Penman-Monteith model. Therefore, ANN models that were derived based on temperature, solar radiation, wind speed and relative humidity as input and the ET_0as output variables. The ANN model results, when using (T, RH, R_s and U_2) from the four essential meteorological variables as input, seldom show the same values of coefficient of determination (R^2). These results prove that the relative humidity has a very low contribution to ET_0 when using ANNs models. The overall accuracies of most models were found to be similar to each other.

Furthermore, the results of the ANN can be significantly influenced by the number of input data which can lead to significant error and deviation. On the other hand, lowering the number of neurons in the input layer to three or even two can give us satisfactory results in the estimation of the reference evapotranspiration. The most critical inputs for accurate estimation of ET_0 using an ANN are temperature and radiation data (Jain et al. 2015). The results showed that the proper choice of ANN architecture allows not only error minimization but also maximizes the relationship between the dependent and the independent variables. The results of the study reveal that temperature and solar radiation as the most influencing variables compared to relative humidity and wind speed for semi-arid climate conditions, as demonstrated in the present study. Given the intense data requirements for applying the Penman-Monteith model, the study employed ANN with minimum input variables such as temperature, and solar radiation. The trained and tested algorithms developed based on empirical models can be valuable tools to predict ET_0 for limited data case studies.

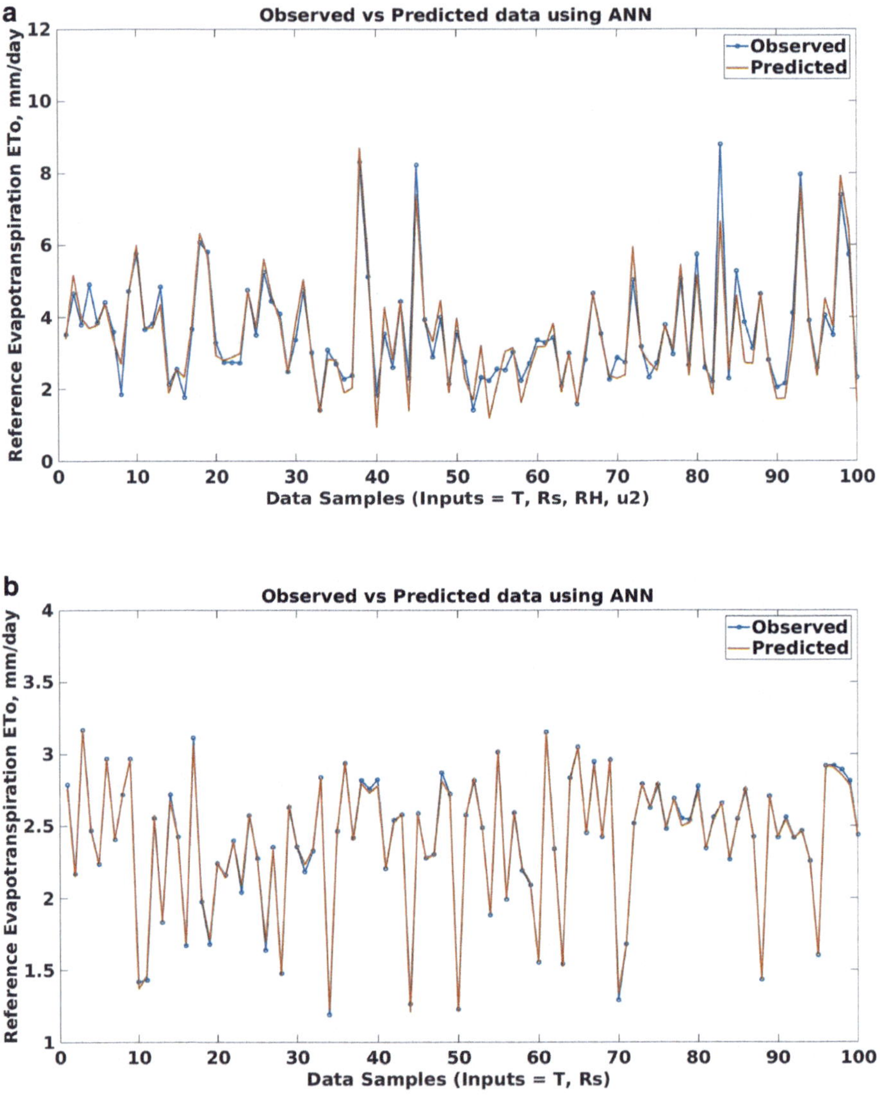

Fig. 3 **a** Variation in reference evapotranspiration (ET_0) from Penman-Monteith for the first 100 data points using ANN. **b** Variation in reference evapotranspiration (ET_0) from Turc for the first 100 data points using ANN. **c** Variation in reference evapotranspiration (ET_0) from Hargreaves for the first 100 data points using ANN. **d** Variation in reference evapotranspiration (ET_0) from Priestley-Taylor for the first 100 data points using ANN

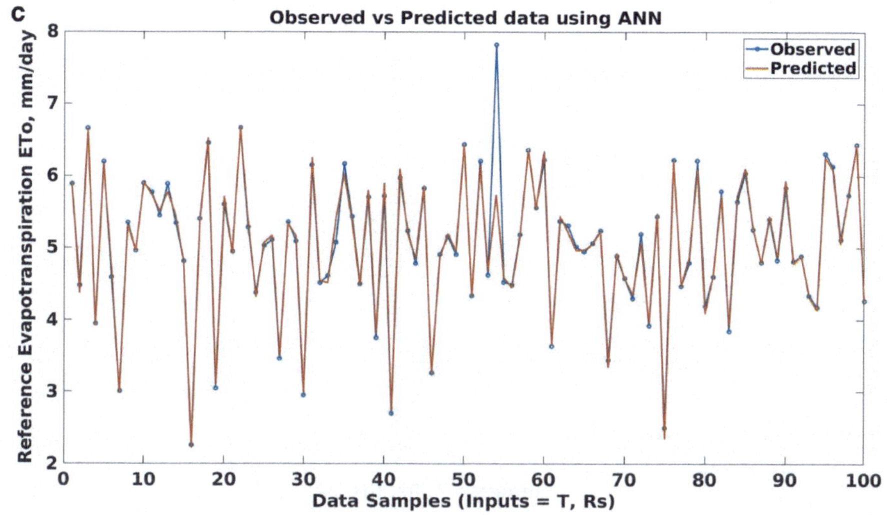

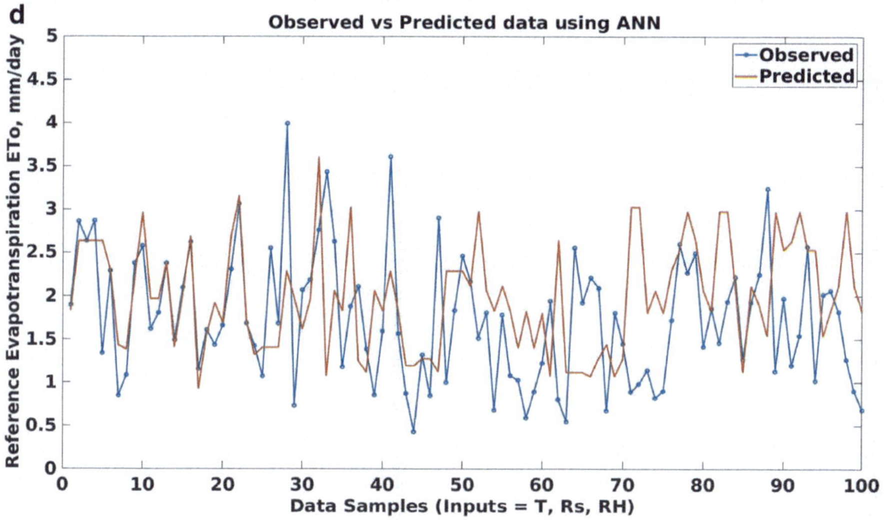

Fig. 3 (continued)

Analysing the sensitivity of each climate variable on ET_0 and testing the statistical dependencies, data pre-processing to acquire relevant information before the development of such data-driven algorithms is of most importance in the implementation. Analysis of compensating accuracies with the inclusion of limited climate input variables in the ET_0 estimates compared to standard empirical models can be a potential area of research.

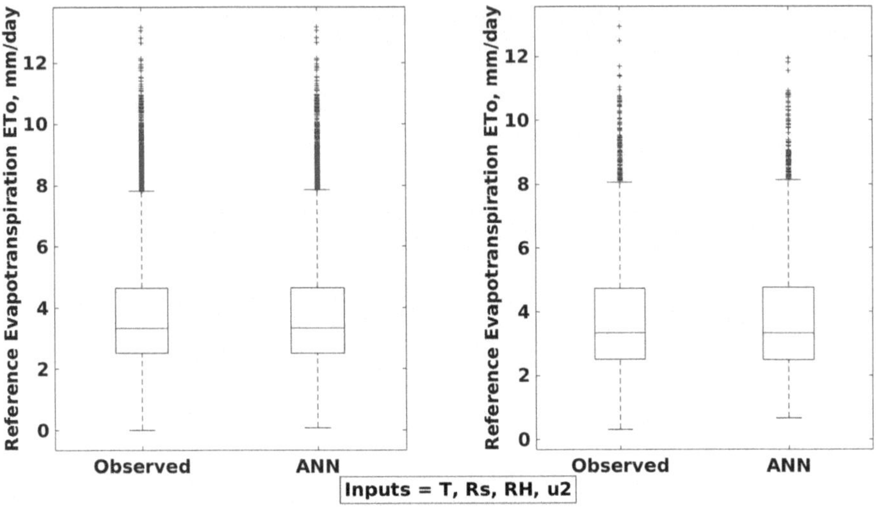

Fig. 4 Comparison of ET_0 predicted by ANN and Penman-Monteith method values for training and testing periods

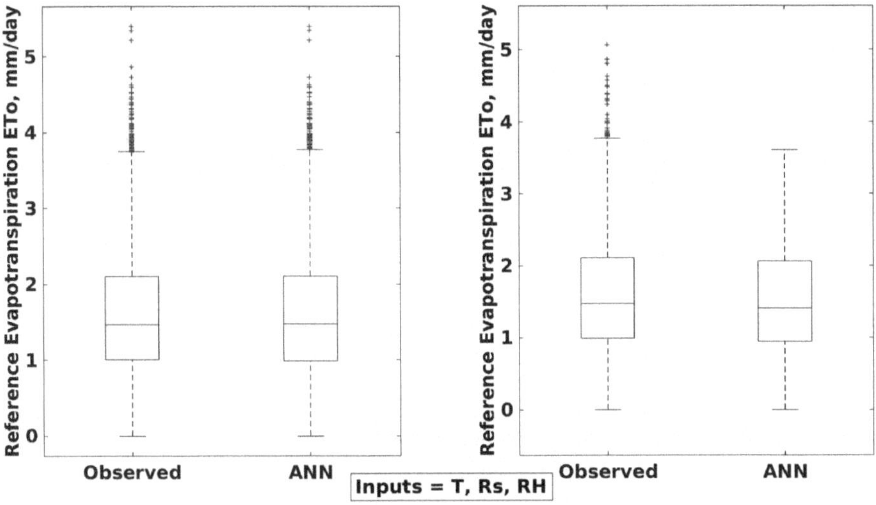

Fig. 5 Comparison of ET_0 predicted by ANN for Priestley-Taylor method values for training and testing periods

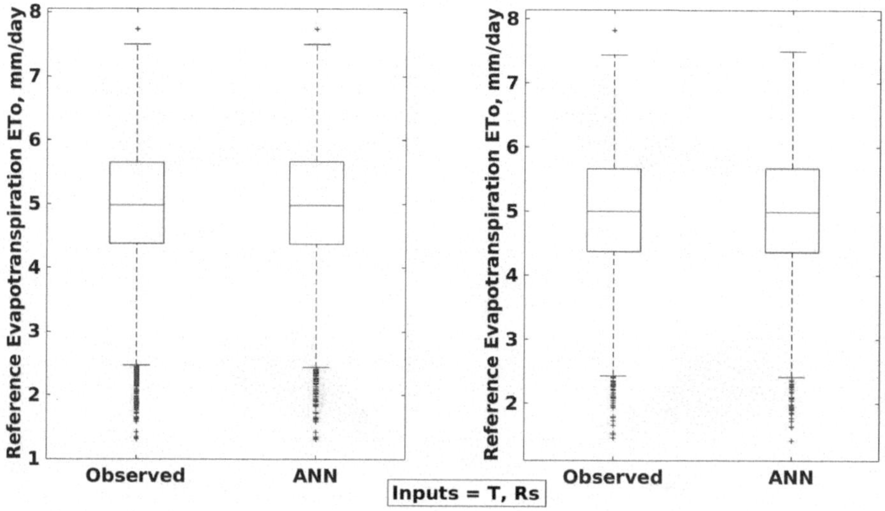

Fig. 6 Comparison of ET_0 predicted by ANN for Hargreaves method values for training and testing periods

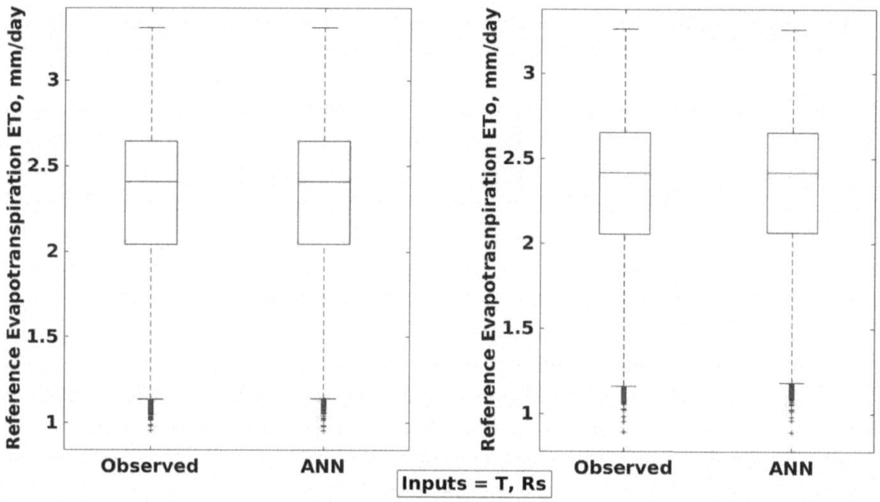

Fig. 7 Comparison of ET_0 predicted by ANN for Turc method values for training and testing periods

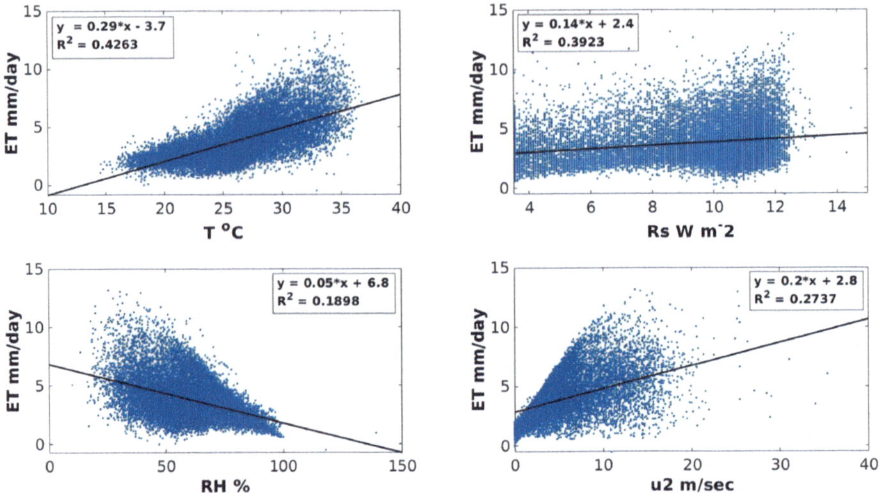

Fig. 8 Correlation of main meteorological parameters such as temperature, relative humidity, solar radiation, wind speed to ET_0

5 Conclusions

The daily reference evapotranspiration over semi-arid climatic conditions over Hyderabad, Telangana, India, were modelled using empirical and data-driven models. The Penman-Monteith model estimates of reference evapotranspiration were considered as standard reference models for various temperature and radiation-based empirical models and also for data-driven models. The daily reference evapotranspiration rates were estimated with ANN modelling technique using four input variables as maximum and minimum air temperatures, relative humidity, solar radiation, and wind speed; three input variables as average air temperature, relative humidity, and solar radiation; two input variables as temperature and solar radiation. The results were discussed with the results of alternative methods of ET_0 calculation, such as the combination-based method of Penman-Monteith, the radiation-based methods of Priestly-Taylor, the temperature-based methods of Hargreaves, and the Turc method. The correlation coefficient values suggest that temperature is the most important factor followed by solar radiation, wind speed, and relative humidity, respectively. ANN with all-climate variables as input was able to simulate ET_0 values estimated using the Penman-Monteith method. Temperature and solar radiation have a maximum correlation with ET_0 estimates of Penman-Monteith models as compared to relative humidity and wind speed. The Turc model uses temperature and solar radiation as input variables and high accuracy with the ANN model. Whereas, the relative humidity has the least correlation with the reference ET_0 estimates. The Priestly-Taylor model considers relative humidity, temperature, and solar radiation as input variables. Due to the lower dependency of relative humidity on the reference ET_0 estimates, the Priestly-Taylor model has lower accuracy with ANN compared to the Turc model. The study concludes that the empirical models work well with data-driven algorithms that consider the climate variables having high dependency with the standard reference ET_0 estimates. Such studies can be implemented for the development of data-driven models statistically dependent with reference model ET estimates. Further, it can be concluded that when a parameter or an input variable with a lower correlation is added to the set of features for training over ANN, the accuracy of prediction will be decreased. The results showed that ANN provides quite good agreement with the ET_0 obtained by the Penman-Monteith method. The study demonstrated that modelling of ET_0 through the use of the ANN technique gave better estimates that proved with their performance criterion, i.e. R^2 as 0.96. The study concludes that the performance of the model varies according to the number of inputs as well as the predicted time step. Overall, results are of significant practical use when limited climate data is available to estimate the reference evapotranspiration.

Acknowledgements The authors sincerely thank Professor Jayashankar Telangana State Agricultural University, Rajendranagar Mandal, Hyderabad, Telangana, for providing meteorological data. The research work presented in the paper is partly funded by Science and Engineering Research Board (SERB), Department of Science and Technology, Government of India through Start-up Grant for Young Scientists (YSS) Project no. YSS/2015/002111 to Dr. S. Rehana.

References

Adamala, S. (2018). Temperature based generalized wavelet-neural network models to estimate evapotranspiration in India. *Information Processing in Agriculture, 5*(1), 149–155. https://doi.org/10.1016/j.inpa.2017.09.004.

Alexandris, S., Kerkides, P., & Liakatas, A. (2006). Daily reference evapotranspiration estimates by the "Copais" approach. *Agricultural Water Management, 82*(3), 371–386. https://doi.org/10.1016/j.agwat.2005.08.001.

Allen, R. G., & Food and Agriculture Organization of the United Nations. (Eds.). (1998). *Crop evapotranspiration: Guidelines for computing crop water requirements*. FAO irrigation and drainage paper. Rome: Food and Agriculture Organization of the United Nations.

Antonopoulos, V. Z., & Antonopoulos, A. V. (2017). Daily reference evapotranspiration estimates by artificial neural networks technique and empirical equations using limited input climate variables. *Computers and Electronics in Agriculture, 132,* 86–96. https://doi.org/10.1016/j.compag.2016.11.011.

Bandyopadhyay, N., Bhuiyan, C., & Saha, A. K. (2020). Drought mitigation: Critical analysis and proposal for a new drought policy with special reference to Gujarat (India). *Progress in Disaster Science, 5,* 100049. https://doi.org/10.1016/j.pdisas.2019.100049.

Castellvi, F., Stockle, C. O., Perez, P. J., & Ibañez, M. (2001). Comparison of methods for applying the Priestley-Taylor equation at a regional scale: Priestley-Taylor equation. *Hydrological Processes, 15*(9), 1609–1620. https://doi.org/10.1002/hyp.227.

Chauhan, S., & Shrivastava, R. K. (2009). Performance evaluation of reference evapotranspiration estimation using climate based methods and artificial neural networks. *Water Resources Management, 23*(5), 825–837. https://doi.org/10.1007/s11269-008-9301-5.

Chiew, F. H. S., & McMahon, T. A. (2002). Modelling the impacts of climate change on Australian streamflow. *Hydrological Processes, 16*(6), 1235–1245. https://doi.org/10.1002/hyp.1059.

Feng, W., Shum, C., Zhong, M., & Pan, Y. (2018). Groundwater storage changes in china from satellite gravity: An overview. *Remote Sensing, 10*(5), 674. https://doi.org/10.3390/rs10050674.

Gafurov, Z., Eltazarov, S., Akramov, B., Yuldashev, T., Djumaboev, K., & Anarbekov, O. (2018). Modifying Hargreaves-Samani equation for estimating reference evapotranspiration in dryland regions of Amudarya River Basin. *Agricultural Science, 09*(10), 1354–1368. https://doi.org/10.4236/as.2018.910094.

Gupta, A. K., & Singh, Y. P. (2011). Analysis of Hamming network and MAXNET of neural network method in the string recognition. *International Conference on Communication Systems and Network Technologies, 2011,* 38–42. https://doi.org/10.1109/CSNT.2011.15.

Hargreaves, D. J., & Bolton, N. (1972). Selecting creativity tests for use in research. *British Journal of Psychology, 63*(3), 451–462. https://doi.org/10.1111/j.2044-8295.1972.tb01295.x.

Hargreaves, G. H. (1983). Closure to "*Estimating potential evapotranspiration*" by Hargreaves, G. H., & Samani, Z. A. (September, 1982). *Journal of Irrigation and Drainage Engineering, 109*(3), 343–344. https://doi.org/10.1061/(asce)0733-9437(1983)109:3(343).

Hargreaves, G. H., & Samani, Z. A. (1985). Reference crop evapotranspiration from temperature. *Applied Engineering in Agriculture, 1*(2), 96–99.

Heddam, S., & Kisi, O. (2018). Modelling daily dissolved oxygen concentration using least square support vector machine, multivariate adaptive regression splines and M5 model tree. *Journal of Hydrology, 559,* 499–509. https://doi.org/10.1016/j.jhydrol.2018.02.061.

Irmak, S., & Haman, D. Z. (2003a). Evapotranspiration: Potential or reference. *IFAS Extension, ABE, 343,* 1–3.

Irmak, S., & Haman, D. Z. (2003b). Evaluation of five methods for estimating class A pan evaporation in a humid climate. *HortTechnology, 13*(3), 500–508. https://doi.org/10.21273/HORTTECH.13.3.0500.

Jain, V. K., Pandey, R. P., Jain, M. K., & Byun, H.-R. (2015). Comparison of drought indices for appraisal of drought characteristics in the Ken River Basin. *Weather and Climate Extremes, 8,* 1–11. https://doi.org/10.1016/j.wace.2015.05.002.

Kisi, O. (2007). Evapotranspiration modelling from climatic data using a neural computing technique. *Hydrological Processes, 21*(14), 1925–1934. https://doi.org/10.1002/hyp.6403.

Krause, P., Boyle, D. P., & Bäse, F. (2005). Comparison of different efficiency criteria for hydrological model assessment. *Advances in Geosciences, 5,* 89–97. https://doi.org/10.5194/adgeo-5-89-2005.

Kumar, M., Raghuwanshi, N. S., Singh, R., Wallender, W. W., & Pruitt, W. O. (2002). Estimating evapotranspiration using artificial neural network. *Journal of Irrigation and Drainage Engineering, 128*(4), 224–233. https://doi.org/10.1061/(ASCE)0733-9437(2002)128:4(224).

Kumar, N., Shankar, V., & Poddar, A. (2020). Investigating the effect of limited climatic data on evapotranspiration-based numerical modeling of soil moisture dynamics in the unsaturated root zone: A case study for potato crop. *Modeling Earth Systems and Environment, 6*(4), 2433–2449. https://doi.org/10.1007/s40808-020-00824-8.

Kurian, C., Sudheer, K. P., Vema, V. K., & Sahoo, D. (2020). Effective flood forecasting at higher lead times through hybrid modelling framework. *Journal of Hydrology, 587,* 124945. https://doi.org/10.1016/j.jhydrol.2020.124945.

Landeras, G., Ortiz-Barredo, A., & López, J. J. (2008). Comparison of artificial neural network models and empirical and semi-empirical equations for daily reference evapotranspiration estimation in the Basque Country (Northern Spain). *Agricultural Water Management, 95*(5), 553–565. https://doi.org/10.1016/j.agwat.2007.12.011.

Legates, D. R., & McCabe, G. J., Jr. (1999). Evaluating the use of "goodness-of-fit" measures in hydrologic and hydroclimatic model validation. *Water Resources Research, 35*(1), 233–241. https://doi.org/10.1029/1998WR900018.

Liu, Y., Hejazi, M., Li, H., Zhang, X., & Leng, G. (2018). A hydrological emulator for global applications—HE v1.0.0. *Geoscientific Model Development, 11*(3), 1077–1092. https://doi.org/10.5194/gmd-11-1077-2018.

Magliulo, V., Andria, R., & Rana, G. (2003). Use of the modified atmometer to estimate reference evapotranspiration in Mediterranean environments. *Agricultural Water Management, 63*(1), 1–14. https://doi.org/10.1016/S0378-3774(03)00098-2.

Naoum, S., & Tsanis, I. K. (2003a). Hydro informatics in evapotranspiration estimation neural networks. *Water Resources Management, 18,* 143–161.

Naoum, S., & Tsanis, I. (2003b). Hydroinformatics in evapotranspiration estimation. *Environmental Modelling and Software, 18*(3), 261–271. https://doi.org/10.1016/S1364-8152(02)00076-2.

Odhiambo, H. O., Ong, C. K., Deans, J. D., Wilson, J., Khan, A. A. H., & Sprent, J. I. (2001a). *Plant and Soil, 235*(2), 221–233. https://doi.org/10.1023/A:1011959805622.

Odhiambo, L. O., Yoder, R. E., Yoder, D. C., & Hines, J. W. (2001b). Optimization of fuzzy evapotranspiration model through neural training with input–output examples. *Transactions of the ASAE, 44*(6), 1625.

Pal, M., & Deswal, S. (2009). M5 model tree based modelling of reference evapotranspiration. *Hydrological Processes: An International Journal, 23*(10), 1437–1443. https://doi.org/10.1002/hyp.7266.

Peng, L., Zeng, Z., Wei, Z., Chen, A., Wood, E. F., & Sheffield, J. (2019). Determinants of the ratio of actual to potential evapotranspiration. *Global Change Biology, 25*(4), 1326–1343. https://doi.org/10.1111/gcb.14577.

Penman, H. L. (1948). Natural evaporation from open water, bare soil and grass. *Proceedings of the Royal Society of London: Series Mathematical, and Physical Sciences, 193*(1032), 120–146. Retrieved from http://www.jstor.org/stable/98151.

Pereira, A. R., & Pruitt, W. O. (2004). Adaptation of the Thornthwaite scheme for estimating daily reference evapotranspiration. *Agricultural Water Management, 66*(3), 251–257. https://doi.org/10.1016/j.agwat.2003.11.003.

Priestley, C. H. B., & Taylor, R. J. (1972). On the assessment of surface heat flux and evaporation using large-scale parameters. *Monthly Weather Review, 100*(2), 81–92. American Meteorological Society. https://doi.org/10.1175/1520-0493(1972)100%3c0081:otaosh%3e2.3.co;2.

Rahimikhoob, A. (2010). Estimating global solar radiation using artificial neural network and air temperature data in a semi-arid environment. *Renewable Energy, 35*(9), 2131–2135. https://doi.org/10.1016/j.renene.2010.01.029.

Rajashekar, N. R. *Estimation of reference evapotranspiration using empirical methods and CROPWAT model for Rangareddy district.* Professor Jayashankar Telangana State Agricultural University.

Rehana, S. (2019). River water temperature modelling under climate change using support vector regression. In S. K. Singh & C. T. Dhanya (Eds.), *Hydrology in a changing world* (Springer water) (pp. 171–183). Cham: Springer International Publishing. https://doi.org/10.1007/978-3-030-02197-9_8.

Rehana, S., Sireesha Naidu, G., Monish, N. T., & Sowjanya, U. (2020). Modeling hydro-climatic changes of evapotranspiration over a semi-arid river basin of India. *Journal of Water and Climate Change.*

Rumelhart, D. E., Hinton, G. E., & Williams, R. J. (1986) Learning representations by back-propagating errors. *Nature, 323*(6088), 533–536. Nature Publishing Group. https://doi.org/10.1038/323533a0.

Saggi, M. K., & Jain, S. (2019). Reference evapotranspiration estimation and modeling of the Punjab Northern India using deep learning. *Computers and Electronics in Agriculture, 156,* 387–398. https://doi.org/10.1016/j.compag.2018.11.031.

Sonali, P., & Nagesh Kumar, D. (2016). Spatio-temporal variability of temperature and potential evapotranspiration over India. *Journal of Water and Climate Change, 7*(4), 810–822. https://doi.org/10.2166/wcc.2016.230.

Sudheer, K. P., Gosain, A. K., & Ramasastri, K. S. (2003). Estimating actual evapotranspiration from limited climatic data using neural computing technique. *Journal of Irrigation and Drainage Engineering, 129*(3), 214–218. https://doi.org/10.1061/(ASCE)0733-9437(2003)129:3(214).

Suryavanshi, S., Pandey, A., Chaube, U. C., & Joshi, N. (2014). Long-term historic changes in climatic variables of Betwa Basin, India. *Theoretical and Applied Climatology, 117*(3–4), 403–418. https://doi.org/10.1007/s00704-013-1013-y.

Tasumi, M. (2019). Estimating evapotranspiration using METRIC model and Landsat data for better understandings of regional hydrology in the western Urmia Lake Basin. *Agricultural Water Management, 226,* 105805. https://doi.org/10.1016/j.agwat.2019.105805.

Thornthwaite, C. W. (1948). An approach toward a rational classification of climate. *Geographical Review, 38*(1), 55. https://doi.org/10.2307/210739.

Trajkovic, S. (2005). Temperature-based approaches for estimating reference evapotranspiration. *Journal of Irrigation and Drainage Engineering ASCE, 131*(4), 316–323. https://doi.org/10.1061/(ASCE)0733-9437(2005)131:4(316).

Turc, L. (1961). Estimation of irrigation water requirements, potential evapotranspiration: A simple climatic formula evolved up to date. *Annals of Agronomy, 12,* 13–49.

Walter, I. A., Allen, R. G., Elliott, R., Jensen, M. E., Itenfisu, D., Mecham, B., et al. (2001). ASCE's standardized reference evapotranspiration equation. *Watershed Management and Operations Management, 2000,* 1–11. Presented at the Watershed Management and Operations Management Conferences 2000, Fort Collins, Colorado, United States: American Society of Civil Engineers. https://doi.org/10.1061/40499(2000)126.

Yao, J., Zhao, Y., Chen, Y., Yu, X., & Zhang, R. (2018). Multi-scale assessments of droughts: A case study in Xinjiang, China. *Science of the Total Environment, 630,* 444–452. https://doi.org/10.1016/j.scitotenv.2018.02.200.

Yates, D. N. (1997). Approaches to continental scale runoff for integrated assessment models. *Journal of Hydrology, 201*(1–4), 289–310. https://doi.org/10.1016/S0022-1694(97)00044-9.

Zanetti, S. S., Sousa, E. F., Oliveira, V. P., Almeida, F. T., & Bernardo, S. (2007). Estimating evapotranspiration using artificial neural network and minimum climatological data. *Journal of Irrigation and Drainage Engineering, 133*(2), 83–89. https://doi.org/10.1061/(ASCE)0733-9437(2007)133:2(83).

Zhang, D., Chen, P., Zhang, Q., & Li, X. (2017). Copula-based probability of concurrent hydrological drought in the Poyang lake-catchment-river system (China) from 1960 to 2013. *Journal of Hydrology, 553,* 773–784. https://doi.org/10.1016/j.jhydrol.2017.08.046.

Verifying Storm Water Drainage System Capacity for Vadodara Airport

Nayana D. Mathasoliya and S. S. Mujumdar

Abstract Vadodara formerly known as Baroda, the Cultural Capital of Gujarat is the third largest city in the Gujarat. Airports play an important role in the development of any city. Flooding of an airport in general and runway specifically can cause serious damage to the connectivity of the city. The recently constructed terminal of the Vadodara airport is soon scheduled to become an international airport. Hence, it is very important to regulate flow of storm water of airport during flooding & other relevant critical situation. Storm water management modeling (SWMM) has been used in thousands of studies of hydraulic and water quality analysis of storm, sanitary and combined sewer systems throughout the world. Major role of it is in flood control and water quality protection. This study attempts to model the Vadodara airport to analyze the discharge carrying capacity of the existing stormwater drains and the possible flood scenario using EPA SWMM _5.1 Software. SWMM model is develop and analyzed for Storm Water drainage system of Vadodara airport for a precipitation with the return period of 1 and 2 year for durations of 15, 30, 45 and 60 min. The simulation results can be used to modify the capacity of the existing storm water drainage of airport and determine the possible changes in drainage network in order to avoid flooding of the airport. INP.PINS tool was used which used to get an areal extent of the flooding for the durations of 1 year and 2 year return periods.

Keywords SWMM · Storm water · Drainage system · Capacity · Hydraulic modeling

N. D. Mathasoliya (✉)
M.E. Hydraulic Structures, Civil Engineering Department, Faculty of Technology & Engineering, MSU, Baroda, Gujarat, India
e-mail: nayna.3001@gmail.com

S. S. Mujumdar
Civil Engineering Department, Faculty of Technology & Engineering, MSU, Baroda, Gujarat, India
e-mail: sanskritimujumdar@yahoo.co.in; ssmujumdar-ced@msubaroda.ac.in

M. Majumder and G. D. Kale (eds.), *Water and Energy Management in India*,
https://doi.org/10.1007/978-3-030-66683-5_8

Abbreviations

AAI	Airport Authority of India
CMS	Cubic Meter per Second
EPA	Environmental Protection Agency
IMD	Indian Meteorological Department
MM	Millimeter
SWMM	Storm Water Management Model
VMSS	Vadodara Mahanagar Seva Sadan

1 Introduction

Importance of storm water management is constantly increasing due to three global developments: rapid urbanization, uncontrolled population growth, and climate change. Storm water management is essential in order to avoid physical injuries to individuals and property from flooding as well as to manage the environmental condition, quality and quantity of our water sources and also helps in groundwater recharge and flood protection. For this purpose, a good number of modeling tools are available around the world for hydrological and hydraulic analysis of urban areas like EPA SWMM, MIKE, SOBEK, Bentley Storm CAD, GeoSWMM. Among all these tools, EPA SWMM is an open source tool to study the urban drainage problems. (Source:-www.niwa.co.nz).

1.1 Flooding of Airports

A well designed airport drainage system is a prime vital for operational safety and efficiency as well as pavement durability. Airports require removal of surface and subsurface water.

Hence, they need integrated drainage system. Removal of water should be done from runways, taxiways, aprons, parking lots etc.

In 2005, Mumbai's airports were shut for more than 30 h due to heavy flooding of the runways, submerged Instrument Landing System equipment and extremely poor visibility. Over 700 flights were cancelled or delayed. (Maharashtra flood July 2005 report).

In June 2013 the amount of rain water, combined with poor drainage in the surrounding area and topography of airport led to back flow of water towards the Terminal 3 of the Delhi airport. (www.indiatoday.intoday.in).

In 2014 Continuous rainfall for over 36 h in Vadodara and surrounding region caused the level of water in Vishwamitri and Mahisagar rivers to tremendous, which flooded several areas in the city as well as in rural areas. It also included flooding of Vadodara airport and areas surrounding airport. (www.indianexpress.com) This phenomenon reoccurred during the July 2019 floods.

In 2015, around 400 passengers are stranded at the Chennai airport, where water has entered the runway, reaching the under carriage of aircraft after two days of heavy and continuous rain. Chennai airport has been shut, after runway and taxiway were flooded in the continuous rain since two days. (www.indianexpress.com).

Due to the recent spell of heavy rains which have caused major floods in Gujarat in July 2017, the runway of the Ahmedabad airport has been damaged, resulting in flights being diverted or cancelled.

1.2 Significance of the Study

A detailed study over the last few periods shows that many cities are facing the problem of flash floods urban flooding most of which can be attributed to disorganized urbanization. Many attribute the change in the rainfall pattern to worldwide climate changes and global warming. This has resulted in flooding in urban areas and causes disturbance in the normal life. A Report On Technical Report Florida Statewide Airport Storm Water Study of 2008 also give inspiration to choose Airport as my case study area. (https://www.stantec.com).

Airports require a large amount of capital investment, and often represent strategic installations required for more functions than only passenger and freight transport, Also, they cannot easily be relocated, nor can they afford to be out of service for any length of time. Open water attracts wildlife, specifically birds which can pose a significant safety hazard to flying aircraft. The impacts of flooding on the operation of an airport take many forms, the most obvious of which is the disruption of air traffic. This in itself can cause severe economic losses in terms of passenger and freight traffic. Flooding of access roads and transport networks surrounding the airport can also lead to down-time for an airport. Closure of an airport has indirect impacts on the operations and finances of the other airports also. (https://www.stantec.com).

The complex of Vadodara airport is spread over 17,500 square meters. It can handle 700 passengers at its peak, 500 domestic and 200 international, that is, 16,800 passengers a day. Built at a cost of Rs. 162 crores, it will be able to handle 500 domestic and 200 international passengers. The runway is 8,100 feet long. The Vadodara city is a fast developing in terms of setting of new industries as well the Airport Authority of India has added the new terminal building keeping this in mind and plans to convert it into an international airport in near future. (https://www.stantec.com).

It is therefore necessary to analyze the existing system of airport and to identify the reasons of this problem so that effective measures can be taken to obtain a remedy to the problem. Application of SWMM performed inside and outside of India. Some of the research articles of Outside of india are given below:

(1) Krebsa et al. (2012) A High Resolution Application of Storm Water Management Model (SWMM) Using Genetic Parameter Optimization

(2) Henri Tikkanen (2013) Hydrological modeling of a large urban catchment using a storm water management model
(3) Biniyam Asfaw (2016) Assessment of Storm Water Drainage Systems in Kemise Town
(4) Sahar Babaei et al. (2018) Urban flood simulation and prioritization of critical urban sub-catchments using SWMM model and PROMETHEE II approach
(5) Yiran Bai et al. (2018) Storm Water Management of Low Impact Development in Urban Areas Based on SWMM.

Detailed Research articles of inside India are given below:

(1) **Swathi et al.** (2014)**Application of Storm Water Management Model to an Urban Catchment**: In this study SWMM has been explored for the catchment of BITS Pilani Hyderabad campus, India. The catchments have been divided into various sub catchments and are modeled for 2006 rainfall event. The study deals with a flexible set of hydraulic modeling capabilities, in particular it is used to assess infiltration using Horton method and flow routing analysis using dynamic wave method. The external inflows through the drainage system network of pipes, channels, storage works and diversion structures were also considered and the critical locations of overflow are identified.
(2) **Bhaniyara et al.** (2015) **Storm Water Drainage Problem of Surat City & Its Solutions Due to Flood in River Tapi**: In this case study by designing some general as well as systematic drainage solution & also appropriate design of storm water drainage system of desire working objective. Aim of the research is begins by examining the performance of current storm water drainage system & the conditions that lead flooding problem at some low lying critical areas of Surat city.
(3) **Ashwini et al.** (2016) **Design of Drainage Network for The Fort Area, Belgavi City Using SWMM**: Due to urbanization, impermeable area expands leading to frequent flooding. Surface overflow generally relies on rainfall, infiltration, slope, percent of imperviousness and land use. The aim of study carried out by the author is to assess the runoff and design drainage network by SWMM.The Soil Service Conservation Service Curve number is being used in the present approach to determine flood magnitudes generated in each zone of the study area.
(4) **Vinay Ashok Rangari et al.** (2018) **Simulation of Urban Drainage System Using a Storm Water Management Model (SWMM)**: Storm water modelling plays an important role in checking issues such as flash floods and urban water-quality problems. In this study a SWMM model is developed to analyze drainage network for the campus of National Institute of Technology, Warangal.The model is simulated for one real storm event and 2-year return period of interval 1-h design storm intensity. Design storm intensity derived from IDF curves for different return periods. GIS methodology is employed for handling spatial data simultaneously. From results, it is observed that some part of campus are commonly affected with flooding.

(5) **Shaik Mohammed Fazal Ahamed et al.** (2019)**Urban Flood Modelling and Management using SWMM for New R.R. Pet Region, Vijayawada, India**: This case study deals with the flexible set of hydraulic modelling capabilities. It is used to assess infiltration using Green Ampt method and flow routing analysis using Dynamic wave method. The study area is represented using AutoCAD map and runoff water flow that can be routed through drainage system like pipes, channels, and outlets are identified. The aim of this study area is to check the runoff from extreme rainfall events and to evaluate the accuracy of the drainage system. The study revealed that storm networks are well planned and more sufficient to cater the simulated Rainfall event.

2 Study Area and Data Collection

The Vadodara city is the 3rd largest and among the most populated cities in Gujarat. Vadodara Airport Situated in the suburb of Harni, around 7 km in the north-east of Vadodara from the city as shown in Fig. 1. It is located at 22°19′46″ N & 73°13′10″ E. The catchment area of airport is 2.2914 sq. km.It is the second green building of India and second largest proposed international airport of Gujarat after the Sardar Vallabhbhai Patel International Airport, Ahmedabad. The new terminal constructed at the cost of Rs.160 Crores is also the second green airport of India after Kochi Airport. The newly integrated terminal building covers an area of 0.018 sq. km. and has the capacity of handling 700 passengers at peak hour. (Source:- VMSS) Table 1 shows the various data collection.

3 Methodology

3.1 Storm Water Management Model (SWMM)

EPA's SWMM is used throughout the world for planning, analysis and design related to storm water runoff, combined and sanitary sewers. SWMM is a dynamic hydrology-hydraulic water quality simulation model. It is used for single event or long-term (continuous) simulation of runoff quantity and quality from primarily urban areas.

Its main governing equation in the routing system relating flow rate, flow depth and bed slope is Manning, while the Hazen-Williams or the Darcy-Weisbach equations are used for pressurized flow. The software can use either kinematic or dynamic wave in routing, therefore, it handles backwater, surcharge, reverse flow and surface ponding. For the purpose of this study the Horton model and Dynamic wave routing method is used. The model is conceptually divided into four major environmental compartments: (i) the atmosphere compartment (ii) the land surface compartment

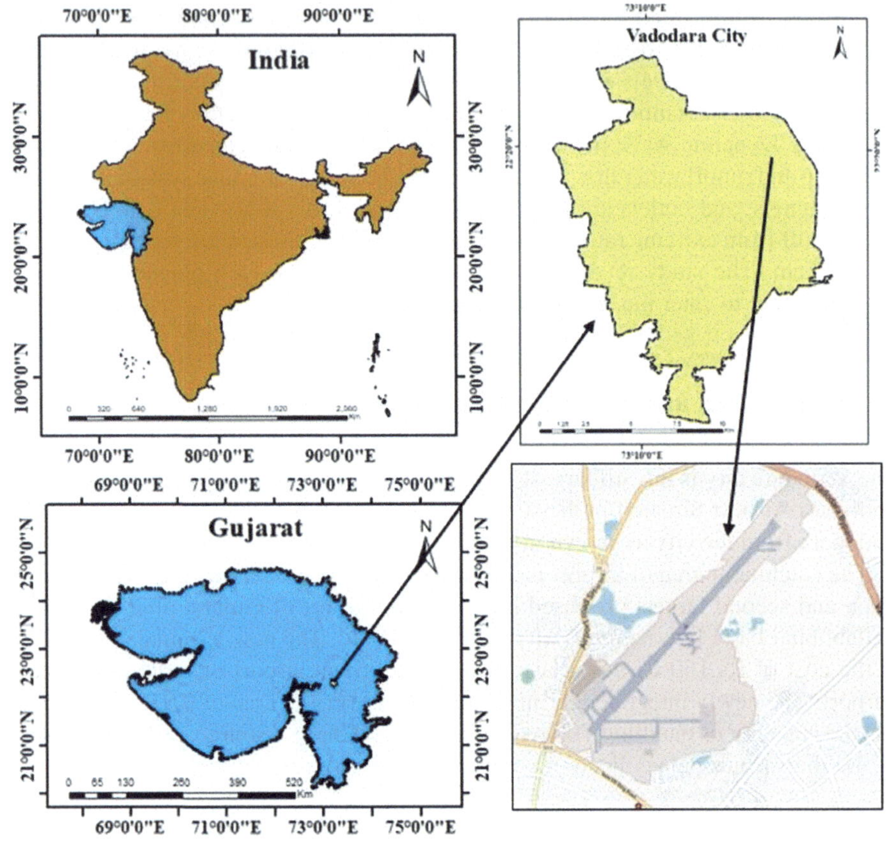

Fig. 1 Study area map of Vadodara Airport (https://openstreetmap.org)

Table 1 Data collection

Sr. No.	Type of data	Name of agency	Duration
1	Daily rainfall data	IMD, Pune	1986–2019
2	Existing storm water drainage layout	AAI, vadodara	

(iii) the transport compartment (iv) the ground-water compartment (Source:- SWMM manual Sect. 2.1.1. (Rossman 2010)).

3.2 Methodology of SWMM Model Development

The actual SWMM model was developed from the ArcGIS data. All the features parameterized had been stored in four separate ESRI Shape files (sub catchments, junctions, conduits, and outfalls). For modeling SWMM project file. Junctions are provided in the drainage network at the points where direction changes. Location of junctions are decided from X & Y co-ordinates obtained from shape file of junction. Two junctions are joined by the conduit according to the flow direction of storm water. Position of sub catchment is decided on the basis of ridge line of contour of areas surrounding the airport. All the sub catchments have connecting nodes which connect to the links and lead to the outlet. Outlets are provided as per the present existing outlet of drain which is connecting to VMSS drain. Rain gauge provided in study area on its actual location on site and gives the time series for rainfall data of various rainfall intensity.

Thus for construction of the model, a total 64 junctions, 70 conduits, 3 outfalls, 11 sub catchment and 1 rain gauge station are placed in the entire network. To create a time series, we are supposed to object and populate it with the help of daily rainfall data. The reporting time step and the dry-weather hydrologic time step were chosen as 20 min & one hours respectively. The hydraulic routing time step was set to 60 s. Figure 2 shows the existing storm water drainage of airport. The first step will be derivation of IDF curve using daily data of rainfall for different storm frequency referring various manuals available for storm water drainage design. The next step will be to Prepare the drainage model for the study area using cross sections of the existing storm water drains, slopes, their catchment areas, land use information and various other parameters in EPA SWMM.

The model is then run for simulation of flow to check the carrying capacity of the drains and identify the links and nodes which fail as shown in Fig. 3. The model was run for the precipitation having a return period of one year and two years for the duration of rainfall as 15, 30, 45 and 60 min to analyze the flooding scenario under various conditions.

4 Result & Analysis

4.1 Results for Sub-Catchments

The summary report generated after running the SWMM model run for the sub-catchments includes the total rainfall, total infiltration, total runoff and rate of peak runoff for each sub catchment. Table 2 shows the Subcatchment runoff summary of 1 year return period for 15,30,45,60 min respectively and Table 3 shows the subcatchment runoff summary for 2 year return period for duration of 15, 30, 45 and 60 min respectively.

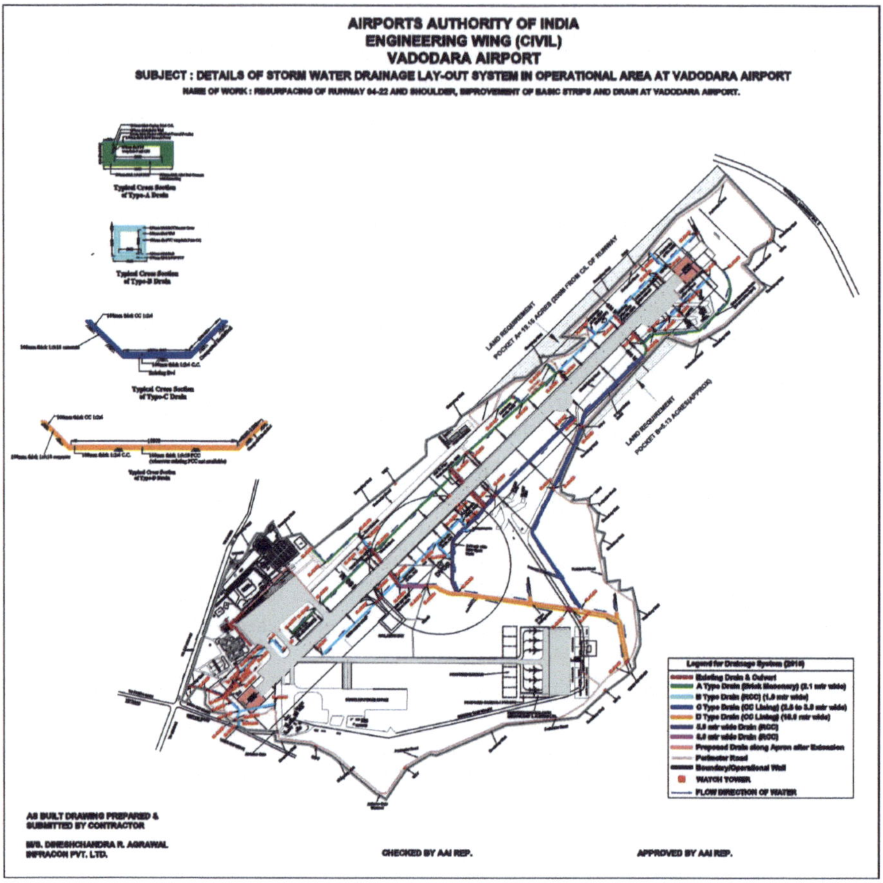

Fig. 2 Existing storm water drainage of Airport (AAI, Vadodara)

4.2 Results of Nodes

The simulation results gave the details of the rate of flow in the nodes for various durations of rainfall. Table 4 shows the rate of flow in the nodes for durations of 15, 30, 45 and 60 min of rainfall for 1 year return period, in which red color indicate node flooding and Table 5 shows the rate of flow in the nodes for durations of 15, 30, 45 and 60 min of rainfall for 2 year return period, in which red colour indicate node flooding.

The simulation results also gave the details of the nodes which are getting flooded for various durations of rainfall and return periods. Table 6 shows the names of the nodes which are flooded for durations of 15, 30, 45 and 60 min of rainfall for 1 and 2 year return period.

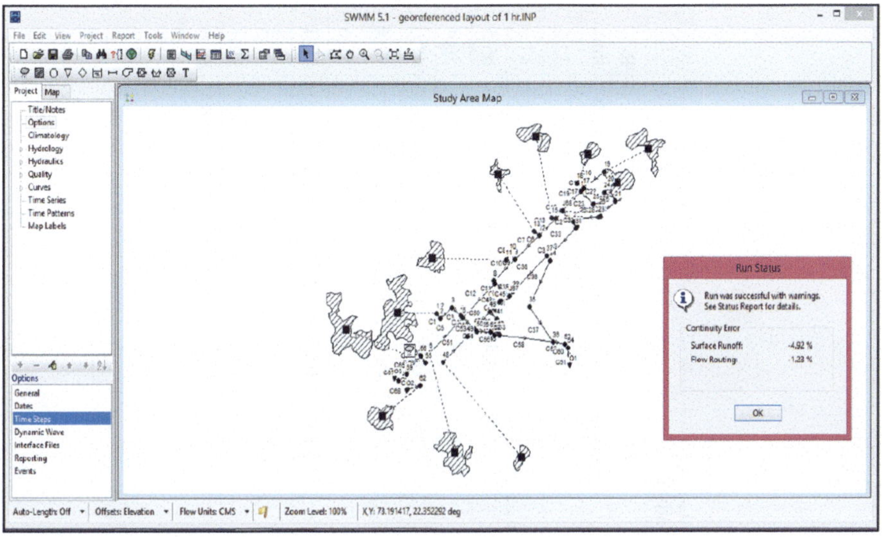

Fig. 3 Run status of model

4.3 *Floodplain Mapping Using INP-PINS*

The SWMM software does not directly generate maps for observing the extent of flooding. For this purpose, INP.PINS tool was used which acts as an interface with SWMM and Google earth to generate flood maps. Figure 4 shows the working diagram of inp-pins.

The areal extent of the flooding will change according to the time of concentration selected for the SWMM model simulation. It will also depend on the conditions of the water level at the outfall, which in our case is the stormwater drain constructed by the Vadodara Mahanagar Seva Sadan.Since the water levels at the outfall are variable and unknown, random plots were created for flood mapping considering the VMSS storm water drain to be empty and the outfall to be in free falling condition. Figures 5 and 6 show the areal extent of the flooding on for the various durations for 1 and 2 year return periods.

5 Conclusion

The study was aimed at assessing the carrying capacity of the existing storm water drains of the Vadodara airport for the 1 and 2 year return periods of 15, 30, 45 and 60 min duration for each return period. The following conclusions are derived from the present study:

Table 2 Sub catchment runoff summary for 1 year return period

Sub catchment	15 min duration			30 min duration			45 min duration			60 min duration		
	P	I	R	P	I	R	P	I	R	P	I	R
	mm	mm	mm	mm	mm	mm	mm	mm	mm	mm	mm	mm
S1	23.1	3.0	8.4	46.2	2.9	21.0	69.3	2.8	31.4	92.5	2.6	32.6
S10	23.1	3.0	11.5	46.2	2.9	29.4	69.3	2.8	44.6	92.5	2.6	47.9
S11	23.1	3.0	14.6	46.2	2.9	36.3	69.3	2.8	56.4	92.5	2.6	65.5
S2	23.1	3.0	9.2	46.2	2.9	23.3	69.3	2.8	34.9	92.5	2.6	36.4
S3	23.1	3.0	10.9	46.2	2.9	27.9	69.3	2.8	42.2	92.5	2.6	45.0
S4	23.1	3.0	10.4	46.2	2.9	26.6	69.3	2.8	40.1	92.5	2.6	42.4
S5	23.1	3.0	12.2	46.2	2.9	31.0	69.3	2.8	47.4	92.5	2.6	51.6
S6	23.1	3.0	10.4	46.2	2.9	26.6	69.3	2.8	40.1	92.5	2.6	42.4
S7	23.1	3.0	9.7	46.2	2.9	24.8	69.3	2.8	37.3	92.5	2.6	39.2
S8	23.1	3.0	10.0	46.2	2.9	25.6	69.3	2.8	38.6	92.5	2.6	40.7
S9	23.1	3.0	13.7	46.2	2.9	34.4	69.3	2.8	53.1	92.5	2.6	59.9

Table 3 Sub catchment runoff summary for 2 year return period

Sub catchment	15 min duration			30 min duration			45 min duration			60 min duration		
	P	I	R	P	I	R	P	I	R	P	I	R
	mm	mm	mm	mm	mm	mm	mm	mm	mm	mm	mm	mm
S1	27.0	3.0	10.7	54.0	2.9	26.2	81.0	2.8	38.7	108.0	2.6	40.2
S10	27.0	3.0	14.7	54.0	2.9	36.2	81.0	2.8	54.6	108.0	2.6	58.8
S11	27.0	3.0	18.4	54.0	2.9	43.9	81.0	2.8	67.8	108.0	2.6	79.1
S2	27.0	3.0	11.7	54.0	2.9	28.9	81.0	2.8	43.0	108.0	2.6	44.9
S3	27.0	3.0	14.0	54.0	2.9	34.5	81.0	2.8	51.8	108.0	2.6	55.3
S4	27.0	3.0	13.3	54.0	2.9	32.9	81.0	2.8	49.3	108.0	2.6	52.2
S5	27.0	3.0	15.5	54.0	2.9	38.1	81.0	2.8	57.8	108.0	2.6	63.1
S6	27.0	3.0	13.3	54.0	2.9	32.9	81.0	2.8	49.3	108.0	2.6	52.2
S7	27.0	3.0	12.4	54.0	2.9	30.8	81.0	2.8	45.9	108.0	2.6	48.3
S8	27.0	3.0	12.8	54.0	2.9	31.8	81.0	2.8	47.5	108.0	2.6	50.1
S9	27.0	3.0	17.3	54.0	2.9	41.9	81.0	2.8	64.2	108.0	2.6	72.8

Table 4 Rate of Inflow in the nodes for various duration and 1 year return period

Node name	15 min	30 min	45 min	60 min
	Total inflow in CMS			
j1	1.43	2.38	3.13	3.57
j10	2.76	2.76	2.75	2.75
j11	3.41	3.41	3.41	3.40
j12	11.02	10.99	10.96	10.93
j13	15.82	15.82	15.82	15.81
j14	15.63	15.64	15.64	15.64
j15	15.63	15.64	15.64	15.64
j16	0.56	0.84	1.12	1.38
j17	0.84	1.51	2.00	2.04
j18	0.41	0.96	1.45	1.47
j19	0.58	1.09	1.52	1.82
j2	1.42	2.36	2.41	2.41
j20	0.40	0.62	0.84	1.01
j21	0.40	0.63	0.85	1.02
j22	0.32	0.55	0.70	0.83
j23	0.15	0.23	0.30	0.37
j24	0.34	0.53	0.71	0.86
j25	0.34	0.53	0.72	0.87
j26	0.55	0.82	0.93	1.01
j27	0.57	1.23	1.44	1.44
j28	0.73	1.23	1.60	1.83
j29	0.27	0.46	0.56	0.60
j3	1.38	2.26	2.26	2.26
j30	0.49	0.79	1.06	1.24
j31	92.44	92.44	92.45	92.45
j32	16.43	12.51	12.51	12.29
j33	15.00	12.37	12.52	12.32
j34	14.54	14.36	14.38	12.41
j35	14.65	12.52	12.55	12.36
j36	0.79	1.31	1.50	1.56
j37	0.94	1.04	1.09	1.05
j38	2.32	2.67	2.36	1.00
j39	12.51	13.21	13.44	13.29
j4	1.29	1.53	1.53	1.53
j40	12.51	13.29	13.43	13.27

(continued)

Table 4 (continued)

Node name	15 min	30 min	45 min	60 min
j41	12.51	13.30	13.47	13.27
j42	11.49	10.30	9.68	10.18
j43	8.30	8.31	8.33	8.35
j44	7.87	7.86	7.84	7.78
j45	6.71	6.71	6.71	6.71
j46	92.44	92.44	92.45	92.45
j47	31.10	31.10	31.10	31.10
j48	31.65	31.42	31.39	31.37
j49	31.80	31.80	31.80	31.80
J5	1.27	1.53	1.53	1.53
j50	14.91	13.41	13.46	13.41
j51	15.72	14.39	14.44	14.49
j52	9.98	9.98	9.97	9.97
j53	92.44	92.44	92.45	92.45
j54	4.00	3.91	3.58	3.52
j55	1.07	1.07	1.07	1.10
j56	0.13	0.19	0.25	0.31
j57	0.21	0.21	0.24	0.30
j58	1.05	1.63	2.21	2.65
j59	1.02	1.28	1.28	1.28
j6	1.82	2.33	2.61	2.87
j60	0.98	1.04	1.04	1.04
j61	0.94	1.03	1.03	1.03
j62	0.92	1.03	1.03	1.03
j63	0.89	1.03	1.04	1.03
j64	1.10	1.63	1.85	2.00
j65	0.39	0.62	0.83	0.99
j7	0.42	0.42	0.43	0.43
j8	1.38	1.38	1.38	1.38
j9	2.26	2.26	2.25	2.25
O1	15.64	15.64	15.64	15.64
O2	1.09	1.63	1.85	2.00

Table 5 Rate of inflow in the nodes various durations of rainfall for 2 year return period

Node name	15 min	30 min	45 min	60 min
	Total inflow in CMS			
j1	1.77	2.96	3.81	4.27
j10	2.76	2.76	2.75	2.75
j11	3.41	3.41	3.41	3.40
j12	11.02	11.00	10.97	10.95
j13	15.82	15.82	15.82	15.81
j14	15.63	15.64	15.64	15.64
j15	15.63	15.64	15.64	15.64
j16	0.69	1.04	1.39	1.69
j17	1.01	1.78	2.04	2.04
j18	0.43	1.23	1.47	1.47
j19	0.73	1.34	1.85	2.20
j2	1.76	2.41	2.41	2.41
j20	0.49	0.77	1.04	1.23
j21	0.49	0.78	1.04	1.23
j22	0.40	0.65	0.84	0.99
j23	0.18	0.28	0.37	0.45
j24	0.42	0.65	0.88	1.05
j25	0.42	0.66	0.88	1.05
j26	0.72	0.89	1.02	1.12
j27	0.74	1.44	1.44	1.44
j28	0.91	1.52	1.95	2.19
j29	0.33	0.56	0.67	0.71
j3	1.69	2.25	2.25	2.25
j30	0.60	0.98	1.29	1.49
j31	107.99	108.00	108.00	108.01
j32	17.05	12.31	12.71	12.27
j33	15.10	12.17	12.65	12.38
j34	14.70	14.31	14.64	12.61
j35	14.74	12.68	12.53	12.42
j36	1.00	1.58	1.78	1.83
j37	0.99	1.07	1.09	1.01
j38	2.40	2.74	1.84	1.10
j39	12.71	13.14	13.28	13.51
j4	1.52	1.53	1.53	1.53
j40	12.72	13.14	13.27	13.41

(continued)

Table 5 (continued)

Node name	15 min	30 min	45 min	60 min
j41	12.73	13.14	13.27	13.45
j42	11.56	10.43	9.82	9.86
j43	8.29	8.33	8.32	8.35
j44	7.87	7.86	7.85	7.83
j45	6.71	6.71	6.71	6.71
j46	107.99	108.00	108.00	108.01
j47	31.10	31.10	31.10	31.10
j48	31.69	31.40	31.40	31.38
j49	31.80	31.80	31.80	31.80
J5	1.53	1.53	1.53	1.53
j50	15.03	13.68	13.44	13.44
j51	15.85	14.44	14.49	14.54
j52	9.98	9.98	9.97	9.97
j53	107.99	108.00	108.00	108.01
j54	4.20	4.07	3.74	3.51
j55	1.07	1.07	1.10	1.15
j56	0.16	0.23	0.31	0.38
j57	0.21	0.22	0.29	0.36
j58	1.29	2.02	2.71	3.22
j59	1.27	1.28	1.28	1.28
j6	2.14	2.51	2.87	3.18
j60	1.04	1.04	1.04	1.04
j61	1.01	1.03	1.03	1.03
j62	1.01	1.03	1.03	1.03
j63	0.99	1.04	1.04	1.03
j64	1.34	1.78	2.03	2.21
j65	0.48	0.77	1.02	1.19
j7	0.42	0.42	0.43	0.43
j8	1.38	1.38	1.38	1.38
j9	2.26	2.26	2.25	2.25
O1	15.64	15.64	15.64	15.64
O2	1.33	1.77	2.03	2.20

Table 6 Node flooded for various Durations & 1 year return period

Node name	15 min	30 min	45 min	60 min	15 min	30 min	45 min	60 min
	Node flooding in CMS				Node flooding in CMS			
j1	0.00	0.00	0.92	1.36	0.00	0.74	1.59	2.06
j17	0.84	1.51	2.00	2.04	1.01	1.78	2.04	2.04
j19	0.00	0.18	0.64	0.93	0.00	0.44	0.96	1.31
j2	0.00	0.50	0.56	0.56	0.00	0.57	0.57	0.57
j27	0.00	0.57	0.78	0.78	0.00	0.78	0.78	0.78
j28	0.00	0.00	0.17	0.39	0.00	0.09	0.51	0.75
j3	0.00	0.73	0.73	0.73	0.19	0.73	0.73	0.73
j31	80.26	80.35	80.34	80.31	95.82	95.86	95.87	95.87
j36	0.00	0.35	0.54	0.60	0.00	0.63	0.82	0.87
j37	0.58	0.81	0.82	0.84	0.80	0.83	0.84	0.84
j41	6.30	7.09	7.26	7.06	6.51	6.93	7.06	7.24
j46	61.30	61.33	61.34	61.34	76.85	76.88	76.89	76.90
j48	1.62	1.28	1.24	1.22	1.66	1.27	1.25	1.23
j49	18.53	18.41	18.41	18.41	18.56	18.41	18.41	18.41
j51	8.72	6.23	5.36	5.43	8.92	7.12	5.41	5.48
j53	89.45	89.47	89.48	89.49	105.00	105.03	105.04	105.05
j58	0.00	0.39	0.96	1.41	0.00	0.78	1.47	1.98
j59	0.00	0.24	0.24	0.24	0.23	0.24	0.24	0.24
j6	1.15	1.73	2.02	2.28	1.52	1.92	2.28	2.59

Fig. 4 Working diagram of inp-pins

1. The nodes j17, j31, j37, j41, j46, j48, j49, j51, j53 & j6 are observed to get flooded for all the durations of rainfall and all return periods for 2-h time of concentration. However, change in the time of concentration can vary the results.

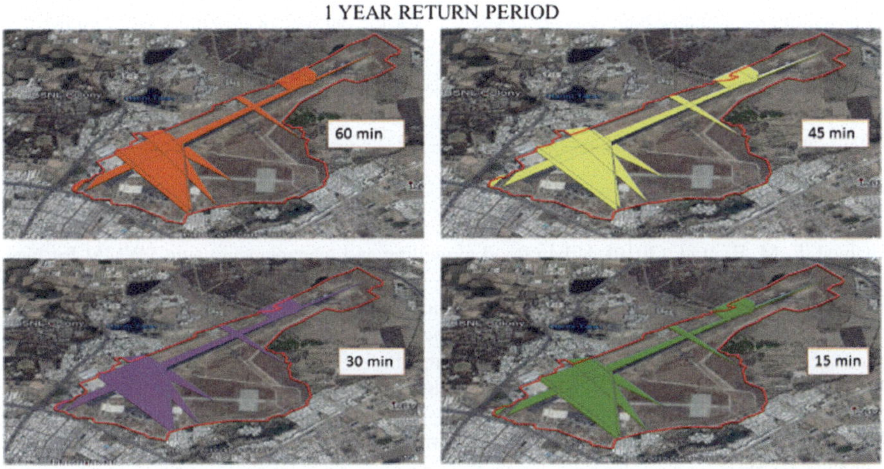

Fig. 5 Flood inundation map of 1 year return period & various durations

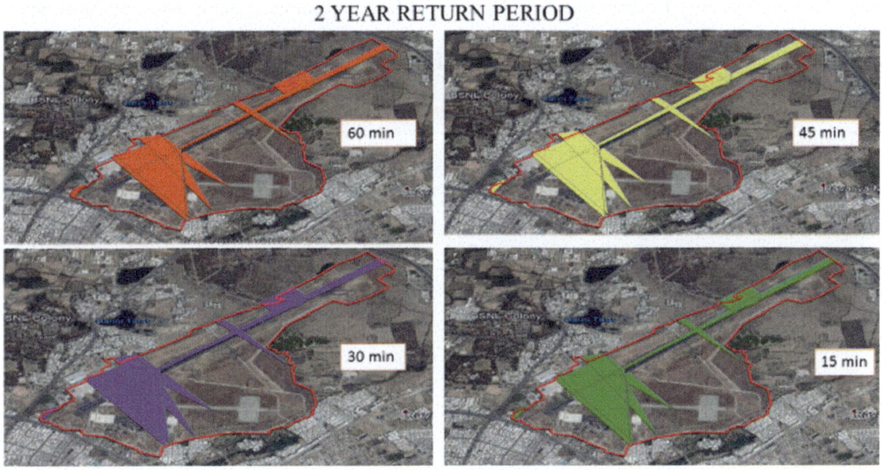

Fig. 6 Flood inundation map of 2 year return period & various durations

2. The storm drainage system of Vadodara airport can be modified using the above results to prevent flooding observed for smaller durations of rainfall and return periods.
3. Regular maintenance of the storm water drains is essential to reduce the flooding of the airport.

Acknowledgements I would like to thanks to **Srivani Acharya,** project assistant ITRA-Water team, IISc Bangalore **& Swathi Vemula**, Ph.D. scholar, BITS Hyderabad for providing support in direction to this work. I am also thankful to **Shri Ravi Shankar, Civil** Engineer, Airport Authority for providing me valuable data & conduct site visit for the purpose of the study. We acknowledge the use of software SWMM, ArcGIS and INP.PINS.

References

Ahamed, S. M. F., et al. (2019). Urban flood modelling and management using SWMM for New R.R. Pet Region, Vijayawada, India.

Asfaw, B. (2016). Assessment of storm water drainage systems in Kemise Town.

Babaei, S., et al. (2018). Urban flood simulation and prioritization of critical urban sub-catchments using SWMM model and PROMETHEE II approach.

Bai, Y., et al. (2018). Storm water management of low impact development in urban areas based on SWMM.

Rangari, V. A., et al. (2018). Simulation of urban drainage system using a storm water management model (SWMM).

Swathi, V., et al. (2014). Application of storm water management model to an urban catchment.

Krebsa, G., et al. (2012). A high resolution application of storm water management model (SWMM) using genetic parameter optimization

Henri Tikkanen (2013) Hydrological modeling of a large urban catchment using a storm water management model

Kishan, J., Bhaniyara et al. (2015) Storm Water Drainage Problem of Surat City & Its Solutions Due to Flood in River Tapi

Ashwini, H. U., et al. (2016). Design of drainage network for the fort area, Belgavi City using SWMM

Energy Management

Optimal Trade-Off Between the Energy—Economy of a Hydropower Plant for Better Management of the Renewable Energy Resources

Priyanka Majumder and Abhijit Saha

Abstract In the present paper, an attempt has been made to identify the optimal trade-off between utilization efficiency and financial liability by two optimization techniques; Firefly Algorithm and Differential Evolution. As utilization efficiency, as well as financial liability of the power plant, depends on multiple parameters, decision-making techniques are utilized to produce an index representing the utilization and financial requirement of the power plant. According to the result, it is found that the Efficiency of turbine and Maintenance and operation cost are the most important indicators of utilization and economic liability of hydropower plants respectively. Also for the present trade-off analysis, Differential Evolution is found to be a better optimization technique compared to FireFly Algorithm.

Keywords Trade-off analysis · MCDM · Optimization · Efficiency of HPP · Financial liability

1 Introduction

The hydropower plants (HPPs) uses water as fuel to generate energy. Such sort of renewable energy power plants has the minimum expenditure requirement and complexity in energy conversion among all the sources of alternative energy. Approximately, 19% of the global energy demand is presently supplied by HPP. The contribution among the renewable energy resources to satisfy global energy demand was found to be 63% (Listverse 2009).

The populace excess alongside mechanical progression has expanded the interest for power everywhere on over the World. The expense of Electricity has been expanded all the while. As the assets of traditional fills are restricted, interchange

P. Majumder (✉) · A. Saha
Department of Basic Science and Humanities, Techno College of Engineering Agartala, Maheshkhola 799004, Tripura, India
e-mail: majumderpriyanka94@yahoo.com

A. Saha
e-mail: abhijit84_mt@yahoo.in

M. Majumder and G. D. Kale (eds.), *Water and Energy Management in India*,
https://doi.org/10.1007/978-3-030-66683-5_9

vitality sources to supplant petroleum derivatives are presently liked to gracefully the overabundance request.

Among all the wellsprings of exchange Energy, hydropower was discovered to be the most dependable yet reasonable wellspring of elective Energy (Berga 2016). In any case, locational suggestions and variety in active vitality of water stream during the storm and non-rainstorm seasons draw in adequate measure of budgetary risk. Accordingly for any hydropower ventures, the money related obligation is assessed before favoring the establishment of the task. The monetary obligation of HPP relies upon certain pointers. The pointers are Maintenance and activity cost, Capital cost, Total creation cost, Project cost, Energy hardware cost, Plant cost, common work cost, Turbine cost, Investment cost, Initial venture cost, Fuel cost and, Labor cost.

In addition, the use productivity of hydropower plants relies upon numerous boundaries like Efficiency of the penstock, Efficiency of the turbine, Efficiency of the generator, Labor effectiveness, Amount of release, Turbulence in water, Turbulence in water, Difference in flexibly and request vitality, the Pressure distinction among bay and force to be reckoned with just as the Pressure contrast between power house and tailrace and numerous other (Majumder et al. 2016; Majumder and Saha 2018a).

Mingyue et al. in (2017) (Pang et al. 2017) utilized compromise between the carbon decrease benefits and biological execution of including carbon catch and capacity (CCS) to the force plant. As indicated by the outcome, it was discovered that leaders should consider both the carbon decrease benefits and the biological expenses in the improvement of biomass power plants (BioCCS) frameworks.

In 2017 Ji et al. (2017) examined compromise data between framework cost and danger for chiefs with various danger inclinations lastly they found that compromise data would be important for the ideal long haul power framework extension arranging while at the same time confronting a future unsure circumstance.

In 2018 Rayamajhee and Joshi distributed a paper (Rayamajhee and Joshi 2018) on Economic compromise examination on hydropower plants considering the compromise between hydroelectricity creation and natural externalities. They utilized an ideal control system to lead near investigations of exogenous and endogenous Vexternality mitigation fund (EMF) comparative with the base case with no EMF. As per the outcome, endogenous EMF, when contrasted with the base case, decreases crop misfortune by 87.5% with a relating vitality creation compromise of 11.8%.

Amini and Almassalkhi contemplated vigor and execution compromises in subsiding skyline control with questionable vitality assets in 2018 (Amini and Almassalkhi 2018). The chance-constrained model predictive control (CCMPC) dispatches powerfully the questionable virtual power plants (VPPs) and ordinary generators while considering financially ideal, secure reference direction for producing resources. Shut circle execution limits the deviation of ordinary generators from their reference direction.

In 2018, Singh and Singal (2018) talked about operational trade-off of little scope hydropower plants utilizing nonlinear constrained optimization algorithm (NLC). In their investigation, they attempted to recognize the most critical choice boundaries to produce greatest vitality with least misfortunes lastly acquired greatest yearly benefit and ideal estimations of the operational boundaries.

In 2019, Majumder and Saha analyzed the efficiency of HPP using a Bootstrap MCDM Approach by Decision-Making Trial and Evaluation Laboratory (DEMATEL) with the Analytic Hierarchy Process (AHP) (Majumder and Saha 2019a). In that year Majumder and Saha also analyze the efficiency of HPP to identify the most significant indicators of performance efficiency of HPP under climate change and urbanization scenario. In this study they also used an hybrid MCDM tool using DEMATEL, AHP and Statistical Process Control (SPC) (Majumder and Saha 2019b).

Majumder et al. (2019) studied in Real time reliability monitoring of HPP. In that study has tried to develop a new methodology for determination of the reliability assessment of hydro power projects with the help of MACBETH MCDM and PNN models. The benefit of this new method will be it can objectively and cognitively analyze reliability of HPP. The model if embedded in a real time monitoring (RTM) can continuously monitor the performance reliability of HPP (Majumder et al. 2019).

In 2020, Majumder et al. proposing an intelligent mitigation measure to control the trade-off with the help of some group of indicators which have the maximum impact on production reliability of a power plant. This significance-based parameter modification entails recognition of the indicators and their significance in controlling reliability of a hydropower plant with the help of objective decision making methods and validating the selection by laboratory based physical models as well as real-life case studies. A number of multi-criteria decision making methods which were popular in the identification of best decision out of many options were utilized in the detection of the significant indicators and their importance where the ensembled output from multiple multi-criteria decision making methods was used to detect the priority indicators and their priority (Majumder et al. 2020a).

To reduce the cost and infrastructural requirements of a smart system Majumder et al. (2020a) represent the plant performance for instant mitigation of system failures by replacing the requirement of multi-indicator tracking by single weighted function monitoring. This monitoring upgradation will reduce the process cost of the system, thereby elevating the profitability of the power plant. The functional tracking will also increase the efficiency of the MAD (monitoring, analysis and decision-making) and minimize the memory requirement of the real-time monitoring as single pointer will be required to be analysed and evaluated before taking a decision (Majumder et al. 2020b).

However, all the studies described in previous paragraphs utilize one or more single input or response parameters for trade-off analysis but the economic performance, as well as the utilization capacity of a HPP, is dependent on multiple parameters that influences both economic liability and operational profitability of the HPP. As a consequence, a single causal and response parameter is not sufficient to represent a viable solution throughout trade-off analysis. But if the contribution of the multiple parameters on the response parameter can be incorporated in a single function then the trade-off can yield a plausible solution that can ensure the reliability of the HPP. That is why, in the present investigation a trade-off analysis is conducted with the help of two multi-parameter indexes; one representing the utilization efficiency and the other one is developed from financial parameters considering the individual contribution of the parameters toward the conclusive impact on the utilization capacity of the HPP.

As a result, the main aim of the current investigation is to find the trade off between the multi-parameter indexes representing the utilization efficiency and financial liability of a HPP. In this aspect, we develop two indexes for representing the performance efficiency and economic liability by an objective and unbiased selection of parameters and their significance in modifying the efficiency and liability responses of the power plant. Aims of the study is to investigate the trade-off for maximization of efficiency and minimization of financial liability.

The present investigation is initiated by the process of determination of priority for the parameters or indicators which can represent the utilization efficiency and financial liability of a HPP. In this aspect, two MCDM tools are selected and their ensembled values of each parameter are selected as the priority value of those parameters which are directly proportional to the significance of the indicator in modifying the performance and expenditures incurred by the power plant. As the objective of the determination of priority of the parameters is fuzzy due to the similarity in the contribution of the indicators, Fuzzy Analytical Hierarchy Process (FAHP) (Chang 1996) and Fuzzy Weighted Sum Method (FWSM) (Sisodia et al. 2018) are used for determination of the priority value (PV) of the alternatives.

Once the priority of the parameters is determined, the optimization algorithms are applied to find the trade-off between the index representing utilization efficiency and the index representing the financial liability of the HPP. The multi-objective optimization is presented by two optimization technique viz Differential Evolution (DE) (Price et al. 2006) and FireFly Algorithm (FFA) (Yang 2009) and the optimization performed in the present investigation is a non-linear and gradient-based optimization which encourages the authors to select these two techniques for identification of the trade-off.

Section 2 describes the methods used and the detailed methodology adopted for development and trade-off analysis of the indexes.

2 Detail Methodology

The methodology adopted in the present investigation can be divided into two subsections. The first section will include the application of MCDM for the determination of priority and the second section will depict the application of OTs in identification of trade-off. Sections 2.1 and 2.2 depicts the procedure adopted for the determination of priority value and identification of the trade-off between the utilization and economic response from the HPP. Figure 1 shows the schematic diagram of detailed methodology.

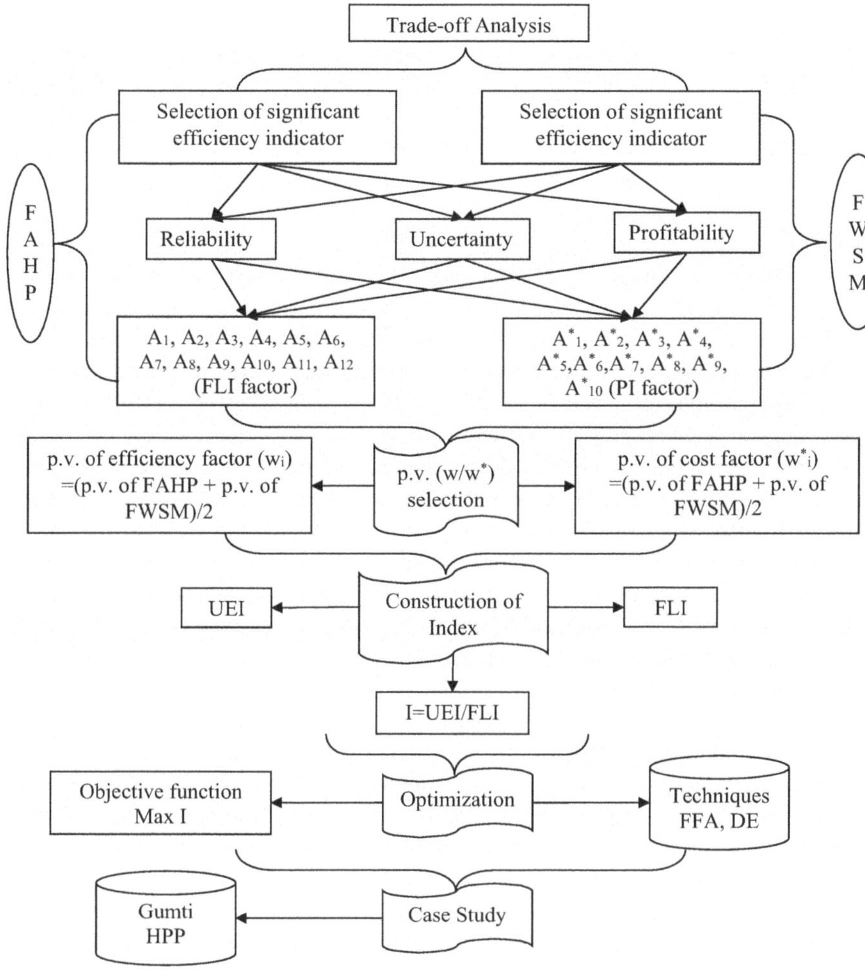

Fig. 1 Schematic of the current methodology

2.1 Determination of Priority Value

The priority value was determined by the application of two different MCDM methods and ensembling their outputs. As per the methodology of MCDMs, the priority value of each alternative is estimated for criteria. The alternatives are evaluated based on their significance either concerning each other based on the criteria (as is the case in FAHP) or with respect to the criteria (as is the case in FWSM). In the present study Reliability, Uncertainty and Profitability of the HPP was selected as the criteria as all the indicators selected for the present investigation will effect; maybe in different scale; all these four response parameters and the performance, as well as economic liability, is affected by all these four criteria. That is why all the four parameters were selected as the criteria followed by the indicators shown in Table 2 as alternatives of the MCDM technique for determination of the priority value of each of the indicators. Table 1 depicts the criteria and their mathematical representations.

The indicators which can represent or modify the utilization efficiency of a HPP is depicted in Tables 2 and 3 respectively for the Utilization Efficiency Index and Financial Liability Index to represent the utilization capacity and financial liability of the HPP.

The Efficiency of Penstock, Turbine and Generator are selected as the indicator of utilization as all these parameters represent the utilization efficiency of the power plant. If the efficiency is more then the utilization of hydropower potential will be more and vice versa. The efficiency of skilled and unskilled manpower will also affect the production of the power plant but will not be as sensitive as the efficiency of the penstock, turbine, and generators.

The power potential of HPP is determined based on the available water head between the inlet and power house which is converted kinetic energy for rotating the turbines which produce energy from the generators. The head between the power house and tailrace is deducted from the total head to find the net head which actually is converted to produce power from the power plant and also determines the power potential of the HPP. Both of these heads are included as an indicator for the representation of utilization efficiency. Distance from the nearest grid and the difference between supply and demand indicates that even after the successful conversion of mechanical potential into kinetic energy and thereby producing electrical energy, the amount of actual utilization of the generated electricity. Transmission loads and deficit in power supply will depict the incomplete utilization of the production from a power plant which is equal to the utilization efficiency of the power plant. That is why both deficits in supply and distance from the grid is included as an indicator.

In case of financial liability of an HPP, O&M cost is the variable cost and cost for hiring skilled and unskilled manpower and procurement of fuels to supply the auxiliary demand for power in the HPP are some of the recurring cost incurred in an HPP. Capital and project cost along with civil, turbine and energy equipment cost (cost for procurement of generators, transformers and many more electrical equipment) are classified under the fixed cost. The Investment Cost and Initial Investment Cost

Table 1 Criteria selection

Criteria	Equation	Abbreviation
Reliability R(t) (Majumder et al. 2019)	$R(t) = e^{-\frac{t}{\theta}}$	$\frac{1}{\theta}$ = Failure rate, t = time
Uncertainty $U_c(y)$ (Majumder and Saha 2019a)	$U_c(y) = \sqrt{\sum_{i=1}^{n} [c_i u(x_i)]^2}$	c_i = Sensitivity coefficient, $u(x_i)$ = Standard uncertainty
Profitability P(i) (Majumder and Saha 2019b)	$P(i) = \frac{NI}{S}$	NI = Net Income, S = Sales

Table 2 Indicator selection of efficiency

Alternatives or indicator	Mathematical representation	Notation
Efficiency of penstock (P Eff) (Majumder et al. 2020a)	$P\ Eff = \frac{H_n}{H_k} \times 100$	H_n = Net head at end of penstock, H_g = Gross head
Efficiency of turbine (T Eff) (Majumder et al. 2020a)	$T\ Eff = 550 \times \frac{P_{Thp}}{(\rho/g_c)Q_m H_T}$	P_{Thp} = turbine output (hp) ρ = density of water (lbm/ft3) g_c = gravitational constant
Efficiency of generator (G Eff) (Majumder et al. 2020a)	$G\ Eff = \frac{O}{I} \times 100$	O = Output, I = Input
Efficiency of skilled and unskilled manpower (L Eff) (Majumder et al. 2020b)	$L\ Eff = \frac{E}{At} \times 100$	E = Expected time, At = Actual time
Amount of discharge (A Dis) (Majumder et al. 2019)	$A\ Dis = Av$	A = cross-sectional area of flow in m^2 or ft^2 v = mean velocity of flow in m/sec or ft/sec
Turbulence in water (T water) (Majumder et al. 2020a)	$T\ Water\ \alpha\ \mathrm{Re} = \frac{L \times U}{v}$	Re = Reynold's number, L = Velocity scale, U = Length scale, v = kinematic viscocity of the fluid
Difference in supply and demand energy (Majumder and Saha 2019a)	$q_t^d(p_t; x_t) = q_t^s(p_t; x_t)$	T = For each market t, p_t = the econometrician can only observe the equilibrium price, q_t = the equilibrium quantity, and x_t = the covariates, but cannot observe either the demand function $q_t^d(\cdot)$ or the supply function $q_t^s(\cdot)$
Pressure difference between inlet and power house (IP Press) (Majumder and Saha 2019b)	$IP\ Press = \rho g\left(H - \frac{v_1^2 - v_2^2}{2g}\right)$	H = The net head, v and 2 v are the water velocities in m/s at the inlet and outlet points, respectively, ρ = The actual water density, g = The value of the acceleration due to earth's gravity at the site
Pressure difference between power house and tailrace (PHT Press) (Majumder and Saha 2019b)	$PHT\ Press = H - \frac{p_1' - p_2''}{\rho g} - \frac{v_1^2 - v_2^2}{2g}$	H = The net head, v and 2 v are the water velocities in m/s at the inlet and outlet points, respectively, ρ = The actual water density, g = The value of the acceleration due to earth's gravity at the site. p_1' and p_2' are the readings of the two pressure transducers in Pa installed at the inlet and outlet points, respectively

(continued)

Table 2 (continued)

Alternatives or indicator	Mathematical representation	Notation
Distance from nearest grid ($D_f(p)$) (Majumder and Saha 2019a)	$D_f(p) = \min_{q \in G}\{d(p, q) + f(q)\}$	G = Regular grid, f = Function from G to set of real number, d = distance between the points p and q, f(q) = q is closure point p, f = small value of some location, D_f = small value at that location and nearby point

Table 3 Cost factor selection

Cost factor	Equation	Notation
Maintenance and operation cost (OM) (Parida and Kumar 2009)	$M\&O\ Cost = \frac{TMC}{TPC}$	TMC = Total maintenance cost, TPM = Total production cost
Capital cost (C) (Steffen 2018)	$C\ Cost = k_d \frac{D}{D+E} + k_e \frac{E}{D+E}$	k_d = The cost of debt, D = Market value of debt, E = Market value of equity, k_e = The cost of equity
Total production cost (TP cost)	$TP\ Cost = DL + DM + FO$	DL = Direct labor cost, DM = Direct material cost, FO = Factory overhead
Project cost (P cost) (Majumder and Saha 2017)	$P\ Cost = \frac{\sum_l k_l}{L}(1 - e)$	L = Cumulative principal cost items percentage, e = Engineering contingency %, k = Item cost per lane Km_l
Energy equipment cost (EE cost) (de Oliveira 2013)	$EE\ Cost = E_P(v_p - v_i) + E_w(v_w - v_i)$	E_p = Energy rate/flow rate performance, v_p = Specific value (m^3/kg); value scale performance, v_i = Specific value (m^3/kg); value scale component, E_w = Energy rate/flow rate Waste

(continued)

Table 3 (continued)

Cost factor	Equation	Notation
Plant cost (P cost)	$P\ Cost = \frac{P \times H}{\beta}\left\{COE - \left(\frac{OM_f}{P \times H} + \mu \times OM_{v,b}\right) - \frac{f}{\eta}\right\}$	β = Levelized carring charge factor, COE = the unit cost of generating electricity, H = Annual operating hours, P = Net rated output (KW), f = Levelized fuel cost (s/KWh [LHV]), η = Net rated efficiency of the combined-cycle plant (LHV), OM_f = Fixed O & M cost ($), $OM_{v,b}$ = Variable O & M cost for baseland operation ($/KWh), μ = Maintenance cost escalation factor (1.0 for baseland operation)
Civil work cost (CW cost) (Bettini et al. 2016)	$CW\ Cost = EAC + ADMC + EXP$	EAC = Equivalent area cost, ADMC = Administration cost, EXP = Extra cost
Turbine cost (T cost) (Aggidis et al. 2010)	$T\ Cost = 15,000 \times (Q \times H)^{0.68}$ (£, 2008) (Kaplan turbine costs) $T\ Cost = 142,000 \times (Q \times H^{0.5})^{0.07}$ (£, 2008) (Francis turbine costs) $T\ Cost = 282,000 \times (Q/H^{0.5})^{0.11}$ (£, 2008) (Francis turbine costs) $T\ Cost = 50,000 \times (Q/H^{0.5})^{0.52}$ (£, 2008)(Francis turbine costs) $T\ Cost = 8,300 \times (Q \times H)^{0.54}$ (£, 2008) (Pelton turbine costs)	Q = The flow rate. In two of the bands the cost of Francis turbine does not change a lot for different values of H

(continued)

are the indicators that represent the future value of the power plant in terms of the present expenditures. All these variable viz. fixed, recurrent and future costs are included to determine the Financial Liability Index which represents the economic status of the power plant.

All the indicators are evaluated for its ability in modifying the criteria considered in the study and fuzzy logic is utilized in representing the evaluation. Similarly the evaluation of alternatives is also conducted with respect to each other based on the ability to modify selected criteria and the evaluation result is depicted by the fuzzy logic. After the rating of the indicators by fuzzy logic, the aggregation is completed

Table 3 (continued)

Cost factor	Equation	Notation
Investment cost (I cost) (Tuhtan 2007)	$I\ Cost = NPV - \sum_{t=1}^{n} \frac{R_t - M_t}{(1+r_t)^t}$	NPV = The resulting net present value, t = number of years from the present, n = total number of years (power plant design life of 30 years), R_t = annual income from energy generation at year t (+), M_t = annual operation and maintenance costs at year t (−), r_t = annual discount rate (here taken as a constant)
Initial investment cost (II cost) (Majumder and Saha 2017)	$II\ cost = \frac{\mu - \tau}{\theta}$	μ = Annual life cycle cost, τ = Annual expenses, θ = capital recovery factor
Fuel cost (F cost)	$F\ cost = \frac{TC \times M}{D}$	TC = Total cost, D = Distance (km), M = Milage (km/l)
Labour cost (L cost) (Lipská et al. 2005)	$F\ cost = \frac{TC \times M}{D}$	TCE = Total nominal cost per employee, RLP = Real labor productivity

with the help of both AHP and WSM. The result of the aggregation for each indicator is ensembled to find the final Priority Value (PV) of the indicators of both UEI and FLI.

The mathematical representation of the Utilization Efficiency Index and Financial Liability Index is depicted in Eqs. 1 and 3 and explained in Sects. 2.1.1 and 2.1.2 respectively.

2.1.1 Development of Utilization Efficiency Index (UEI)

After the PV is determined by the ensemble output from FAHP and FWSM, an index (Eq. 1) is developed with the help of the PV and the magnitude of the indicators which are utilized for representation of utilization efficiency of a HPP. The indicator is made in such a way that it become directly proportional with respect to the utilization efficiency of the power plant (Majumder and Saha 2018a).

$$UEI = \frac{\sum_{i=1}^{10} w_i^* A_i^*}{\sum_{i=1}^{10} w_i^*} \quad (1)$$

where,

$$p.v.\ of\ UEI\ indicator\left(w_i^*\right) = \frac{p.v.\ of\ FAHP + p.v.\ of\ FWSM}{2} \tag{2}$$

and A_1^* denotes the magnitude of Efficiency of Penstock, A_2^* denotes the magnitude of Efficiency of Turbine, A_3^* denotes the magnitude of Efficiency of Generator, A_4^* denotes the magnitude of Efficiency of Transformer, A_5^* denotes the magnitude of Amount of discharge, A_6^* denotes the magnitude of Turbulence in water, A_7^* denotes the magnitude of Difference in supply and demand energy, A_8^* denotes the magnitude of Pressure difference between inlet and power house, A_9^* denotes the magnitude of the Pressure difference between power house and tailrace, A_{10}^* denotes the magnitude of Distance from the nearest grid, $w_i^* = PV$ of the indicators (Eq. 2).

2.1.2 Development of Financial Liability Index (FLI)

After the weightage of significance is resolved a Index (Eq. 3) is created with the assistance of the weightage and the size of the cost parameters. The weighted normal of the apparent multitude of boundaries is proposed as the list for $FLI\ \alpha$ Liability Index (Majumder and Saha 2018b).

$$FLI = \frac{\sum_{i=1}^{12} w_i A_i}{\sum_{i=1}^{12} w_i} \tag{3}$$

where

$$p.v.\ of\ FLI\ indicator(w_i) = \frac{p.v.\ of\ FAHP + p.v.\ of\ FWSW}{2} \tag{4}$$

and A_1 denotes the magnitude of Maintenance and operation cost, A_2 denotes the magnitude of Capital cost, A_3 denotes the magnitude of Total production cost, A_4 denotes the magnitude of Project cost, A_5 denotes the magnitude of Energy equipment cost, A_6 denotes the magnitude of Plant cost, A_7 denotes the magnitude of Civil work cost, A_8 denotes the magnitude of Turbine cost, A_9 denotes the magnitude of Investment cost, A_{10} denotes the magnitude of Initial investment cost, A_{11} denotes the magnitude of Fuel cost, A_{12} denotes the magnitude of Labour cost, w_i denotes the PV of the FLI indicator (Eq. 4).

2.1.3 Development of Trade-off Index (I)

After construction of the UEI (Eq. 1) and FLI (Eq. 3), the index for the representation of trade-off which is also used as the objective equation of the optimization problem is developed where UEI is placed in the numerator and FLI is located in the denominator.

It means that the objective equation TI is made to get optimality of the power for increasing values of TI where efficiency is compared to the financial liability.

Equation 5 depicts the formula for TI which is used in the present study as the objective for maximization.

$$TI = \frac{UEI}{FLI} = \frac{\frac{\sum_{i=1}^{10} w_i^* A_i^*}{\sum_{i=1}^{10} w_i^*}}{\frac{\sum_{i=1}^{12} w_i A_i}{\sum_{i=1}^{12} w_i}} = \frac{\sum_{i=1}^{10} w_i^* A_i^*}{\sum_{i=1}^{12} w_i A_i} \times \frac{\sum_{i=1}^{12} w_i}{\sum_{i=1}^{10} w_i^*}$$

$$= \frac{\sum_{i=1}^{10} w_i^* A_i^*}{\sum_{i=1}^{12} w_i A_i} \quad (5)$$

$$\left[Since \sum_{i=1}^{12} w_i = \sum_{i=1}^{10} w_i^* = 1 \right]$$

2.2 *Application of Optimization Techniques (OT) for Identification of Optimal Trade-Off*

Equation 6 depicts the mathematical representation of the optimization problem solved in the present investigation. Here TI is maximized by changing the magnitude of the indicators which is also designated as the design variable for the present optimization problem. The constraint of the design variables is depicted in Eqs. 7a and 7b. Here it is to be noted that the normalized magnitude of the design variables are used for optimization such that the inference due to scale difference can be avoided.

$$Maximize\ TI$$

$$= \frac{\left[w_1^* w_2^* w_3^* w_4^* w_5^* w_6^* w_7^* w_8^* w_9^* w_{10}^*\right] \times \left[A_1^* A_2^* A_3^* A_4^* A_5^* A_6^* A_7^* A_8^* A_9^* A_{10}^*\right]^T}{\left[w_1 w_2 w_3 w_4 w_5 w_6 w_7 w_8 w_9 w_{10} w_{11} w_{12}\right] \times \left[A_1 A_2 A_3 A_4 A_5 A_6 A_7 A_8 A_9 A_{10} A_{11} A_{12}\right]^T} \quad (6)$$

$$Subject\ to$$

$$[0000000000] < \left[A_1^* A_2^* A_3^* A_4^* A_5^* A_6^* A_7^* A_8^* A_9^* A_{10}^*\right] < [1111111111] \quad (7a)$$

$$[0000000000] < [A_1 A_2 A_3 A_4 A_5 A_6 A_7 A_8 A_9 A_{10} A_{11} A_{12}] < [1111111111] \quad (7b)$$

As depicted in Sect. 2.1.3 two different optimization techniques are utilized to maximize the TI function. The results are selected from the technique which has yielded a maximum value of TI as its best, average, and worst value among all the iterations compared to the same value estimated by the other optimization algorithm.

The magnitudes of the indicators are collected from the Gumati HPP in Tripura. Section 2.2.1 depicts a description of the study area.

2.2.1 Study Area

Gumti is one of the bigger streams in Tripura, India which streams toward the west and releases into Bangladesh. Because of the development of a dam for the hydropower plant, an enormous supply was made, known as Gumti repository. This repository is in the upper catchment of the Gumti River. The capacity limit of the repository is a 23570-ha m. The lowered territory at F.R.L of 92.05 m and M.W.L. of 95.25 m was discovered to be 46.34 and 74.86 km^2 individually. With the assistance of this repository, Gumti Hydro Power plant produces capacity to moderate the emergency of intensity in Tripura. The plan limit of this hydropower plant was 15 MW. It has three units. The first and second units were appointed in 1976 and the rearward in 1984. Be that as it may, out of a 15 MW limit, at present just 8 MW–9 MW force is created from Gumti HPP during the stormy season. During the dry season, creation diminishes to 0.5 MW. Figure 2 demonstrating the Location of Gumti HPP.

The results from the priority determination phase and trade-off identification phase are delineated in Sect. 3.

Fig. 2 Location of Gumti HPP

3 Results and Discussions

The results from the present investigation can be sub-divided into two parts, namely the results of the MCDM method to evaluate the priority value of the indicators and results of the optimization.

3.1 Results from MCDM

The comparison of one alternative with respect to the criteria or with respect to another alternative based on the criteria helped to estimate the PV for each of the alternatives from the two different MCDM methods. Table 4 depicts the result and the ensembled value of the result which was selected as the PV for the respective indicator t, the time of optimization. According to the result, it is found that the efficiency of turbine and maintenance and operation cost are the most significant indicators for utilization efficiency and financial liability of hydropower plants respectively. The studies depicted in 35, 41 and 42 also shows the same indicators as the most significant when performance efficiency of HPP is required to be estimated.

Table 4 PV of the indicators

FLI indicator	FAHP	FWSM	Ensembled PV	UEI indicator	FAHP	FWSM	Ensembled PV
A_1^*	0.0841	0.1198	0.1019	A_1	0.1923	0.1364	0.1644
A_2^*	0.1908	0.1461	0.1684	A_2	0.0774	0.0768	0.0771
A_3^*	0.163	0.1299	0.1464	A_3	0.0657	0.0866	0.0761
A_4^*	0.0871	0.1246	0.1058	A_4	0.0437	0.0698	0.0568
A_5^*	0.0842	0.1035	0.0938	A_5	0.0901	0.108	0.099
A_6^*	0.0582	0.0647	0.0615	A_6	0.0973	0.0857	0.0915
A_7^*	0.1209	0.1244	0.1227	A_7	0.0867	0.1029	0.0948
A_8^*	0.0798	0.0711	0.0754	A_8	0.0536	0.0597	0.0567
A_9^*	0.0788	0.0711	0.0749	A_9	0.0699	0.0846	0.0773
A_{10}^*	0.0511	0.0449	0.048	A_{10}	0.0567	0.0324	0.0445
				A_{11}	0.066	0.0667	0.0664
				A_{12}	0.1013	0.0904	0.0959

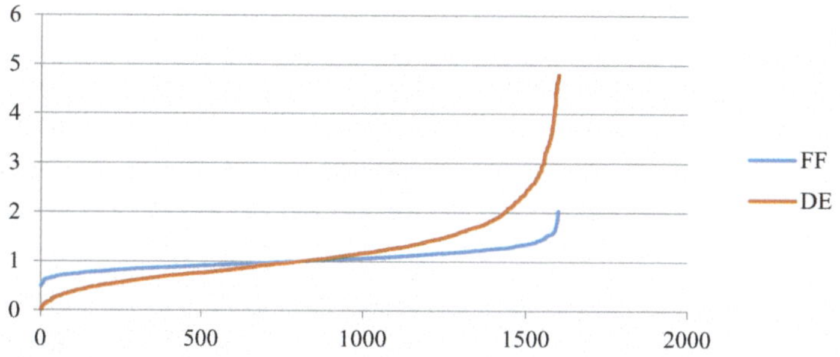

Fig. 3 Result of optimization

Table 5 Result from Optimization

Name of OT	Worst	Average	Best	Time of convergence (s)
FFA	0.495153	1.02826	2.036845	15.32
DE	0.035552	1.170957	4.806858	11.25

3.2 *Identification of Trade-Off Between the UEI and FLI*

Figure 3 depicts the magnitude of TI for each of the iterations as estimated by the FFA and DE. Table 5 shows the best, average and worst value of TI as determined by the FFA and DE algorithms. From Table 5, it is clear that the results from DE are better than that of FFA. Also, the iteration time or rate of convergence is lesser in DE compared to FFA. So the results estimated by DE are identified as the optimal Trade-off which is equal to 4.806858. Table 6 depicts the magnitude of the design variables for which the optimal value of TI is achieved.

Figure 4 depicts the trade-off between UEI and FLI. In the present study, trade-off analysis is applied to Gumti HPP in Tripura. The region of optimality is shown in the figure.

4 Conclusion

In the present investigation, MCDM and OT are applied for trade-off analysis between PI and FLI. This trade-off analysis is applied for HPP of Tripura. According to the result, it is found that the Efficiency of turbine and Maintenance and operation cost are the most important indicators of performance efficiency and economic liability of hydropower plant respectively. Also, DE is the best OT for that trade-off analysis. If the number of criteria is increased or decreased then the importance of criteria may or may not be changed but their priority value must be changed, which is the major disadvantage of this analysis. The Magnitude value of the indicator of PI

Table 6 The magnitude value of the factors PI and FLI

FLI		UEI	
Indicator	PV	Indicator	PV
A_1	0.164364	A_1^*	0.101941
A_2	0.077092	A_2^*	0.168448
A_3	0.076134	A_3^*	0.146449
A_4	0.056778	A_4^*	0.105849
A_5	0.099034	A_5^*	0.093842
A_6	0.091539	A_6^*	0.06148
A_7	0.094772	A_7^*	0.122682
A_8	0.056669	A_8^*	0.075432
A_9	0.07727	A_9^*	0.074944
A_{10}	0.044549	A_{10}^*	0.047996
A_{11}	0.066352		
A_{12}	0.095867		

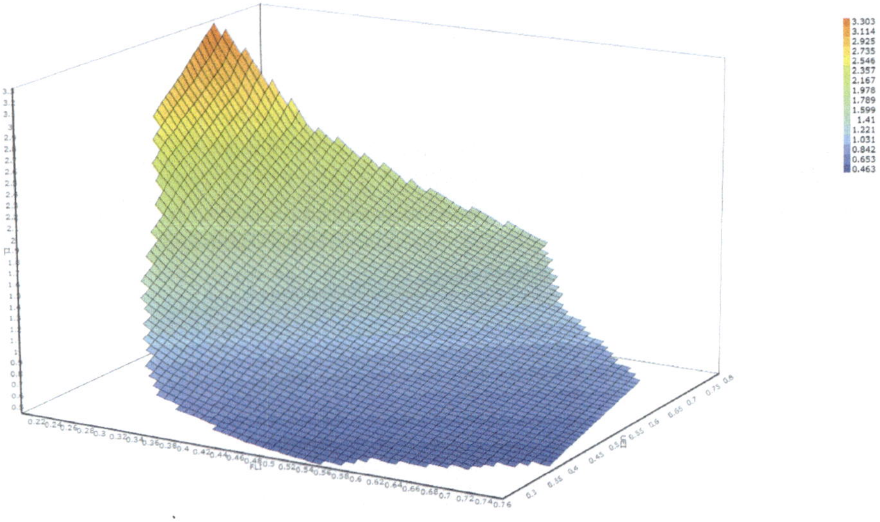

Fig. 4 Trade-off between UEI and FLI

and FLI in optimal condition is the advantage of this study. This model is applied for trade-off between hydroelectricity production and environmental externalities, trade-off between efficiency and financial feasibility of HPP, trade-offs between the economic and environmental performance of an autonomous energy system utilizing an existing Small hydro powerplant while improving its future reliability. This model can be applied for trade-off between the carbon reduction benefits and ecological performance of adding CCS to the power plant.

References

Aggidis, G. A., Luchinskaya, E., Rothschild, R., & Howard, D. C. (2010). The costs of small-scale hydro power production: Impact on the development of existing potential. *Renewable Energy, 35*(12), 2632–2638. https://doi.org/10.1016/j.renene.2010.04.008.

Amini, M., & Almassalkhi, M. (2018). Trading off robustness and performance in receding horizon control with uncertain energy resources. In *2018 Power systems computation conference (PSCC)* (pp. 1–7). IEEE. https://doi.org/10.23919/pscc.2018.8442985.

Berga, L. (2016). The role of hydropower in climate change mitigation and adaptation: a review. *Engineering, 2*(3), 313–318. https://doi.org/10.1016/J.ENG.2016.03.004.

Bettini, C. R., Longo, O. C., Alcoforado, L. F., & Maia, A. C. G. (2016). Method for estimating of construction cost of a building based on previous experiences. *Open Journal of Civil Engineering, 6*(5), 749–763. https://doi.org/10.4236/ojce.2016.65060.

Chang, D. Y. (1996). Applications of the extent analysis method on fuzzy AHP. *European journal of operational research, 95*(3), 649–655. https://www.expertchoice.ir/wp-content/uploads/2017/08/FAHP-Chang-1996.pdf.

de Oliveira, S. (2013). Exergy, exergy costing, and renewability analysis of energy conversion processes. In *Exergy* (pp. 5–53). Springer London. https://doi.org/10.1007/978-1-4471-4165-5_2.

Ji, L., Huang, G. H., Xie, Y. L., Niu, D. X., & Song, Y. H. (2017). Explicit cost-risk tradeoff for renewable portfolio standard constrained regional power system expansion: A case study of Guangdong Province, China. *Energy, 131,* 125–136. https://doi.org/10.1016/j.energy.2017.05.017.

Lipská, E., Vlnková, M., & Macková, I. (2005). Unit labour costs. *BIATEC, 13*(1), 8–12. http://www-ext.nbs.sk/_img/Documents/BIATEC/BIA01_05/8_12.pdf.

Listverse, L. V. (2009). *Top 10 renewable energy sources.* http://listverse.com/2009/05/01/top-10-renewable-energy-sources/. Accessed 9 Jan 2016.

Majumder, P., & Saha, A. K. (2017). Development of financial liability index for hydropower plant with MCDM and neuro-genetic models. In *Application of geographical information systems and soft computation techniques in water and water based renewable energy problems* (pp. 71–105). Springer, Singapore. https://doi.org/10.1007/978-981-10-6205-6_4.

Majumder, P., & Saha, A. K. (2018a). Efficiency assignment of hydropower plants by DEMATEL-MAPPAC approach. *Water Conservation Science and Engineering, 3*(2), 91–97. https://doi.org/10.1007/s41101-018-0041-y.

Majumder, P., & Saha, A. K. (2018). Development of financial liability index for hydropower plant with MCDM and neuro-genetic models. In *Application of geographical information systems and soft computation techniques in water and water based renewable energy problems* (pp. 71–105). Springer, Singapore. https://doi.org/10.1007/978-981-10-6205-6_4.

Majumder, P., & Saha, A. K. (2019a). Ranking of indicators for estimation of plant efficiency in hydropower plants by a bootstrap MCDM approach. *International Journal of Energy Optimization and Engineering (IJEOE), 8*(3), 69–92. https://doi.org/10.4018/IJEOE.2019070104.

Majumder, P., & Saha, A. K. (2019b). Identification of most significant parameter of impact of climate change and urbanization on operational efficiency of hydropower plant. *International Journal of Energy Optimization and Engineering (IJEOE), 8*(3), 43–68. https://doi.org/10.4018/IJEOE.2019070103.

Majumder, P., Majumder, M., & Saha, A. K. (2016). Application of decision making for optimal condition method to analyze operational efficiency of hydropower plants. *International Journal of Control Theory and Applications, 9*(42), 79–94. https://serialsjournals.com/abstract/73561_cha-10.pdf.

Majumder, P., Majumder, M., Saha, A. K., Sarkar, K., & Nath, S. (2019). Real time reliability monitoring of hydro-power plant by combined cognitive decision-making technique. *International Journal of Energy Research, 43*(9), 4912–4939. https://doi.org/10.1002/er.4530.

Majumder, P., Majumder, M., Saha, A. K., & Nath, S. (2020a). Selection of features for analysis of reliability of performance in hydropower plants: A multi-criteria decision making approach. *Environment, Development and Sustainability, 22*(4), 3239–3265. https://doi.org/10.1007/s10668-019-00343-2.

Majumder, P., Majumder, M., & Saha, A. K. (2020b). Real-time monitoring of power production in modular hydropower plant: Most significant parameter approach. *Environment, Development and Sustainability, 22*(5), 4025–4042. https://doi.org/10.1007/s10668-019-00369-6.

Pang, M., Zhang, L., Liang, S., Liu, G., Wang, C., Hao, Y., et al. (2017). Trade-off between carbon reduction benefits and ecological costs of biomass-based power plants with carbon capture and storage (CCS) in China. *Journal of Cleaner Production, 144,* 279–286. https://doi.org/10.1016/j.jclepro.2017.01.034.

Parida, A., & Kumar, U. (2009). Maintenance productivity and performance measurement. In *Handbook of maintenance management and engineering* (pp. 17–41). Springer, London. https://doi.org/10.1007/978-1-84882-472-0_2.

Price, K., Storn, R. M., & Lampinen, J. A. (2006). *Differential evolution: a practical approach to global optimization.* Springer Science & Business Media. https://books.google.co.in/books?hl=en&lr=&id=hakXI-dEhTkC&oi=fnd&pg=PR7&dq=17.%09Price,+K.,+Storn,+R.+M.,+%26+Lampinen,+J.+A.+(2006).+Differential+evolution:+a+practical+approach+to+global+optimization.+Springer+Science+%26+Business+Media.&ots=c0YDDPNcb1&sig=pPJEzP9xI9PwaevB2XbsciEbnEo&redir_esc=y#v=onepage&q&f=false.

Rayamajhee, V., & Joshi, A. (2018). Economic trade-offs between hydroelectricity production and environmental externalities: A case for local externality mitigation fund. *Renewable Energy, 129,* 237–244. https://doi.org/10.1016/j.renene.2018.06.009.

Singh, V. K., & Singal, S. K. (2018). Optimal operation of run of river small hydro power plant. *Biophysical Economics and Resource Quality, 3*(3), 10. https://doi.org/10.1007/s41247-018-0045-4.

Sisodia, G., Sharma, K., & Gupta, S. (2018). Intuitionistic fuzzy weighted sum and product method for electronic service quality selection problem. *International Journal of Modern Education & Computer Science, 10*(9). http://mecs-press.net/ijmecs/ijmecs-v10-n9/IJMECS-V10-N9-5.pdf.

Steffen, B. (2018). The importance of project finance for renewable energy projects. *Energy Economics, 69,* 280–294. https://doi.org/10.1016/j.eneco.2017.11.006.

Tuhtan, J. A. (2007). Cost optimization of small hydropower. *Mini & Micro Hydro Power Generation EBARA Hatakeyama Memorial Fund Tokyo, Japan.* https://d1wqtxts1xzle7.cloudfront.net/38455513/Tuhtan_Thesis.pdf?1439387896=&response-content-disposition=inline%3B+filename%3DSmall_Hydro_Power.pdf&Expires=1602944448&Signature=CXlx0RNkayma6E2oBKS~atTTmXGVfU0uQ~r4YiMlxl9e~77D684VZr6OEPUfU4RkVdDpRajF0eKYLxD0SDp4h0k~cwwhjATE4Ywa746yGj67d3E3zjmLH404P5GENGK7yr6i6cwN67FAHb7jYF~rnmFsiKrtQZSXw6e7BSS0yWsRQVbKIqV99TuI73UXEAqEYQfxGn2fpyHvczxmurTrn9pjYew-z8HbOCtJ4I9xLzrKSipyEd2HyxHtFlkT6qiR4aw16Uo4HmbQH2e2w6GFeV8k6WYmJyg0N3c9rkS8kkeORZhL66IwdgriZjwYlABPUNwD6U9bf43PlNPYhZaNGg_&Key-Pair-Id=APKAJLOHF5GGSLRBV4ZA.

Yang, X. S. (2009). Firefly algorithms for multimodal optimization. In *International symposium on stochastic algorithms* (pp. 169–178). Springer, Berlin, Heidelberg. https://doi.org/10.1007/978-3-642-04944-6_14.

Impact Analysis of Water, Energy, and Climatic Variables on Performance of Surface Water Treatment Plants

Sudipa Choudhury and Apu K. Saha

Abstract In India, water shortages and poor water quality continue to be major challenges in both domestic and industrial sectors. The methods that aim to optimize Water Treatment Systems can go some way towards addressing these pressing challenges. The change in climate and large scale urbanization has imbibed vulnerabilities in water and energy use of Surface Water Treatment Plants (SWTP). As SWTPs take water from surface sources like rivers, lakes, etc., the change in the climate of a location can also impact the plant's performance. The external factors like climate change and urbanization have created an imbalance in the hydrologic cycle, which in turn compromised both the quality and quantity of surface water used in the SWTP for processing and treatment. The imbalance has also influenced the time taken for the completion of the treatment operations in an SWTP. The overuse of the electrical machinery is another effect of this natural in-equilibrium. That is why; there is a necessity for the analysis of the impact of climatic changes, and because of the change, the impact of variation in water, and energy used by the SWTP and duty cycle of the treatment processes followed in an SWTP. As such, factors will directly influence the performance of SWTP and the quality of the treated water. The present study tries to apply cognitive and objective decision-making tools to develop a framework for identifying the most significant factor which will be most affected by the uncertainty and its impact on the performance efficiency of the SWTP. Data derived from water treatment plants in Tripura of North Eastern India has been utilized to demonstrate the reliability of the proposed method.

Keywords Water treatment plant · Analytical network process · Polynomial neural network

S. Choudhury (✉) · A. K. Saha
Department of Mathematics, National Institute of Technology, Agartala, Tripura, India
e-mail: sudipachoudhury032@gmail.com

A. K. Saha
e-mail: apusaha_nita@gmail.com

M. Majumder and G. D. Kale (eds.), *Water and Energy Management in India*,
https://doi.org/10.1007/978-3-030-66683-5_10

1 Introduction

The process of water treatment plants (WTP) is often dynamic. It can also become capital intensive, and as such, optimization methods are essential in making projects commercially viable, technically efficient to maintain quality standards. Primarily as the performance of each treatment unit affects the efficiency of the subsequent groups (Li et al. 2020; Dorussen and Wassenberg 1997).

The purpose of the system of municipality water delivery is to convey fresh water from the water treatment plants to the consumers, for the purpose of drinking, cooking, and other works. Water supply is also essential for businesses and industries that operate in the same municipal environment (Khan et al. 2017). Since most WTPs work using rule-of-thumb approaches, it becomes imperative to carry out an efficiency analysis process based on several performance indicators, mainly as such performance factors have a notable impact on the quality and efficiency of the processes observed (Kumar 2010).

In recent years, the uncontrolled expansion of urban areas and the extraction of natural resources have imbibed modifications in the regular pattern of climate in most of the places of the world. The reason for the change is attributed to global warming, but it has impacted the regular functionality of many natural systems and processes (Raseman et al. 2017). This change also influences the performance of the SWTP. The water and energy consumed by an SWTP depend on the quantity and quality of water collected for treatment. Change in climate has altered the quantity and large scale urbanization has modified the quality of water all over the World (Fant et al. 2017). As a result, many SWTP has been forced to change its regular operational protocols for adjusting with the current uncertainties due to the degradation in quality and variations in the quantity of water intake. The change in the intake will change the quantity as well as quality of the output water. Not only the quantity and quality of the treated water will be altered but also the time required to complete the operation will change because of these vulnerabilities.

There are many available simulation and predictive models for estimation of performance efficiency for SWTP (Chen et al. 2020) but none among them included the impact of the climatic and urbanization uncertainties. The exclusion of such uncertainties has compromised model performance and therefore, yielded erroneous evaluation about the SWTP performance.

This book chapter will attempt to develop a framework that will estimate plant performance including the impact of climate change. The impact of these changes will be encoded in the model in terms of the water and energy used and the duty cycle of the operations as both are mostly affected as a consequence of the change in the regular pattern of the climate. Specifically, it will do so through the application of Analytic Network Process (ANP) (Saaty and Vargas 2013). The analytic network process. In Decision making with the analytic network process (pp. 1–40). Springer, Boston, MA., that take into account the conclusions from the related kinds of literature looking after the execution of the WTP after evaluation by experts and stakeholders to identify the most significant parameter which controls the performance efficiency of

SWTP followed by the use of Polynomial Neural Networks (PNN) (Oh and Pedrycz 2002; Ivakhnenko and Ivakhnenko 1995) in the estimation of an indicator which was made to be directly proportional to the performance efficiency of Surface Water Treatment Plant (SWTP).

1.1 Study Objective

The objective of the present investigation is to examine the impact of the change in climate on the surface water treatment plants in terms of water and energy used and the duty cycle of the operations. In this aspect, the most significant parameter is required to be identified first. After the most important parameter or MIP was identified the next aim is to develop an indicator-based model which can represent the performance of SWTP in terms of the related parameters comprised according to their importance as decided concerning the impact of change in water and energy consumed by the plant and change in the duty cycle of the operations for the treatment of the intake water. The impact of the alteration in the climatic parameters on the performance strictures of the SWTP is also encompassed as per their influence on the plant routine and being influenced due to the variation in the climatic pattern.

1.2 Literature Reviews

Water plays a crucial role in promoting and protecting human health and the creation of sustainable ecosystems (Yan et al. 2016). As the population, urbanization, industrialization and consumption are increasing, the resources of freshwater is also demanding (Bagatin et al. 2014). Several diseases like Cholera, Typhoid, and Dysentery, etc. are the diseases caused by infected water. Near about 80% of the diseases in the developing countries are associated with water infectivity (Sundara et al. 2010). To decrease the percentage of these diseases and to take care of human health, Water Treatment Plants (WTPs) are innovated to supply potable water to the people.

Vasquez (Vasquez, 2017) highlighted the importance of technological advances and trade agreements concerning company relocation. The importance of adapting the globalization while relocating was identified in this study. Akbari et al., (Akbari et al. 2017) applied multi-criteria decision-making methods for site selection of low-speed wind forms with respect to investment returns. Due to the inclusion of randomness in the decision making, distortion of information during the normalization process, and exclusion of managers from criteria rating, the reliability of existing MCDM is not satisfactory. That is why the authors used cloud-based decision-making methods with the 2-tuple cloud-based linguistic environment. In some of those works, synthesis and optimization of wastewater treatment with single and multiple contaminants have been proposed (Tsiakis & Papageorgiou 2005; Skiborowski et al.

2012; Castro and Teles 2013). Several articles were published on wastewater treatment (Dong et al. 2008; Ahmetović and Grossmann 2011; Galán and Grossmann 2011; Rojas-Torres et al. 2013; Ibrić et al. 2014; Yang and Grossmann 2013). A review article analysed the various contributions made to Water Network Synthesis (Ahmetović et al. 2011).

2 Methods Used

In the present investigation, the ANP Multi-Criteria Decision Making method was used for feature selection and the PNN model was used to estimate the indicator from the selected factors. Following Section tries to highlight the methodology, applications, strengths, and weaknesses of the methods and why it was used to achieve the objective of the present investigation.

2.1 Multiple Criteria Decision-Making Methods (MCDM)

In this study, one MCDM methods that are well suited to this line of inquiry are the Analytical Network Process (ANP) was used and outlined briefly in Sect. 2.1.1.

2.1.1 Analytic Network Process (ANP)

Like the highly popular Analytical Hierarchy Process (AHP), the Analytical Network Process is a more generalized version of the AHP (Hernández et al. 2010). That is, whilst the AHP regards parameters independently according to the given criteria and then allocates them into hierarchies, the ANP allows for the interdependency of parameters with respect to the criteria (Cambron and Evans 1991). The AHP method if employed would consider each of the factors independently with respect to the criteria (Saaty 1980). The ANP method however would allow for interdependency, for instance, if the cost of building a new school was very like that of building a new road, then the importance of cost in the criteria would be reduced.

The main advantage of using the ANP method is that often, it is impossible to structure a decision-making problem hierarchically, since they involve interaction and connections between the highest and lowest elements. To take into account the complexities of real-world problems, Saaty in 2005 (Saaty 2005) proposed the ANP model as a deliberate modification of the AHP method.

It practices the ANP provides the solution of such decision-making problems which cannot be structured hierarchically by connecting the criteria and factors by a network. It preferred a pairwise comparison to create the connections within the structure. The evaluation process of this method is composed of four core principles, viz.; the structure development of the problem, the pairwise comparison, super matrix creation, and the priority value (Saaty and Vargas 2006). The initial stage of this analysis is to subdivide the objective into clusters which includes various nodes and alternatives to form a network. Relationships must then be established between the different sections of the network. As per the process of AHP, ANP also produces pairwise comparisons to evaluate the relative importance of the various factors with respect to the other component belonging to the network. Another similarity is that the ANP also uses Saaty's Fundamental Scale to quantify network comparisons.

The ANP MCDM was used in decision support systems developed for various objectives related to water treatment plant performance and efficiency improvement (Özdemir et al. 2020; Zolfaghary et al. 2020).

2.1.2 Polynomial Neural Network (PNN)

PNN networks are algorithms that emulate the biological structure of neurons. In nature, neurons operate by transmitting an impulse only when a certain threshold has been surpassed, once an impulse in initiated it then moves between different neurons to initiate a reaction within the organism (Oh et al. 2003). Similarly, PNN networks consist of a set of input characteristics that are processed by "hidden" functions, which then map them to outputs. They are particularly well adapted to approximating complex decision making situations as "hidden" functions work to identify the most significant factors, factoring this into the outputs generated (Atashkari et al. 2007). This aspect of the PNN has contributed to its wide applicability in a range of real-time applications (Majumder et al. 2020), including, functional approximation (Mohammadi et al. 2020), or regression analysis, as well as time series prediction and modelling (Pham et al. 2020) etc.

However, the PNN procedure does have some drawbacks. For one, it is not a daily life general-purpose problem solver, especially as there is no structured methodology available in PNN. There is also no single standardized paradigm for PNN development (Xu et al. 2020).

The main advantage of the PNN method (Ghodsi and Khanjani 2020) is it auto-select the topology from the data and information fed to the model about the input and output variables and the method is also capable of self-estimation of parameters important for reliable functionality of the model (Jiang and Guo 2013). The parameters like type of activation function and weight of connections in between input and hidden layers and hidden and output layers and the number of hidden layers ideal for generation of a reliable framework. The PNN model utilizes 100 s of algorithms to estimate all the parameters and based on some metrics identify the best algorithm which features optimal configurations (Pusat and Akkaya 2020; Ivakhnenko and Ivakhnenko 1995).

3 Methodology

In this study, the ANP has been used to a problem regarding a surface water treatment plant. When these methods are applied in this manner, the consultation of experts is often beneficial as they can improve the decision with their opinions on this and how best to solve it. For this reason, this study sought the expertise of five researchers who have studied the sector of wastewater treatment extensively.

3.1 Application of MCDM

The application of ANP involves the selection of criteria, alternatives, and methods as detailed below. A team of experts was selected to evaluate the of the water treatment plant. Then the corresponding importance of the factors has been calculated by the survey of Literature, Hazard potential, consumer's feedback, and Engineer's feedback. The relative weights of importance are then estimated by the application of ANP methods.

3.1.1 Selection of Criteria

Some criteria have to be identified for which the alternatives will be compared and the difference in importance can be determined. In this regard, the following factors are considered as Criteria:

1. *Impact on Time of Operation (TO)*

The parameters which are generally used to evaluate the performance of the SWTP are affected by the change in the duty cycle of operation (Deng et al. 2020). But not all the parameters are affected equally. The degree of influence will be different for different parameters. The change in climate greatly modifies the duty cycle as, under the uncertainty, both quality and quantity of the intake water will change and when quality and quantity changes the duty cycle will also change. For larger quantity and degraded quality of water duty cycle will be elongated and vice versa. The impact of change in the duty cycle was considered when the priority value of the priority parameters is evaluated (Deng et al. 2020).

2. *Impact of Climatic Variables (CV)*

Climate change will enforce alteration in the regular pattern of climate observed in a location. SWTP collects water from surface sources and such resources depend on climatic parameters like Precipitation, Evapotranspiration, etc. When the pattern of this climatic parameter changes the quantity of water in the surface sources will also change (Dutta et al. 2020). The study has included the influence of change in climatic parameters at the time the MIP is identified, and the weight of importance is being estimated by using the parameter as a criterion (Debnath et al. 2015).

3. *Impact on Energy Use (EU)*

The change in climate modifies the amount of water in the surface resources. The large-scale urbanization increases the demand for treated water. The exponential increase in industrialization and agriculture to satisfy the need of the burgeoning population has compromised the quality of the surface water. As a result, the characteristic of the intake water has changed. The regular operation procedure must be adapted to mitigate the uncertainty caused by the change. Therefore, the energy consumed by the plant will also vary. The intake water is carried throughout the plant either by gravity or using auxiliary pumps. The clari-flocculators and aerators are run by electric motors. When the duty cycle of the processes gets elongated the running time of the motors will also upsurge. As a result, the consumption of energy will also rise. The effect of this excess use of energy is also considered at the time of calculation of weightage of the parameters used in the performance evaluation of SWTPD (De and Majumder 2019).

4. *Impact on Water Use (WU)*

The amount of water used by the plant for smoothly carrying out the treatment operations also depends on the quality and quantity of intake water. When the quantity is more the operation time will be more and as a result, more amount of water will be required to maintain the efficiency of the operations. The filtration requires additional water to clear the suspended and residual contaminants (Rodríguez et al. 2010). If the quality of the water is detreated, the amount of water to clear the residuals will surge. The influence endowed by this additional water is also considered at the time of evaluation for the identification of MIP.

3.1.2 Selection of Alternative

The evaluation parameters for performance evaluation of SWTP were identified with the help of government reports, published literature, and logbooks of water treatment plants. After the initial selection, a group of experts preselected the seven most important, independent, and coherent parameters and treated them as alternatives of the multi-criteria decision-making problem used to identify the priority value of the parameters which will, in turn, identify the MIP.

Every SWTP under the state or national government produces an annual report about the performance of an SWTP it maintains. The parameters discussed in those reports are important in terms of evaluating the performance of the SWTP.

The logbooks are maintained to store the daily functionality of an SWTP. The parameters which re-logged daily is certainly significant enough to represent the performance of the SWTP.

The literature published in reputed journals and conferences is a reliable source for the identification of parameters utilized to estimate the operational efficiency of SWTP.

The parameters collected in this aspect were evaluated by the experts and among all the selected parameters, eight most significant parameters were used for comparing their importance for the selected criteria. All the eight parameters were ranked by the experts and the common rank of the parameters was determined by considering the ranks given by each of the experts to each of the parameters with respect to the selected criteria. At last, the common rank was complemented to find the score of the importance of the parameters with respect to the selected criteria. The scores were utilized in the pairwise comparison step to rate the parameters for each of the other parameters based on the impact of the selected criteria and the goal of the decision-making problem.

Once the eight parameters and their score were identified the aggregation methods like ANP were applied to find the relative weights of importance or the priority value (PV) of the parameters. The PV of the parameters along with the magnitude of the variables were used to find the value of the index for the SWTP to estimate the performance of an SWTP. If annual values of the parameters were used to calculate the index then annual performance will be depicted and if monthly or daily values are used then monthly and daily indicators will be calculated to represent monthly and daily performance of the SWTP in terms of the 'climatic impact' and change in 'water' and 'energy' use of the SWTP and the change in 'duty cycle' of the plant operations.

Section 3.2 shows the way the indicator was developed.

3.2 Development of Index

After the weightage of importance was determined, an index was developed with the help of the PV and the magnitude of the priority parameters. The Eq. 1 is proposed as the index which represents the performance of the SWTP in terms of the change in climate and urbanization impacts.

$$P = \text{Sin}(a^z) \quad (1)$$

where,

$$\mathbf{Z} = \frac{\sum_{i=1}^{n} w_i * l_i}{\sum_{i=1}^{n} w_i} \quad (2)$$

where, w_i is the weightage of the importance of the parameters and l_i is the *ith* priority parameters as determined in Sect. 3.1.

A polynomial neural network model was developed where all the selected priority parameters were used as input and the index (Eq. 1) was used as output. The development of the model is described in Sect. 3.2.

3.3 Application of Polynomial Neural Network

The present study aims to develop an automated framework for estimation of the index value with the help of the priority parameters and their influence on plant performance. Although the Artificial Neural Networks are widely used for prediction, simulation, and optimization purposes there are three major drawbacks of the modeling algorithms. The weightage of the connections is one of the major model parameters which is a medium for enabling equalization between the desired and predicted output. In ANN, the weightage is estimated by using various algorithms one at a time and based on most optimal training performance the model selects the weights determined by an algorithm which have the most optimal metrics. But as trial and error method is followed and no standard recommendation is there for determination of the weights, the time taken for identification of the best training algorithm degrades the electability of the procedure where quick prediction is required. Similarly, the determination of the other parameters like type of activation function and number of hidden layers is performed with the help of trial and error or some search algorithms both of which are time-consuming and computationally intensive.

But on the other hand, the PNN method can self-identify the optimal topology and uses 100 s or more algorithms at a time to determine the weightage of the network. The capability of parameter estimation at a time has enforced the authors to use the PNN based algorithms in the estimation of SWTP performance.

In the present investigation, eight inputs and one output was used to develop the model. The same output was predicted by the PNN model where Group Method of Data Handling (GMDH) was used as the training procedure which comprises the application of 100 algorithms to estimate the weight of the connections in between input and hidden and hidden and output layer. The number of hidden layers for optimal prediction is also estimated by the GMDH algorithm. Similarly, Quick Combinatorial (QC) algorithm was also used to determine the weight and follow the same architecture of the PNN model. The difference between the GMDH and QC lies in the type of transfer functions used by the procedures.

The output parameter was also used in two different formats. In one case, the normal value of the output is used and in the other, its arc tangent transformation was taken as the predictor. The advantage of using arc tangent transformation is a further reduction of scale influences which generally reduce the learning time and increases the accuracy of the developed model.

In this manner, the same output was predicted with the help of the same eight parameters but by using four different configurations. The predictions from each configuration were compared with the help of simple metrics like Root Mean Square

Error, Mean Absolute Error, and Correlation Coefficient between the desired and predicted output from the model.

The developed model was used to estimate the annual performance of Barjala Treatment Plant located in Agartala, a peri-urban city in North-East India. The plant intake is 4MLD and supplies to nearly 1 crore consumer of both industrial and domestic types.

The result of the methodology is depicted in Sect. 4.

4 Result and Discussion

The priority value of both criteria and alternatives with respect to alternatives and criteria is depicted in Table 1. The priority value of the parameters for each of the criteria and criteria weightage with respect to alternatives is shown in Figs. 1 and 2. The comparison of the PV determined by AHP and ANP method is given in Fig. 3. The performance metrics of the developed different configurations of the PNN models are shown in Table 2. The performance analysis treatment plant in the case study area from 2010 to 2014 is given in Fig. 4.

Section 4.1 discussed the result of the ANP method and 4.2 about the PNN method. Section 4.3 highlights the results of the case study.

4.1 Results from the ANP Method

These results show that the CV is the most important criterion followed by the EU, TO and WU. Among the alternatives, IE was found to be the MIP followed by ME and CE as per the priority value assigned to the parameters in the ANP method.

4.2 Results from the PNN Method

The configuration which used Arc Tangent transformation and GMDH procedure respectively to transform output data and train the model was found to have a higher CC of 0.999 and lower RMSE of 0.0002 and MAE 0f 0.0003 compared to that of the other configurations in the training phase of the model. The RMSE of the G2A model in the training phase was found to be 0.6 times less compared to the G1N Training and G2A testing phase model. The MAE of the G2A model Training phase was found to be 25% smaller compared to the G1A model training phase which has the second-best MAE among all the eight models. In the case of CC, both G1A Training and testing G2A training and testing have the highest value compared to the other models which were trained by the QC algorithm. Even, G4A model training phase, which used the arc tangent transformation of output, as model output and trained by QC algorithm has 4 and 1.67 times more RMSE and MAE along with 0.99 times less

Table 1 Super matrix showing the PV of selected criteria and alternatives after comparison with each other with respect to the alternative and criteria respectively

	TO	CV	EU	WU	LE	IE	CE	ME	FE	OC	MC	LC	Final
TO					0.364	0.364	0.100	0.364	0.333	0.200	0.200	0.364	0.100
CV					0.091	0.091	0.400	0.091	0.167	0.100	0.100	0.091	0.400
EU					0.364	0.364	0.100	0.364	0.167	0.400	0.400	0.273	0.300
WU					0.182	0.182	0.400	0.182	0.333	0.300	0.300	0.273	0.200
LE	0.111	0.054	0.103	0.105	0.102								0.102
IE	0.222	0.162	0.205	0.184		0.204							0.204
CE	0.083	0.108	0.077	0.211			0.143						0.143
ME	0.194	0.162	0.205	0.184				0.194					0.194
FE	0.056	0.081	0.077	0.132					0.089				0.089
OC	0.167	0.216	0.154	0.079						0.140			0.140
MC	0.139	0.189	0.154	0.079							0.132		0.132
LC	0.028	0.027	0.026	0.026								0.027	0.027

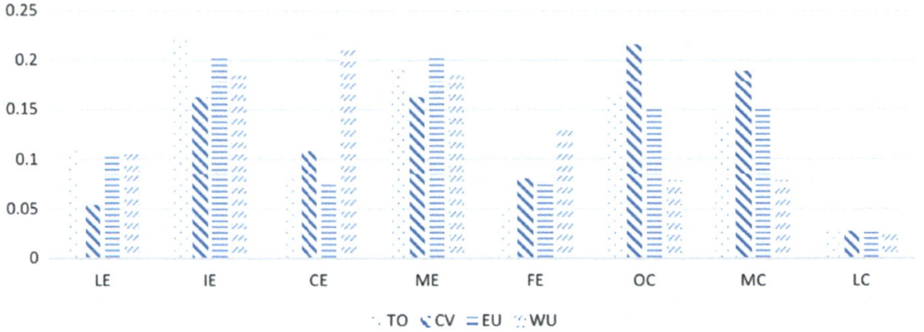

Fig. 1 The PV of the parameters after comparing with each other with respect to the selected criteria

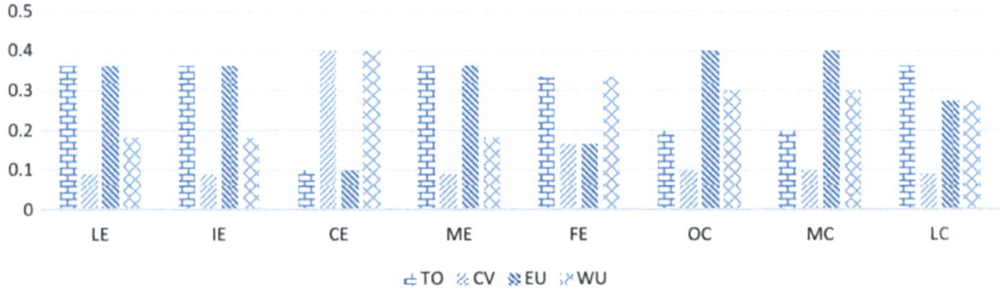

Fig. 2 The PV of the Criteria after comparing with each other with respect to the selected alternatives

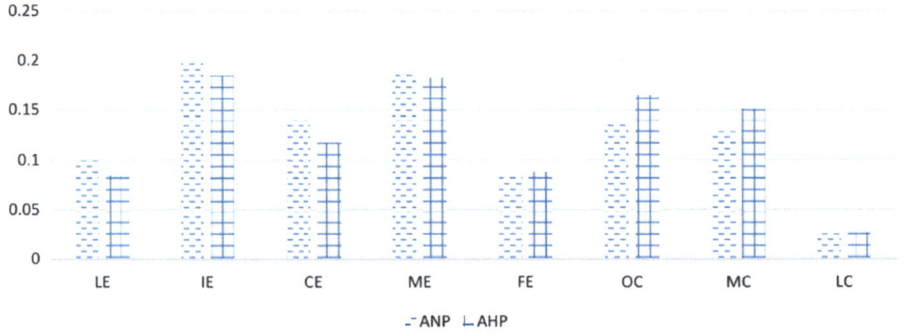

Fig. 3 The final PV value after the application of ANP and AHP aggregation method

CC compared to the G2A model training phase. That is why the model trained with GMDH and which uses arc tangent to transform output data was used to estimate the performance of the SWTP. The model generated equation is given in Eq. 2 which was used to determine the performance of the Barjala Water Treatment Plant for the year 2010–2015.

Table 2 The performance metrics of the four different configurations to develop the PNN model

Model name	Phase	Training procedure	Output transformation	RMSE	MAE	CC
G1N	Training	GMDH	None	0.005	0.004	0.999
G1N	Testing	GMDH	None	0.007	0.006	0.999
G2A	Training	GMDH	Arc Tangent	0.002	0.003	0.999
G2A	Testing	GMDH	Arc Tangent	0.005	0.005	0.999
Q1N	Training	QC	None	0.046	0.017	0.986
Q1N	Testing	QC	None	0.056	0.019	0.976
Q2A	Training	QC	Arc Tangent	0.008	0.005	0.998
Q2A	Testing	QC	Arc Tangent	0.011	0.001	0.997

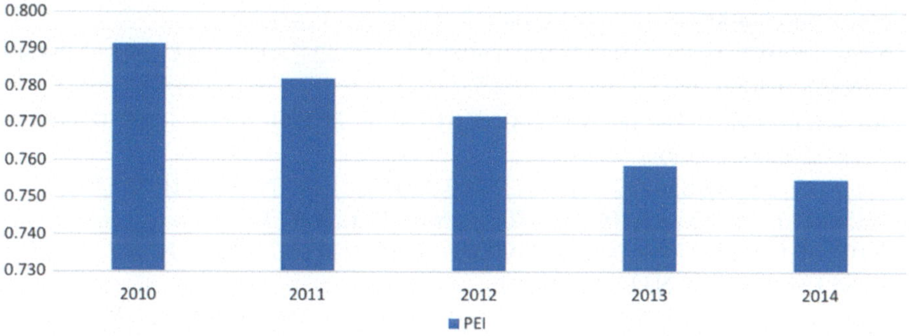

Fig. 4 The performance of the Barjala SWTP for the year 2010–14 as estimated by the PEI

4.3 *Results from the Case Study*

Based on the selected model, the annual performance of Barjala SWTP was estimated by Eq. 2 and is given for five consecutive years from 2010 to 2014.

According to the results, the performance of the SWTP in Barjala for the year 2014 was found to be 1.04 times inferior compared to that of 2010. The reason can be attributed to the large-scale urbanization that has taken place in the Agartala for the last five years. The urban area of the city has increased compared to that in 2010. As a result, the demand for treated water has increased. The increase in the urban population implies that the area of deforestation will also increase and as a result, the quality of the surface water will degrade which has increased the pressure on the SWTP. The indicator was successful in the representation of the SWTP performance with respect to the impact of climate change and corresponding consequences.

$$\begin{aligned}\mathbf{Y1} &= -\mathbf{0.000158213} - \mathbf{N190} * \mathbf{0.0470396} - \mathbf{N190} * \mathbf{N2} * \mathbf{0.11437} \\ &\quad + \mathbf{N190}^2 * \mathbf{0.113561} + \mathbf{N2} * \mathbf{1.04777}\end{aligned} \tag{3}$$

Note: See Annexure 1 for the value of N190 and N2.

Annexure 1

The value of the parameters Nn where n = 1 to 546, which implies that 546 number of sub-model was prepared to estimate the output.

N2 = −4.35114e-06 + N286*N3*0.805475 − N286^2*0.40569 + N3*1.00054 − N3^2*0.400614
N3 = 0.00361389 − N507*0.0271919 − N507*N4*0.0324194 + N507^2*0.0491858 + N4*1.0092 + N4^2*0.0051021
N4 = −5.25143e-05 − N157*N5*2.74272 + N157^2*1.37767 + N5*0.999803 + N5^2*1.36527
N5 = −0.0137968 + N535*0.0732828 + N535*N6*0.0223251 − N535^2*0.0957644 + N6*0.990607
N6 = −0.0109597 + N514*0.0642941 + N514*N7*0.0497147 − N514^2*0.096555 + N7*0.985109 − N7^2*0.00811556
N7 = −8.95764e-05 + N342*N8*0.608784 − N342^2*0.300953 + N8*1.00099 − N8^2*0.309011
N8 = 0.000288154 − IE*N9*0.00602018 + IE^2*0.00270669 + N9*0.997848 + N9^2*0.00577336
N9 = −0.000836897 + N352*0.050116 + N352*N10*0.63067 − N352^2*0.372585 + N10*0.954707 − N10^2*0.263973
N10 = 0.000274375 − N55*0.449027 − N55*N11*6.24051 + N55^2*3.64213 + N11*1.44751 + N11^2*2.60019
N11 = −0.00864384 + N468*0.0381293 − N468*N12*0.0751688 + N12*0.998376 + N12^2*0.0369512 N12 = −0.000644025 + N109*0.114227 − N109*N13*8.91966 + N109^2*4.28922 + N13*0.887827 + N13^2*4.62841
N13 = −0.014036 + N474*0.0891585 + N474*N14*0.0464754 − N474^2*0.125914 + N14*0.97554 + N14^2*0.00608043
N14 = −0.000976324 − N51*0.882412 + N51*N15*26.4945 − N51^2*12.2561 + N15*1.88731 − N15^2*14.2435
N15 = 0.0557807 − N541*0.318108 − N541*N16*0.202622 + N541^2*0.473507 + N16*1.05341 + N16^2*0.0400011
N16 = 0.0239975 − N486*0.155858 − N486*N17*0.103061 + N486^2*0.234665 + N17*1.04264
N17 = 0.00067865 − N315*0.0436705 + N18*1.04214
N18 = −0.0112054 + N472*0.128727 + N472*N19*0.363663 − N472^2*0.332952 + N19*0.926366 − N19^2*0.0954475
N19 = 0.0730708 − N522*0.403297 − N522*N23*0.205558 + N522^2*0.53677 + N23*1.08325 + N23^2*0.0154941
N23 = 0.00139764 − LE*0.00630617 − LE*N26*0.0234872 + LE^2*0.017214 + N26*0.992125 + N26^2*0.0234394
N26 = −0.00178969 + LC*0.0178568 + LC*N42*0.0102531 − LC^2*0.0236901 + N42*0.99665
N42 = 0.083059 − N509*0.36254 + N509^2*0.360679 + N54*1.01368
N54 = 0.00127719 + N82*0.991557 − N82*N83*1.06668 + N83^2*1.07871
N83 = −0.0211799 + N386*0.257882 − N386^2*0.0448513 + N158*0.810405

N158 = −0.0105328 + N278*0.634579 + N337*0.389229

N278 = −0.880796 + N429*2.26836 − N429*N488*1.47984 − N429^2*0.737464 + N488*1.85417 − N488^2*0.319865

N488 = −0.00435897 + N522*N545*2.28281

N429 = −0.615573 + N498*1.82818 − N498^2*0.949933 + N514*0.988258

N386 = −1.5273 + N449*3.04572 − N449*N529*2.91014 − N449^2*0.783989 + N529*3.66071 − N529^2*1.3795

N529 = 0.114718 + N539*N543*1.67153

N315 = −0.268079 + N424*1.30611 − N424*N499*1.21302 + N424^2*0.213542 + N499^2*1.65527

N499 = −0.13039 + N536*2.94979 − N536^2*2.18074 − N545*2.4907 + N545^2*4.07033

N424 = −0.915167 + N483*2.16785 − N483*N514*2.78551 + N514*2.13438

N483 = 0.311843 + ME*0.101757 + ME^2*0.0303621 + FE*0.189757 − FE^2*0.0784192

N541 = 0.0482325 + N546*N547*2.01331

N547 = 0.406532 + LE*0.119069 − LE^2*0.0676594

N51 = −0.00116357 + N82*0.581435 + N111*0.421195

N111 = 0.000852128 + N157*0.98843 − N157*N201*1.02622 + N201^2*1.04696

N201 = −0.127346 + N335*0.843406 − N335*N409*0.133544 + N409*0.630276 − N409^2*0.277272

N409 = −0.942898 + N470*0.979493 + N533*3.34512 − N533^2*2.6798

N470 = 0.00178486 + N509*N528*2.25131

N474 = −0.0513266 + ME*0.09137 + ME^2*0.0427421 + N516*0.977257

N109 = −0.00268512 + N242*1.00225 − N242^2*0.635445 + N194^2*0.643709

N194 = −0.0918785 + N337*0.849754 − N337*N404*0.279123 + N337^2*0.057818 + N404*0.45755

N404 = −1.33558 + N472*1.8672 − N472*N533*1.99306 + N533*4.23067 − N533^2*2.68812

N533 = 0.032233 + N544*N546*2.09513

N544 = 0.395184 + OC*0.0967102

N472 = −0.0607508 + ME*0.160032 − ME^2*0.0273872 + N514*0.974332

N337 = −0.725901 + N453*1.27485 − N453*N466*1.05546 + N466*2.41636 − N466^2*1.2901

N466 = 0.0105541 + N509*N516*2.20597

N509 = 0.340809 + LE*0.113182 − LE^2*0.0622609 + ME*0.117534 + ME^2*0.019951

N453 = −0.985615 + N500*2.22706 − N500*N524*2.76067 + N524*2.22208

N242 = −0.698257 + N425*2.35571 − N425*N494*1.80084 − N425^2*0.67488 + N494*0.913494 + N494^2*0.919567

N494 = −0.903791 + N534*2.18701 − N534*N536*2.97117 + N536*2.17165

N425 = −0.693308 + N491*2.19279 − N491^2*1.36445 + N514*0.985333

N55 = 0.00177621 + N82*0.988737 − N82*N105*0.968159 + N105^2*0.983907

N105 = 0.0669965 + N522*N193*1.06519 − N522^2*0.769448 + N193*0.845233 − N193^2*0.282968

N193 = −0.0103703 + N290*0.604365 + N335*0.419075

N335 = −0.88497 + N463*2.29295 − N463*N468*1.44732 − N463^2*0.913638 + N468*2.06145 − N468^2*0.665222

N468 = −0.383989 + N514*0.935538 + N516*0.932398

N463 = 0.0237337 + N492*N524*2.138

N492 = 0.315205 + ME*0.12814 − ME*MC*0.0667144 + ME^2*0.044814 + MC*0.125061

N290 = −0.91018 + N421*2.2033 − N421*N496*1.46376 − N421^2*0.690019 + N496*2.11821 − N496^2*0.678333

N496 = −0.190352 + N534^2*1.08005 + N537*0.949251

N421 = −0.4394 + N498*0.996601 + N500*0.996582

N82 = 0.00448029 − LE*0.0533156 + LE^2*0.0622887 + N145*1.00288

N145 = 0.0558503 + N522*N266*0.903129 − N522^2*0.773407 + N266*0.947021 − N266^2*0.288659 N266 = −1.00658 + N416*1.41177 − N416*N520*1.54519 + N416^2*0.318871 + N520*3.10425 − N520^2*1.57343

N416 = −1.15242 + N486*2.91695 − N486*N498*1.71136 − N486^2*1.35811 + N498*2.41022 − N498^2*0.796024

N486 = 0.324185 + IE*0.126925 + FE*0.110457

N522 = 0.333516 + FE*0.140349 − FE*MC*0.0548683 + MC*0.105483

N352 = −0.972664 + N491*2.7401 − N491*N457*1.75145 − N491^2*1.1413 + N457*1.74489

N457 = −0.325861 + N500*1.60679 − N500*N515*1.53545 + N515^2*1.81299

N500 = 0.324102 + IE*0.140114 + MC*0.0949479

N342 = −0.549263 + N415*1.84039 − N415*N489*1.42734 − N415^2*0.351965 + N489*0.976161 + N489^2*0.486085

N489 = −0.459776 + N536*1.01703 + N537*1.02221

N537 = 0.347702 + LE*0.156324 − LE^2*0.0961625 + CE*0.0972907

N415 = −0.848669 + N482*1.23947 − N482^2*0.331539 + N516*2.76859 − N516^2*2.107

N516 = 0.354148 + CE^2*0.0835151 + FE*0.120053

N535 = 0.365766 + CE^2*0.0869876 + MC*0.0919093

N157 = −0.0613612 + N465*0.331828 + N465*N270*0.721188 − N465^2*0.5203 + N270*0.865073 − N270^2*0.324565

N270 = −0.745174 + N449*2.29828 − N449*N487*1.88317 − N449^2*0.530701 + N487*1.06987 + N487^2*0.868139

N487 = −0.644125 + N527*1.60598 − N527*N536*1.66751 + N536*1.58849

N536 = 0.335849 + OC*0.132186 − OC*MC*0.080884 + MC*0.123088

N527 = 0.342024 + LE*0.118368 − LE*FE*0.0561952 − LE^2*0.044503 + FE*0.143457

N449 = −0.711395 + N491*2.31589 − N491^2*1.5131 + N528*0.969693

N528 = 0.366651 + IE*0.134984 + LC*0.0170788
N491 = 0.325926 + CE*0.0967273 + ME*0.105905 + ME^2*0.0370063
N465 = −0.481991 + N514*1.04075 + N524*1.0487
N524 = 0.337054 + CE*0.101628 + OC*0.108863
N514 = 0.351541 + LE*0.0410637 + LE*IE*0.0562627 + IE*0.115624
N507 = −0.320373 + N539*1.12971 + N545^2*1.33895
N539 = 0.361884 + LE*0.115066 − LE^2*0.0668289 + OC*0.0948431
N286 = −0.629476 + N408*2.20287 − N408*N510*1.70777 − N408^2*0.55311 + N510*0.790064 + N510^2*0.977505
N510 = −1.65816 + N534*3.88046 − N534*N543*6.70986 + N543*3.83786
N408 = −0.748837 + N498*2.18485 − N498^2*1.26511 + N504*1.07384
N504 = 0.331422 + IE*0.133903 + CE*0.0857028
N498 = 0.333875 + ME*0.130903 + OC*0.0855882
N190 = 0.261635 − N518*0.86643 + N518*N275*1.39802 + N275*0.850835 − N275^2*0.439447
N275 = −0.963417 + N410*1.36521 − N410*N520*1.4853 + N410^2*0.335547 + N520*2.97481 − N520^2*1.47374
N520 = −0.0239965 + N543*N545*2.38345
N545 = 0.391001 + CE*0.0885287 − LC*0.00735221 + LC^2*0.0297561
N543 = 0.372105 + LE*0.0923278 − LE^2*0.0502149 + MC*0.082214
N410 = −0.802709 + N482*2.06143 − N482*N515*1.91384 − N482^2*0.305739 + N515*1.73795
N515 = 0.323404 + FE*0.184557 − FE*OC*0.0557973 − FE^2*0.0437479 + OC*0.114325
N482 = 0.308152 + IE*0.10002 + IE^2*0.0323773 + ME*0.164063 − ME^2*0.0292263
N518 = 0.0427538 + N534*N546*2.0409
N546 = 0.397435 + MC*0.0885541
N534 = 0.366726 + FE*0.122349 + LC*0.0311747

The study identifies the MIP by using ANP and then with the help of the priority value and the selected priority parameters the performance of SWTP is estimated. At the time of estimation of PV, the impact of change in water and energy use along with the time taken by an operation of the SWTP to produce treated water is considered. The influences due to the change in the regular pattern of the climatic variables are also encoded in the priority value. Then the PNN model has developed a model for estimation of plant performance.

5 Conclusion

The present investigation tries to propose a novel method to estimate the performance of SWTP in terms of the change in consumed water and energy and change in the duty cycle of the treatment operations followed in the plant due to the change in the regular climatic pattern. The study at first tried to identify the MIP by using the ANP multi-criteria decision-making method and then with the help of the priority value and the selected priority parameters the performance of SWTP is estimated. At the time of estimation of PV, the impact of change in water and energy use along with the time taken by an operation of the SWTP to produce treated water is considered. The influences due to the change in the regular pattern of the climatic variables are also encoded in the priority value. The PNN model was used with the GMDH training procedure to develop the automated framework for the estimation of plant performance. The case study of a peri-urban city of North East India was used to show the applicability of the new method. As per the result, the EU criteria and IE alternative were found to have the most significance, and if the five-year performance of the SWTP is compared to a regular trend of degradation was observed. It can also be attributed to the impact of climate change and urbanization observed in the peri-urban city of Agartala in the last five years. The result shows that the CV is the most important criterion and among the alternatives, IE was found to be the MIP followed by ME and CE as per the priority value assigned to the parameters in the ANP method. Then a model is generated by the PNN method along with GMDH by which the performance of the plants can be determined.

This study has illustrated some insights as to the advantages of the ANP method also. It has shown that this method can be applied in a manner that allows it to adapt to situations where interdependency of factors influencing output is prevalent. Indeed, the ANP can be particularly effective at considering resource allocations in public service provision, such as water treatment plants where competing options for allocation may share similar significance with respect to particular criteria. For instance, if the cost of implementing training for staff was similar to the cost of implanting specialized pumps that could increase technical efficiency then the relative significance of cost as a criterion for judgment can be reduced using the ANN method. That is, supplementary criteria that can better differentiate between two competing factors can be applied to help decision-makers choose between alternative courses of action that may display similar characteristics. In this present work, the application of the ANP analysis indicates that it is suitable to deal with complex decision-making problems because they allow several criteria to be considered and compared systemically. The final ranking of the alternatives, derived from its application, shows that the gap among the scores of the options increases as the level of complexity of the technique increases. This key result suggests that greater complexity implies greater overlap in terms of evaluation criteria that further supports the suitability of the ANP model. The limitation of this work is the weights obtained in ANP depend on decision-makers. Based on decision-makers thinking, the criteria and the alternatives are ranked. So, the priority value will be changed.

Note: This study tries to investigate the impact of climate change on SWTP. That is why; MCDMs are utilized to obtain the importance of the parameter as decided with respect to the impact of climate change and energy consumed by the plant. After that, a neural Network model (PNN) is used to develop an index.

References

Ahmetović, E., & Grossmann, I. E. (2011). Global superstructure optimization for the design of integrated process water networks. *AIChE Journal, 57*(2), 434–457.

Akbari, N., Irawan, C. A., Jones, D. F., & Menachof, D. (2017). A multi-criteria port suitability assessment for developments in the offshore wind industry. *Renewable Energy, 102,* 118–133.

Atashkari, K., Nariman-Zadeh, N., Gölcü, M., Khalkhali, A., & Jamali, A. J. E. C. (2007). Modelling and multi-objective optimization of a variable valve-timing spark-ignition engine using polynomial neural networks and evolutionary algorithms. *Energy Conversion and Management, 48*(3), 1029–1041.

Bagatin, R., Klemeš, J. J., Reverberi, A. P., & Huisingh, D. (2014). Conservation and improvements in water resource management: A global challenge. *Journal of Cleaner Production, 77,* 1–9.

Cambron, K. E., & Evans, G. W. (1991). Layout design using the analytic hierarchy process. *Computers & IE, 20,* 221–229.

Castro, P. M., & Teles, J. P. (2013). Comparison of global optimization algorithms for the design of water-using networks. *Computers & Chemical Engineering, 52,* 249–261.

Chen, K., Chen, H., Zhou, C., Huang, Y., Qi, X., Shen, R., & Zhang, Y. (2020). Comparative analysis of surface water quality prediction performance and identification of key water parameters using different machine learning models based on big data. *Water Research, 171,* 115454.

Debnath, A., Majumder, M., & Pal, M. (2015). A cognitive approach in selection of source for water treatment plant based on climatic impact. *Water Resources Management, 29*(6), 1907–1919.

De, P., & Majumder, M. (2019). *Allocation of energy in surface water treatment plants for maximum energy conservation* (pp. 1–24). Development and Sustainability: Environment.

Deng, W., Lai, Z., Hu, M., Han, X., & Su, Y. (2020). Effects of frequency and duty cycle of pulsating direct current on the electro-dewatering performance of sewage sludge. *Chemosphere, 243,* 125372.

Dong, H. G., Lin, C. Y., & Chang, C. T. (2008). Simultaneous optimization approach for integrated water-allocation and heat-exchange networks. *Chemical Engineering Science, 63*(14), 3664–3678.

Dorussen, H. L., & Wassenberg, W. B. A. (1997). Feasibility of treatment of low polluted wastewater in municipal waste water treatment plants. *Water Science and Technology, 35,* 73–78.

Dutta, N., Ghosh, A., Debnath, B., & Ghosh, S. K. (2020). Climate change in hilly regions of India: Issues and challenges in waste management. In *Sustainable Waste Management: Policies and Case Studies* (pp. 657–669). Springer, Singapore.

Fant, C., et al. (2017). Climate change impacts on US water quality using two models: HAWQS and US basins. *Water, 9*(2), 118.

Galán, B., & Grossmann, I. E. (2011). Optimal design of real world industrial wastewater treatment networks. In *Computer Aided Chemical Engineering* (Vol. 29, pp. 1251–1255). Elsevier.

Ghodsi, H., & Khanjani, M. J. (2020). Application of improved GMDH models to predict local scour depth at complex bridge piers. *Civil Engineering Journal, 6*(1), 69–84.

Hernández, C. T., Marins, F. A. S., Rocha, P., & Duran, J. A. R. (2010). Using AHP and ANP to evaluate the relation between reverse logistics and corporate performance in Brazilian industry. *Brazilian Journal of Operations & Production Management, 7*(2), 47–62.

Ibrić, N., Ahmetović, E., & Kravanja, Z. (2014). Two-step mathematical programming synthesis of pinched and threshold heat-integrated water networks. *Journal of Cleaner Production, 77,* 116–139.

Ivakhnenko, A. G., & Ivakhnenko, G. A. (1995). The review of problems solvable by algorithms of the group method of data handling (GMDH). *Pattern Recognition and Image Analysis C/C of Raspoznavaniye Obrazov I Analiz Izobrazhenii, 5,* 527–535.

Jiang, B., & Guo, H. (2013). Permutation invariant polynomial neural network approach to fitting potential energy surfaces. *The Journal of Chemical Physics, 139*(5), 054112.

Khan, Z. I., et al. (2017). Health risk assessment of heavy metals in wheat using different water qualities: Implication for human health. *Environmental Science and Pollution Research, 24*(1), 947–955.

Sundara, K. K., et al. (2010). Performance evaluation of waste water treatment plant. *International Journal of Engineering Science and Technology, 2*(12), 7785–7796.

Li, L., Rong, S., Wang, R., & Yu, S. (2020). Recent advances in artificial intelligence and machine learning for nonlinear relationship analysis and process control in drinking water treatment: A review. *Chemical Engineering Journal, 405,* 126673.

Majumder, P., Majumder, M., & Saha, A. K. (2020). Real-time monitoring of power production in modular hydropower plant: Most significant parameter approach. *Environment, Development and Sustainability, 22*(5), 4025–4042.

Mohammadi, D., Mikaeil, R., & Abdollahi-Sharif, J. (2020). Implementation of an optimized binary classification by GMDH-type neural network algorithm for predicting the blast produced ground vibration. *Expert Systems, 37*(5), e12563.

Oh, S. K., & Pedrycz, W. (2002). The design of self-organizing polynomial neural networks. *Information Sciences, 141*(3–4), 237–258.

Oh, S. K., Pedrycz, W., & Park, B. J. (2003). Polynomial neural networks architecture: Analysis and design. *Computers & Electrical Engineering, 29*(6), 703–725.

Özdemir, A., Özkan, A., Günkaya, Z., & Banar, M. (2020). Decision-making for the selection of different leachate treatment/management methods: The ANP and PROMETHEE approaches. *Environmental Science and Pollution Research,* 1–12.

Pusat, S., & Akkaya, A. V. (2020). Explicit equation derivation for predicting coal moisture content in convective drying process by GMDH-type neural network. *International Journal of Coal Preparation and Utilization,* 1–13.

Raseman, W. J., et al. (2017). Emerging Investigators series: A critical review of decision support systems for water treatment: Making the case for incorporating climate change and climate extremes. *Environmental Science: Water Research & Technology, 3*(1), 18–36.

Rodríguez, N. H., Ramírez, S. M., Varela, M. B., Guillem, M., Puig, J., Larrotcha, E., & Flores, J. (2010). Re-use of drinking water treatment plant (DWTP) sludge: Characterization and technological behaviour of cement mortars with atomized sludge additions. *Cement and Concrete Research, 40*(5), 778–786.

Rojas-Torres, M. G., Ponce-Ortega, J. M., Serna-González, M., Nápoles-Rivera, F., & El-Halwagi, M. M. (2013). Synthesis of water networks involving temperature-based property operators and thermal effects. *Industrial & Engineering Chemistry Research, 52*(1), 442–461.

Saaty, T. L. (1980). The analytic hierarchy process. *McGraw Hill International* (1980).

Saaty, T. L. (2005). *Theory and applications of the analytic network process, Pittsburgh.* PA: RWS Publications.

Saaty, T. L., & Vargas, L. G. (2006). *Decision making with the analytic network process: Economic, political, social and technological applications with benefits, opportunities, costs and risks.* New York: Springer.

Saaty, T. L., & Vargas, L. G. (2013). The analytic network process. In *Decision Making with the Analytic Network Process* (pp. 1–40). Springer, Boston, MA.

Skiborowski, M., Mhamdi, A., Kraemer, K., & Marquardt, W. (2012). Model-based structural optimization of seawater desalination plants. *Desalination, 292,* 30–44.

Tsiakis, P., & Papageorgiou, L. G. (2005). Optimal design of an electrodialysis brackish water desalination plant. *Desalination, 173*(2), 173–186.

Vasquez, V. M. (2017). Critical literacy. In *Oxford Research Encyclopedia of Education.*

Xu, L., Wang, X., Bai, L., Xiao, J., Liu, Q., Chen, E., & Luo, B. (2020). Probabilistic SVM classifier ensemble selection based on GMDH-type neural network. *Pattern Recognition, 106,* 107373.

Yan, Y., Xia, B. Y., Zhao, B., & Wang, X. (2016). A review on noble-metal-free bifunctional heterogeneous catalysts for overall electrochemical water splitting. *Journal of Materials Chemistry A, 4*(45), 17587–17603.

Yang, L., & Grossmann, I. E. (2013). Water targeting models for simultaneous flowsheet optimization. *Industrial & Engineering Chemistry Research, 52*(9), 3209–3224.

Zolfaghary, P., Zakerinia, M., & Kazemi, H. A. Model for the use of urban treated wastewater in agriculture using multiple criteria decision making (MCDM) and geographic information system (GIS). *Agricultural Water Management,* 243, 106490.

Kumar, N., & Sinha, D. K. (2010). An approach to river water quality management through correlation study among various water quality parameters. Int J Environ Sci, 1(2), 253-259

Pham, B. T., Singh°, S. K., & Ly, H. B. (2020). Using Artificial Neural Network (ANN) for prediction of soil. Vietnam Journal of Earth Sciences, 42(4), 311-319.

Power Allocation in an Educational Institute in India: A Fuzzy-GMDH Approach

Apu K. Saha, Deepjyoti Deb, and Prachi D. Khobragade

Abstract The ever-increasing population and their energy demand have raised the stress on conventional energy sources manifold. As the quantity of power supply is receding, the supplied energy consumption is required to be optimized for maximum utilization. The distribution of available energy must be so that the need of the consumer can be satisfied. But if the resources cannot be accessed in an adequate quantity to satisfy consumers' demand, customers have to be preferred based on their socio-economic impact. Various sections of consumers have to be provided with convenient weightage based on their importance. Cognitive optimization of the allotment may be carried out so that even in the days of exiguity, the socio-economic influence for not fulfilling the consumers' needs can be curtailed. In this aspect, the present investigation used the advantage of Fuzzy Logic to accredit weightage to in-house customers of an educational campus and then tried for optimal allocation by applying the competence of non-linearity mapping of Neuro-Genetic algorithms. The output from the algorithm was compared with similar work from Particle Swarm Optimization and Differential Evolution Algorithms. The study results provided an option for optimal exertion of available energy based on consumers' socio-economic importance.

A. K. Saha
Department of Mathematics, National Institute of Technology Agartala, Agartala, Tripura, India
e-mail: apu_saha@gmail.com

D. Deb · P. D. Khobragade (✉)
Hydroinformatics Engineering, Department of Civil, National Institute of Technology Agartala, Agartala, Tripura, India
e-mail: prachikhobragade111@gmail.com

D. Deb
e-mail: ddeb210@gmail.com

M. Majumder and G. D. Kale (eds.), *Water and Energy Management in India*,
https://doi.org/10.1007/978-3-030-66683-5_11

1 Introduction

The optimal allocation of energy to different consumers according to their needs and the socio-economic impact can solve the recent uncertainties in the energy region due to rising demand but scarcity in supply (Li and Guo 2014).

Consumers' social and economic impression can be evaluated as per the income generated; investments consumed, including social and environmental benefits and hazards imbibed by the recipients. Proper weightage can be assigned, which coherently represent the socio-economic effect of the consumers. Then according to the need of the recipient, available energy can be distributed so that grievance from the consumers can be minimized.

1.1 Global Energy Scenario

Energy utilization is expected to mount by 56% globally from 2010 to 2040 as per International Energy Outlook 2013. Further, the report states that there will be a significant increment in the overall energy consumption, which is estimated to be around 820 quadrillion Thermal Units in 2040 compared to 630 quadrillion Thermal Units in 2020. Most of the energy sources emit Carbon dioxide in massive amounts, which have an adverse impact on climate change (IEA 2013). The reduction in energy-related carbon dioxide ejaculation can be achieved by applying energy transformation based on energy effectualness and renewable energy. International Renewable Energy Agency report 2018 suggested that renewable energy must hold around two-thirds of the overall primitive energy supply by 2050, which was only 15% in 2015. Member countries in the Organization for Economic Cooperation and Development have shown a rise in energy consumption by 17%. In non-OECD countries, there is a substantial growth in energy use, which increases up to 90% (IRENA 2019).

As per the U.S. Energy Information Administration report, two of the fastest expanding energy sources in the world are Renewable Energy and Nuclear power, each of which is accounting for an increment of 2.5% every year. Despite that, fossil fuels still monopolize the global energy market by yielding around 80% of the energy supply. Global natural gas being the most rapidly augmenting fossil fuel proliferates its consumption by almost 1.7% per year. The utility of natural gas is projected to increase owing to an extensive supply of coal mine methane, shale gas, and tight gas (Sorensen et al. 2000). Coal industries are still flourishing faster than petroleum and other liquid fuel industries as china is rapidly escalating coal utilization. As per ExxonMobil's energy outlook report of 2017, the energy demand will increase by around 25% by 2040 because of the extensive expansion in the industrial sector and economy (ExxonMobil 2017).

1.2 Present Consumption Pattern

The industrial sector still holds the most significant stake in delivered energy utilization; the world industrial sector still consumes over half of the global returned energy in 2040. Ejaculation of Carbon Dioxide from energy-related activities is estimated to hike to 45 billion metric tons by 2040 from 36 billion metric tons in 2020, which is almost an increment of 46%. In 1999–2000, 49% of the available energy resource was consumed by the Industries, followed by the Transportation (22%) and Residential (10%) sector. The Irrigation sector at that time consumed only 5% of the total global energy resources. In 2012 also the Industrial sector was found to be the largest consumer of energy (45%), followed by the Domestic (22%) and Agriculture (17%) sector. The Transportation sector consumed only 11% of the overall energy available (Ramachandra et al. 2015).

1.3 Problem Areas

Global Energy is facing three major problems. The deficit of supply to the consumers is a result of the ever-increasing urbanization and technological development. As the resource becomes scarce, the allocation of the same becomes another problem. The conventional energy sources contribute considerably to the yielding of greenhouse gases, which is attributed to be among the prime factors of change in the climatic pattern.

Among the three problems discussed, the second problem is among the pivotal issues plaguing the energy industry. The non-optimal allocation of energy becomes the cause of many uncertain situations, which has resulted in hostility among consumers. That is why it is utmost necessary to find a methodology for allocating the available energy in an optimal manner where each consumer can satisfy their basic need, and any sort of favor can be avoided. Weighted distribution of power among various customer sections based upon their condition, impact on society, and few other factors have become the only solution to reduce the energy sector's growing hostility.

1.4 Indian Energy Scenario

Due to an increase in population, climate change, economic growth, India faces a challenge to meet its energy requirement and to fulfill the need for the desired quality of energy to different users. India has a vast amount of coal reserves, and most of the electricity is generated with its help. The thermal power plant generates 41% of electricity in India. Oil and natural gas meet 44% of India's energy requirement.

Nuclear and hydel energy would form 2.5 and 3.5%. According to the 2011 census, 55.3% of rural areas have electricity (Duffy et al. 2018).

The main aim is to study the role of various energy technology options under different scenarios, mainly baseline scenario, high nuclear scenario, high renewable scenario, low growth, and high growth rate scenario. This has been achieved with the help of the model for energy Supply Strategy Alternatives and their General Environmental Impact (MESSAGE) model, which elaborates the different energy supply methods with proper management of fuel distribution and availability.

1.5 Proposed Solution

The available energy can be scarce, but if the same can be optimally allocated, the consumers' hostility can be reduced. In this regard, prioritizing different types of consumers has to be performed to separate the consumers according to their needs and impact on society. A weightage can be assigned to each type of consumer, proportional to the importance of the same. Cognitive methods like Fuzzy Logic can be utilized to determine the ideal weights assigned to diverse users, which can depict the need and impact of the consumer on the socio-economic developments. After proper weightage is set, the available energy can be distributed into different sections of customers depending upon the weightage and nature-based optimization algorithms like Particle Swarm Optimization (PSO), Differential Evolution (DE), or Neuro-Genetic (NG) Optimization.

1.6 Prioritization of Energy Consumers

The prioritization of the consumer can be performed with the help of ranking method where each consumer can be ranked according to its importance due to various factors like Annual Demand, Socio-Economic, and Impact that can be caused by the consumers if supply doesn't meet the demand and Environmental Impact due to the consumer can be considered as the factors which can segregate the consumers according to their importance. Good rank representing their differences can be assigned for each of the elements. After the ranking has been made, Fuzzy Logic can be enforced to determine the consumers' weightage.

1.7 Optimization of Energy Allocation

Once the consumers are rated and separated by the weightage assigned to them based on their importance as determined from the ranking method, the available energy can be distributed with nature-based algorithms. The nature-based algorithms that are competent in mapping the non-linearity and cognitive approach towards the problem's solution are widely used to solve similar issues. That is why the present investigation also utilized the advantages of three different nature-based algorithms to determine the optimal allocation of the available resource among the consumers so that their minimum demand can be fulfilled, and proper justification can be given for not supplying according to their needs. In that case, the grievances of the consumers can be reasonably reduced.

1.8 Global Water Scenario

Six countries, namely Brazil, Canada, China, Columbia, Indonesia, and Russia, contain half of the world's total renewable freshwater resources, which indicates the global water crisis for the rest of the world. The ever-increasing population and the rising demand for water for livelihood, industrial, and agricultural purposes have put massive stress on the water resource system (Mogelgaard 2011). World Water Development Report 2019, given by the United Nations, has stated that there will be an increase of around 30% in the level of present water use rate till 2050, primarily because of industrial and domestic sectors. Water pollution is increasing with each passing day, affecting not only the surface water sources but also the surrounding environment.

World Water Assessment Program led by UNESCO has mentioned that in developing countries, more than 80% of sewage effluent is being discharged in the water bodies without any proper treatment, which degrades and pollutes the rivers, lakes, and coastal areas. Water availability is not the appropriate solution to the global water crisis if its quality remains poor enough to intake. Liu et al. (2020) had undergone a comprehensive multi-territorial input–output study to systematically assess the overall water usage, incorporating energy yield, requirement, and international business in a cooperative structure. The investigation's outcome reveals that China had withdrawn 117 billion cubic meter^{s} of direct water in 2011, which was utilized by several crucial energy sectors such as coal, petroleum, oil, gas, etc.

Taikan and Rose (2020) studies and analyzes physical and economic inadequacy of water at 50 km resolution with the help of high-image processing precipitation, accessible freshwater, and exodus data sets in which they found that around 41 million people are living in areas with concurrent severe economic and water-scarcity constraints. These areas are mostly in semi-arid parts of Sub-Saharan Africa, the Middle East, and Central Asia. Qadri et al. (2019) studied the feasibility of the freshwater ecosystem by applying eco-friendly reclamation and management techniques. They outlined the importance of adopting these methods for the imperishability of life on the planet earth.

1.9 Indian Water Scenario

India is the third-largest groundwater shipper globally, but within very few years, 21 cities' groundwater will get depleted entirely. As per the state of food security and nutrition in the world 2020, the per capita availability of fresh water varies spatially and geographically in various regions of the country. Thus, based on the requirement, water is transferred or shared from one state to another. For example, Haryana is sharing water of Yamuna with Rajasthan from Tajewala headworks.

As per the Business-As-Usual (BAU) scenario, India's total water demand will increase by 22% and 32% by 2025 and 2050, respectively. By 2050, Industrial and household sectors together will account for 85% of the additional demand. Many river basins will be facing water scarcity soon, and the groundwater resources in many subsequent areas will be severely overexploited. Blakeslee et al. (2020) have adopted a quasi-random technique to study and assess the consequences of rising water inadequacy in India, taking topographically based differences concerning groundwater into consideration. The authors concluded a precipitous and persistent decline in farm income and wealth due to dry wells, with little affirmation of agricultural adaptation.

Kumari and Kumar (2020), in their study, focuses on the pathetic condition of almost every river and wetlands of India and draws a comparison between the water readiness in the country in the past and present situation to reflect water crisis and figure out the possible reasons for this crisis. Moreover, water resource management's purpose due to the shortage of water is explained in their study. Kumar et al. (2020) examines the usage of the present urban water system in the urban areas of India and discuss a better way for flexible investigation in which they proposed that data analysis on the available amount of chemical and biological contamination in water is the first pressing requirement for this approach. The authors noticed menace to the availability of edible drinking water, which has been forecasted based on their scientific observations on the degradation of water quality and emphasized the advancement of sustainable water management.

1.10 The Objective of the Present Investigation

In the present investigation, a new method for allocating available energy among the different consumer sections was proposed. In this method, the importance of the consumers was at first determined with the help of Fuzzy Logic. After the customers' segregation as per their preference, the available energy was distributed among the various users with different cognitive algorithms.

The study wants to propose an intelligent allocation methodology that considers the consumers' real needs so that the minimum requirement of each type of consumer can be satisfied. Still, grievances can be prevented for not supplying the full demand. The study also verifies the potential of different nature-based algorithms in optimizing energy resources among different types of consumers.

2 Methods Used

2.1 Particle Swarm Optimization

PSO is a community-based assumptive technique for determining continuous and discrete optimization problems (Wang et al. 2017). The basic ideas on particle swarms of Kennedy and Eberhart were derived from the computer imitation of the navigation of a group of birds or fishes, which mostly depicts a standard action when they look for a specific object such as food than completely individual subjective abilities. The first facsimile was persuaded by the work of Heppner and Grenander (1990), which implicated a correlative approach of a group of birds looking out for corn. These soon developed (Kennedy and Eberhart 1995; Eberhart and Kennedy 1995; Eberhardt et al. 1996) into a robust optimization method.

Tharwat et al. (2019) had applied a topsy-turvy Particle Swarm Optimization technique to achieve the optimal control points of the Bézier curve, with which the flawless serene path that minimizes the distance between starting and ending points is found out. From the study, they have found that the algorithm can find the optimal way. Sengupta et al. (2019) made a study on past and latest advancements with assimilation context making use of PSO. Their study emphasized deployment, development, and improvements of the basic and its most recent state of art implementation. Deng et al. (2019) had introduced the upgraded PSO to investigate the flaws of the motor bearing, which the current fault investigation techniques could not recognize efficiently. They observed that the applied enhanced optimization algorithm could efficaciously upgrade the classification preciseness of least-square support-vector-machine and concluded that their designed fault diagnosis method executes better than other techniques.

- *Strength*:

    1. In academic, industrial, and commercial fields, the PSO has many applications and is widely used, such as operations research, fuel, energy, medicine, biology, chemistry, and control system.
    2. PSO is currently used by many researchers to solve the problems related to electrical, biological, mechanical engineering.
    3. PSO helps in reducing the complexity of the problems and finally gives us an approaching solution.
    4. In medical engineering, the PSO model is constructed to determine the diseases which supposed to be very helpful for the treatment field.

- *Weakness*:

    1. PSO has high computational complexity, slow convergence, sensitivity to some parameters.
    2. PSO does not handle a good relationship between local search and global search.

3. PSO cannot employ the crossover operator as it is used in GA, and hence the distribution of the facts between contenders is not appropriate.

2.2 Differential Evolution

Differential evolution is a progression-based theoretical hierarchical optimization approach introduced to obtain the best solution over continuous search spaces (Rogalsky et al. 2000). DE exercises a very smooth and productive process based on a broad selection of continuous random problems that do not require differentiation. The reproduction operant of DE comprises an algebraic anomaly to construct a primary vector, which develops an offspring when utilized by the crossover operator. The weighted variance between randomly chosen individual vectors represents the mutation step sizes. Thus, the DE algorithm applies only to continuous optimization problems because of the vector arithmetic.

Many practical, real-life problems such as organizing issues, conducting, or dispatching problems involve assigning or aligning discrete components. DE approach is not convenient to solve such problems. However, at present, few researchers are working on developing discrete renditions of DE. For example, one technique established for working out arranging problems by evolution algorithm depends on characterizing an array of z components with a group of z real numbers.

Das and Suganthan (2011) compiled and coordinated the data on the recent developments on Differential evolution. They progressed with the different adaptation of DE in various optimization schemes, the amalgamation of DE with alternative developers, and the future research issues on DE. Wang et al. (2019) came up with a unique automated niching Differential Evolution method based on Affinity Propagation Clustering to elucidate multi-quintessential optimization problems. The authors forecasted the conceivable peak's rough bearings that will aid the speed to converge at a faster rate. Opara and Arabas (2018) surveyed the theoretical results obtained so far for DE. A detailed perspective on the current-day understanding of DE's underlying mechanisms was complemented with a list of research directions.

- *Strength*:

  1. DE is much simpler and straight forward to implement in a field.
  2. The code for DE is simple, as it helps practitioners from other fields communicate with the algorithm and get a solution for their problem.
  3. The performance of DE is reasonable as compared to PSO.

- *Weakness*:

  1. The output solution in DE relies on the unreliable contingent probability to achieve an improved result. Thus, mostly, the output of DE is ambiguous.
  2. The standard output value in Differential Evolution is obtained by reiterating 100 times each of the aggregate simulation.

2.3 Neuro-genetic Optimization

Artificial Neural Networks (ANN) are computational methodologies that perform multifactorial analyses. Layers of smooth calculating nodes sustaining in the artificial neural network models (influenced by the human nervous system) serve as disordered computing mechanism. Graded connection lines opulently correlate these nodes, and throughout the training phase, whenever inputs are passed to the network, the weights get accustomed. Neural Networks can execute tasks like yield value forecasting, object analysis, function proximity, pattern identification in conglomerate data, and recognized pattern finalization and other works when the training is lucrative.

ANNs can solve an extensive series of optimization problems such as lateral computing, matrix calculus, and handling of signals. This soft computing technique is also used in several military operations comprising of detecting an automated object, upgrading engine disturbances, identifying flaws, and regulating flying airship in the complicated engineering structure. Tumors can be seen from the medical images. The classification of dead and general cells can be performed in the genetic evaluation, both of which are retrieved through the system incorporating medical image analysis. Time-series predictions have been conducted with neural networks, including the prediction of irregular and chaotic sequences. While formulating and computing neural network problems, the primitive step is to pick the desired topology for the structure followed by the training stage and testing state, respectively. Sakshin and Kumar (2018) proposed a novel technique for optimizing ANNs weights, which combines pruning and Genetic Algorithm (GA). They have observed that ANNs trained with GA optimized values manifest higher convergence with lower execution time and a higher success rate. Abbasi et al. (2019) studies the scheduling of the flow updates with a hybrid genetic algorithm flow scheduler to curtail the culmination time of rejuvenating the network by searching the solution space in software-defined networking. Moreover, the authors observed that the Genetic Algorithm flow scheduler improves network performance when combined with other existing flow scheduling methods.

- *Strength*:

    1. GA always provides a solution to the problem, which eventually gets improved over time.
    2. It enhances multi-objective arithmetic and various interminable and distinct functions.
    3. It preferably provides a blend of reasonable solutions rather than a single solution.
    4. Compared to other conventional approaches, GA is quicker and more competent than other traditional methods.

- *Weakness*:

    1. GA does not work for problems having derivative orientation and much simpler in the pattern.

2. It can be computationally uneconomical in case of few problems as various iterative calculations are required to determine the most robust value.
3. There is no assurance on the accuracy and worthiness of the output value as the GA method is assumptive.
4. If not appropriately implemented, the GA may not converge to the optimal solution.

2.4 Fuzzy Logic

Lotfi Zadeh is the father of Fuzzy Logic. It is a superior version of Boolean Logic, which is the accustomed way of portrayal. In Boolean Logic, representation is in the form of either true or false. But Fuzzy Logic includes a range of possibilities appearing as continuous values represented in degree or intensity. Fuzzy Logic is a way of dealing with uncertainties. Let us take the example of tap water temperature, which can be described as either hot or cold in Boolean Logic. But in fuzzy Logic, there can have a gradient from hot to cold so that we could have something like 'lukewarm,' 'very hot,' 'slightly cold,' etc. Fuzzy Logic is widely used in the aerospace field for height control of spacecraft, speed and traffic control in automotive systems, Natural Language Processing, and various intensive applications in Artificial Intelligence.

Nagi and Tripathy (2020) study on detecting infection in the crop plants, classifying the disease post-infection, and appraising the contamination's grimness with the fuzzy logic utilization. The authors proposed that the approach will help the farmers recognize the disease class early. Al-Dmour et al. (2019) had employed a fuzzy logic approach to model and accomplish a correlative warning structure based on a warning score to classify patients' condition and austerity of the ailment. Thus, the fuzzy logic-based system suggested by the author will help hospital personnel in providing immediate medical treatment to the patients whose health condition will be very severe and will prevent further worsening of their health. Andrew and Kumanan (2020), in their work, focused on fuzzy logic application in decision making for maintenance planning of the manufacturing equipment using main inputs such as mean time between failures, mean time to repair, availability of spares, and the age of the equipment which will ultimately help the maintenance and plant engineers to plan their activities constructively.

- *Strength*:

  1. Fuzzy Logic describes systems in terms of a combination of arithmetic and phonological (symbolic), which has advantages over pure mathematical approaches or pure symbolic approaches as most often, system knowledge nowadays is available in such a combination.
  2. Fuzzy algorithms are sturdy as that they are not very susceptible to changing environments and erroneous rules.

3. Computing power is preserved as the reasoning process is often elementary when compared to computationally scrupulous systems. This is a very advantageous feature in real-time systems.

- *Weakness*:

1. Fuzzy systems don't have the aptitude for both machine learning and neural network type pattern identification.
2. Extensive experimentation with hardware is essential for Validating and Verifying a fuzzy knowledge-based method.
3. Accuracy of Fuzzy Logic cannot always be relentless, as a result of which it cannot be accepted widely because the results are anticipated based on assumptions.

2.5 Group Method of Data Handling

Group Method of Data Handling (GMDH) networks was originated in 1968 by Prof Alexey G. Ivakhnenko. GMDH is very useful in time series forecasting, mining data, deep learning, optimization, and recognition of patterns. This soft computing technique is a series of numerous algorithms used to work out diverse problems. It incorporates equivalent compounding, conglomeration, interpolation, and feasible algorithms. GMDH sorting algorithms are relatively simple for software development.

Hemmati-Sarapardeh et al. (2019) studied the various confining and amplifying aspects of natural gas under contrasting situations such as fluctuating temperatures and pressures by modeling the consolidating factor, determined using the GMDH approach. Moreover, the validation of the technique is processed by various analytical and graphical studies. The authors concluded that the GMDH model imparts satisfactory efficiency in terms of correlations and equations of states. Nariman-Zadeh et al. (2003) proposed a composite genetic construction of GMDH based neural system. Designing and forecasting meaningful cutting measures is carried out by implementing a unique value dissipation mechanism. The authors inferred that the applied tool gave excellent results in figuring out the bearings of equilateral subexpressions' coefficients that exhibits in such GMDH based networks. Srinivasan (2008) adopted the GMDH systems for computing and assessing the energy demand. This technique was immensely compelling in bringing forth the predictions that are more meticulous with minimal labor requirements compared to various other conventional regression and time series-based models.

- *Strength*:

1. GMDH algorithm requires fewer computations. It throws out detrimental information (variables that do not correlate positively with the checking set), making the assessments practical and helps make the systems of ordinary equations well adapted.

2. Models created by GMDH Shell using their unique algorithm are simple, dissolute, and entirely precise in forecasting, which allows evading construction of overcomplicated models that drain CPU resources and do not accord authentic prophetic advantages.
3. GMDH algorithm has successfully overcome other algorithms' incompetency to explicitly identify the network weights by using standard ordinary least squares fitting, which helps to find out the values in a single step.

- *Weakness*:

1. The significant complexity of the GMDH algorithm can be the overfitting issue and flawed inference. The final model's extensive criteria could create overfitting problems, and therefore techniques, which eradicate the number of parameters, should be adopted.
2. In applications of long-range predictions, the GMDH was often found inaccurate. Possible causes are the deficient functional variety of the model candidates, unreasonable use of a series of external criteria for choosing the best complexity, and individual models' overcomplication.
3. The majority of GMDH algorithms are valid for continuous variables and cannot be applied to binary or discrete problems.

3 Case Studies

3.1 About the Study Area

The National Institute of Technology, Agartala (NIT Agartala) is an Educational Institute of national importance situated 20 km away from the city Agartala in the Princely state of Tripura and the north-eastern part of India. The latitude, longitude, and elevation of the study area are 23.84 N, 91.42 E, and 12.80 m above sea level. Figure 1 shows the satellite image of the study area collected from Bhuvan, India Geo-Platform of ISRO.

3.2 Justification

The present investigation required applying the new methodology in an area where the power supply can be controlled. The study needed a place with different types of consumers having different consumption patterns, demand, and impact on the area's socio-economic characteristics.

NIT Agartala is an educational institute of national repute. It has all types of consumers having distinct consumption patterns, demand, and socio-economic impacts.

Fig. 1 Location of study area

For example, the institute has the following types of consumers:

1. Residential: The residence of the campus, like students, teachers, and staff, have a distinct consumption pattern and demand.
2. Hospital: The Hospital will always have a more significant amount of social impact but its demand and consumption much smaller than the other users.
3. Workshops: The institute develops its products and machines in this arena. Its economic impact concerning the institute is more than any other user, and thus the demand from this user is also much more than the other consumers.
4. Administrative: The Administrative is the section where all the institute's official works are carried out. Thus, its demand is substantial in-office time only.

4 Methodology in Detail

The Fuzzy logic decision-making process is adopted to allocate power resources optimally in an educational institute among various sections of consumers such as residential, hospital, workshop, and administrative. Power supply will be provided to these users based upon their importance, which depends on various parameters like demand, consumption patterns, and socio-economic aspects. At first, different kinds of consumers are ranked based on their significance in the four respective parameters. After that, a fuzzy scale of importance ratings was used in the four pairwise comparison matrices of the consumers concerning different criteria. As far as deriving the weightage of the consumers' importance is concerned, The Fuzzy ratings need to be changed into a numerical value. Accordingly, their Crisp Values are computed from a Triangular Membership Function, which Zadeh suggested, and these values replace the fuzzy ratings. Simultaneously, each row's average was calculated to find the weighted importance of each type of consumer-based upon each of the criteria. The first phase of the work is done. In the second phase, the objective is to optimize the cognitive allocation of power such that the difference between the standard and weighted distribution is minimum. The objective equation

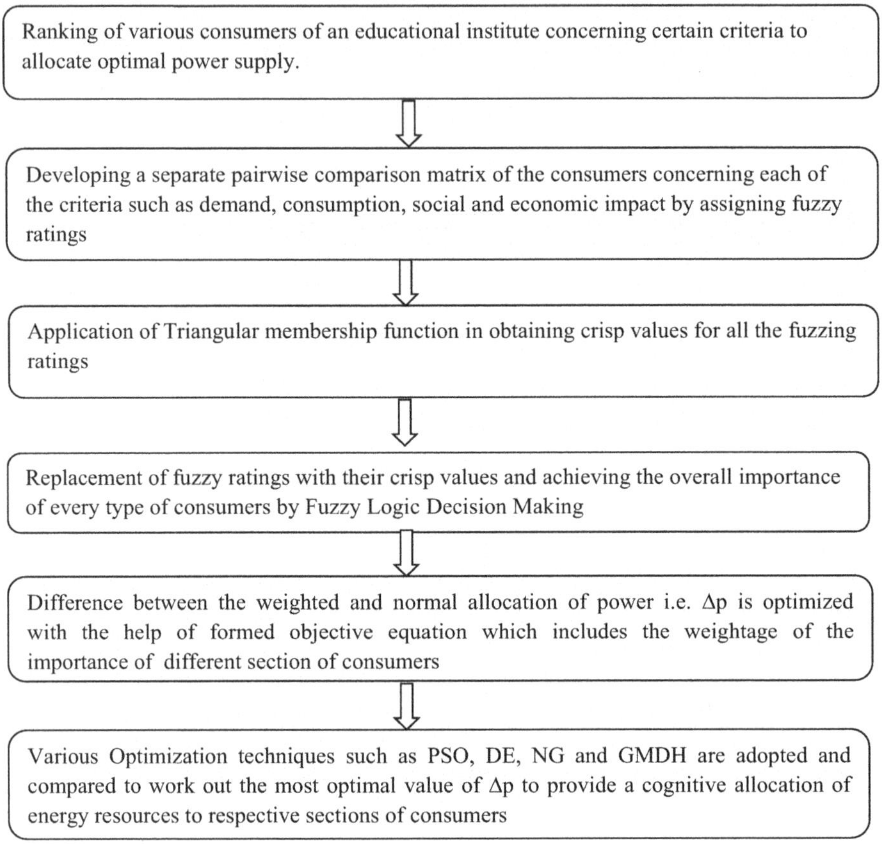

Fig. 2 Schematic Diagram of Detailed methodology of the study

is developed and is utilized in the four applied optimization techniques, namely, Particle Swarm Optimization, Differential Evolution, Neuro Genetic, and GMDH. The competence of these four algorithms is compared, and accordingly, the best algorithm for optimization is selected (as depicted in Fig. 2).

5 Results and Discussions

5.1 Prioritizations of Consumers by Fuzzy Logic

The concepts of Fuzzy Logic Decision Making were applied to prioritize the different consumers of the institute. The consumers' importance based on their demand, consumption pattern, and socio-economic impact, as found from the institute, is

Table 1 Importance of each type of consumers based upon four considered parameters

	Demand	Consumption	Social	Economic
Residential	3	3	2	3
Hospital	4	4	1	4
Workshop	1	1	3	1
Administrative	2	2	4	2

Table 2 Pairwise comparison matrix of the consumers concerning demand

	Residential	Hospital	Workshop	Administrative
Residential	–	SH	VL	VL
Hospital	SL	–	EL	EL
Workshop	VH	EH	–	SH
Administrative	VH	EH	SL	–

shown in Table 1, whereas Tables 2, 3, 4 and 5 depicts the pairwise comparison matrix of the different users based on its importance. Table 6 describes the Fuzzy

Table 3 Pairwise comparison matrix of the consumers concerning consumption pattern

	Residential	Hospital	Workshop	Administrative
Residential	–	SH	VL	VL
Hospital	SL	–	EL	EL
Workshop	VH	EH	–	SH
Administrative	VH	EH	SL	–

Table 4 Pairwise comparison matrix of the consumers concerning social impact

	Residential	Hospital	Workshop	Administrative
Residential	–	SL	VH	VH
Hospital	SH	–	EH	EH
Workshop	VL	EL	–	SL
Administrative	VL	EL	SH	–

Table 5 Pairwise comparison matrix of the consumers concerning economic impact

	Residential	Hospital	Workshop	Administrative
Residential	–	SH	VL	L
Hospital	SL	–	EL	EL
Workshop	VH	EH	–	SH
Administrative	H	EH	SL	–

Table 6 Fuzzy scale of importance rating and its crisp value from a triangular membership function as proposed by Zadeh

	Crisp value
Too high importance (EH)	0.97
Very high importance (VH)	0.85
High importance (H)	0.76
Semi-high importance (SH)	0.67
Neither high nor low importance (NHNL)	0.54
Semi low importance (SL)	0.45
Low importance (L)	0.34
Very low importance (VL)	0.24
Extremely low importance (EL)	0.14

Scale of Importance, which was applied in rating each consumer with the other in the Pairwise comparison Matrix depicted in Tables 2, 3, 4 and 5.

The crisp values replaced the fuzzy values, and the average of each row was calculated to find the importance of each type of consumer-based on each of the criteria. Table 7 depicts the average importance of the consumers for each measure and the overall importance, which was used as the weightage of significance at the time of optimization.

At the time of optimization (1) was used as an objective function.

$$\begin{aligned}\Delta p = {} & \{(p * m_1) + (T - (p * m_1)) * m_2 + (T - (T - (p * m_1)) * m_2) * m_3 + (T - (T - (T - (p * m_1)) * m_2) * m_3) * m_4\} \\ & - \{(p * n_1) + (T - (p * n_1)) * m_2 + (T - (T - (p * n_1)) * n_2) * m_3 + (T - (T - (T - (p * n_1)) * n_2) * n_3) * n_4\} \quad (1)\end{aligned}$$

where

p is the amount of power supplied in peak hour to the consumer having the highest weightage of importance.

$m_{1\text{-}4}$ is the weightage of importance as determined by the Fuzzy Logic where m_1 is the highest and m_4 is the lowest weightage of importance i.e., $m_1 > m_2 > m_3 > m_4$.

$n_{1\text{-}4}$ is the equal weightage which is assigned when the consumers are not separated based on their importance, i.e., $n_1 = n_2 = n_3 = n_4$.

T is the total amount of power supplied to the institute.

Table 7 Average importance of each type of consumers based upon each of the criteria and the consumers' overall importance

	Demand	Consumption	Social impact	Economic impact	Overall importance
Residential	0.383	0.383	0.717	0.417	0.475
Hospital	0.243	0.243	0.870	0.243	0.400
Workshop	0.830	0.830	0.277	0.830	0.692
Administrative	0.757	0.757	0.350	0.727	0.648

Table 8 Performance metrics of the optimization from three different mathematical programming techniques

Performance metrics	PSO	DE	NG	GMDH
Number of iterations	100,000	120,000	200,000	100,000
The magnitude of the optimal value	0.340646	0.341258	0.458392	0.403763

The objective of the optimization was to minimize Δp the difference between the weighted and standard allocation.

The amount of power supplied to each of the consumers, i.e., p was varied between the maximum and minimum energy typically provided to consumers.

The iteration was performed by PSO, DE, NG, and GMDH models.

The optimization from these three methods was compared based on the number of iterations it takes to attain the optimal value and the magnitude of the optimal amount. Table 8 depicts the comparison of the performance of the three algorithms.

The results from the performance analysis, as depicted in Table 8, conclude that PSO is a better option for the optimization problem solved in the present investigation. The optimal allocation value was also realistic and can be adopted without any modifications in the infrastructure or supply pattern.

6 Conclusion

The present investigation attempted to optimize energy resource allocation between different kinds of consumers based on the consumer's demand, consumption pattern, social and economic impact. At first, the study area consumers were differentiated based on their importance concerning their demand, consumption, social, and economic impact. In this aspect, the advantage of fuzzy Logic was utilized. In the next step, the allocation was optimized with the help of the weightage determined in the previous step. The cognitive distribution was optimized to minimize the difference between the weighted and normal distribution. The optimal condition was achieved in PSO within 1,00,000th iteration, whereas DE took 1,20,000 and NG 2,00,000 iterations to complete the optimal state. Again, the magnitude of the objective function was 0.340646 in PSO, but in the case of DE and NG, it was 0.341258 and 0.458392, respectively. The results clearly show that PSO was the better algorithm to optimize. That is why the optimal configuration found from the PSO was utilized to find the optimal pattern of allocation. The study results show how the loss of energy can be minimized after ensuring the consumers' satisfaction. Another conclusion that can be derived from this study is that PSO's competency and soaring performance were established for this kind of problem. The course was adopted in a small area with a few different types of consumers to experiment with the allocation procedure's authorization in a controlled environment. The study can be repeated for a larger area with a more significant number of different types of consumers to verify the investigation's allocation procedure's sensitivity.

References

Andrew, A., & Kumanan, S. (2020). Development of an intelligent decision-making tool for maintenance planning using fuzzy Logic and dynamic scheduling. *International Journal of Information Technology, 12,* 27–36. https://doi.org/10.1007/s41870-019-00384-w.

Abbasi, M. R., Guleria, A., & Devi, M. S. (2019). A genetic algorithm-based flow update scheduler for software-defined networks. *International Journal of Communication Systems*, e4188. https://doi.org/10.1002/dac.4188.

Al-Dmour, J. A., Sagahyroon, A., Al-Ali, A. R., et al. (2019). A fuzzy logic-based warning system for patients classification. *Health Informatics Journal, 25*(3), 1004–1024. https://doi.org/10.1177/1460458217735674.

Blakeslee, D., Fishman, R., & Srinivasan, V. (2020). Way down in the hole: Adaptation to long term water loss in rural India. *American Economic Review, 110*(1), 200–224. https://doi.org/10.1257/aer.20180976.

Das, S., & Suganthan, P. N. (2011). Differential evolution: A survey of the state-of-the-art. *IEEE Transactions on Evolutionary Computation, 15*(1), 4–31. https://doi.org/10.1109/TEVC.2010.2059031.

Duffy, P., Fitzpatrick, C., Conway, T., et al. (2018). Energy sources and supply grids—The growing need for storage. In: *Energy storage options and their environmental impact. Issues in environmental science and technology* (pp. 1–41). eISBN: 978-1-78801-553-0.

Deng, W., Yao, R., Zhao, H., et al. (2019). A novel intelligent diagnosis method using optimal LS-SVM with improved PSO algorithm. *Soft Computing, 23,* 2445–2462. https://doi.org/10.1007/s00500-017-2940-9.

ExxonMobil. (2017). *Outlook for energy: a view to 2040* [online]. Retrieved August 28, 2019, from https://corporate.exxonmobil.com/Energy-and-environment/Looking-forward/Outlook-for-Energy.

Eberhart, R., & Kennedy, J. (1995). A new optimizer using particle swarm theory. In: *Proceedings of the Sixth International Symposium on Micromachine and Human Science*, Nagoya, Japan (pp. 39–43). https://doi.org/10.1109/MHS.1995.494215.

Eberhardt, S., Richter, G., Gimbel, W., et al. (1996). Cloning, sequencing, mapping, and hyperexpression of the ribC gene coding for riboflavin synthase of Escherichia coli. *European Journal of Biochemistry, 242,* 712–719. https://doi.org/10.1111/j.1432-1033.1996.0712r.x.

Heppner, F., & Grenander, U. (1990). A stochastic nonlinear model for coordinated bird flocks. In: S. Krasner (Ed.), *The ubiquity of chaos* (pp. 233–238). Washington, DC: AAAS Publications. Pub. 89-15S.

Hemmati-Sarapardeh, A., Hajirezaie, S., Soltanian, M. R., et al. (2019). Modeling natural gas compressibility factor using a hybrid group method of data handling. *Engineering Applications of Computational Fluid Mechanics, 14*(1), 27–37. https://doi.org/10.1080/19942060.2019.1679668.

International Energy Agency. (2013). *World energy outlook 2013* [online]. Paris: IEA. https://www.iea.org/reports/world-energy-outlook-2013.

International Renewable Energy Agency. (2019). *Renewable power generation costs in 2018* [online]. Abu Dhabi: International Renewable Energy Agency. ISBN 978-92-9260-126-3. https://www.irena.org/publications/2019/May/Renewable-power-generation-costs-in-2018.

Kennedy, J., & Eberhart, R. (1995). Particle swarm optimization. In: *Proceedings of IEEE International Conference on Neural Network* (pp. 1942–1948). Piscataway, NJ. https://doi.org/10.1109/ICNN.1995.488968.

Kumari, O., & Kumar, M. (2020). Water governance: A pragmatic debate of 21st century; An Indian perspective. In: M. Kumar, D. Snow, & R. Honda (Eds.), *Emerging issues in the water environment during anthropocene*. Springer Transactions in Civil and Environmental Engineering. Singapore: Springer. https://doi.org/10.1007/978-981-32-9771-5_19.

Kumar, M., Ram, B., Sewwandi, H., et al. (2020). Treatment enhances the prevalence of antibiotic-resistant bacteria and antibiotic resistance genes in Sri Lanka and India's wastewater. *Environmental Research*, 109179. https://doi.org/10.1016/j.envres.2020.109179.

Li, M., & Guo, P. (2014). A multi-objective optimal allocation model for irrigation water resources under multiple uncertainties. *Applied Mathematical Modelling, 38*(19–20), 4897–4911.

Liu, Y., Chen, B., Wei, W., et al. (2020). Global water use associated with energy supply, demand, and international trade of China. *Applied Energy, 257,* 113992. https://doi.org/10.1016/j.apenergy.2019.113992.

Mogelgaard, K. (2011). *Why population matters to water resources*. Population Action International [online]. https://pai.org/wp-content/uploads/2012/04/PAI-1293-WATER-4PG.pdf.

Nagi, R., & Tripathy, S. S. (2020). Application of fuzzy logic in plant disease management. In: A. Kumar & M. Kalpana (Ed.), *Fuzzy expert systems and applications in agricultural diagnosis* (pp. 261–302). IGI Global.

Nariman-Zadeh, N., Darvizeh, A., & Ahmad-Zadeh, G. R. (2003). Hybrid genetic design of GMDH-type neural networks using singular value decomposition for modeling and prediction of the explosive cutting process. *Proceedings of the Institution of Mechanical Engineers, Part B: Journal of Engineering Manufacture, 217*(6), 779–790. https://doi.org/10.1243/09544050360673161.

Opara, K. R., & Arabas, J. (2018). Differential evolution: A survey of theoretical analyses. *Swarm and Evolutionary Computation, 44,* 546–558.

Ramachandra, T. V., Aithal, B. H., & Sreejith, K. (2015). GHG footprint of major cities in India. *Renewable and Sustainable Energy Reviews, 44,* 473–495. https://doi.org/10.1016/j.rser.2014.12.036.

Rogalsky, T., Derksen, R. W., et al. (2000). Differential evolution in aerodynamic optimization. *Canadian Aeronautics and Space Journal, 46*.

Srinivasan, D. (2008). Energy demand prediction using GMDH networks. *Neurocomputing, 72*(1–3), 625–629. https://doi.org/10.1016/j.neucom.2008.08.006.

Sakshin, S., & Kumar, R. (2018). A neuro-genetic technique for pruning and optimization of ANN weights. *Applied Artificial Intelligence, 33*(1), 1–26. https://doi.org/10.1080/08839514.2018.1525524.

Sengupta, S., Basak, S., & Peters, R. A. (2019). Particle swarm optimization: A survey of historical and recent developments with hybridization perspectives. *Machine Learning and Knowledge Extraction, 1*(1), 157–191.

Taikan, O., & Rose, E. Q. (2020). Economically challenged and water-scarce: Identification of global populations most vulnerable to water crises. *International Journal of Water Resources Development, 36*(2–3), 416–428. https://doi.org/10.1080/07900627.2019.1698413.

Tharwat, A., Elhoseny, M., Hassanien, A. E., et al. (2019). Intelligent Bézier curve-based path planning model using Chaotic Particle Swarm Optimization algorithm. *Cluster Computing, 22,* 4745–4766. https://doi.org/10.1007/s10586-018-2360-3.

Wang, D., Tan, D., & Liu, L. (2017). Particle swarm optimization algorithm: An overview. *Soft Computing, 22*(2), 387–408. https://doi.org/10.1007/s00500-016-2474-6.

Application of New Convergent Point Decision Making Method in Estimation of Vulnerability Index for Hydro Power Reservoirs

Tilottama Chakraborty

Abstract The ever-developing interest for extravagance has expanded the weight on conventional energy resources and urges researchers and designers to search for other choices. Hydropower is one of the most economical and reliable source of energy which is regarded to have the ability to fill in for customary energy sources. The hydropower plants (HPP) is providing the power of 1106 TWh. The main concern regarding the storage based HPP is the efficiency of its Hydro Power Reservoirs (HPR) and its productivity relies upon different elements which are variable with respect to climatic, water powered and financial boundaries. Every one of these boundaries again depends on other factors such as, locational impedance and nature of the machine introduced. As there are various factors having various degrees of effect on the productivity of HPRs, Multi-Criteria Decision Making (MCDM) was applied to develop an indicator which can indicate the performance status of the power plant. MCDM generally estimates priority values (P.V.) for normal conditions, but the present study utilizes the method to determine the contribution of the parameters for optimal conditions only. Measuring Attractiveness by a Categorical Based Evaluation Technique (MACBETH) and Analytical Network Process (ANP) were used to determine the constraints and Group Method of Data Handling (GMDH) was utilized to find the P.V. at optimal condition.

1 Introduction

Progress in sustainable power source boots energy protection tends to natural issues and atmosphere exchange, just as adding to different purposes of social improvement (Flavin and Aeck 2005; IEA 2012). There is the potential for sustainable power source to meet the world's more than 33% of overall energy demand (GEA 2012). Hydropower is the main renewable source for power generation all around, providing 76% of all inexhaustible power (Nautiyal and Goel 2020). All out introduced limit

T. Chakraborty (✉)
Hydro-Informatics Engineering, National Institute of Technology, Agartala, Barjala, Jirania 799046, Tripura, India
e-mail: tilottama86@gmail.com

M. Majumder and G. D. Kale (eds.), *Water and Energy Management in India*,
https://doi.org/10.1007/978-3-030-66683-5_12

was 1000 GW in 2013. Hydropower produces 16.4% of the world's power from all sources (WER 2015). Among renewable resources, hydropower possesses a superior role in the sustainable power source market and leads the path for reliable, sustainable and clean energy (Wang et al. 2014).

The efficiency of HPR relies upon different climatic, pressure driven and financial factors/parameters. As per the review of literature, the factors like Hydraulic feature, Hydrologic feature, Design stability, measure of discharge, turbulence in water, contrast in supply and demand energy, pressure difference among the inlet and powerhouse, pressure difference between powerhouse and tailrace and distance from nearest grid etc. control the overall efficiency of HPP (Majumder et al. 2020a, b; Jadoon et al. 2020; Hatata et al. 2019; Yang et al. 2019; Pasalli and Rehiara 2014; Paish 2002; Kling et al. 2014; Johnson et al. 2014; Locatelli et al. 2015; Rahi and Kumar 2016; Moriarty and Honnery 2012). Although different studies have been led to propose a technique to understand the efficiency of HPP, answers to the accompanying issues stay uncertain:

i. It is not possible to analyze the performance using so many parameters. Therefore, it is necessary to identify the most important parameters (MIP).
ii. Any indicator/media, which represents the performance of HPP must not be biased and must consider the input parameters as per their contribution to the plant O/P.
iii. Any analysis regarding performance must be conducted in view of the optimal scenario, not for the normal conditions.

1.1 Objective

This study tries to identify a solution to these problems by proposing a new MCDM method: Decision Making for Optimal Condition (DMO) which generates a decision for an optimal condition only. The indicator was made in such a way that it does not require human inferences and determines contributions based on the importance of the parameter in influencing the goal of the decision.

Multi-Criteria Decision Making is widely applied for estimation of priority value for a specific decision objective and helps to select an optimal option from the various available options. This type of methods uses objective equations to separate the better option from the multiple options available with the help of either a weight value known as Priority Value (PV) or by ranking of the available alternatives.

The present investigation aims to use the advantages of the MCDM method like Analytical Network Process (ANP) and Measuring Attractiveness by a Categorical Based Evaluation Technique (MACBETH) was used to address the most influential parameter for regulating the performance efficiency of HPP (Solution to Problem i). The priorities of the parameters which is used as the alternative to the MCDM method is determined with the help of statistical control charts which identifies the most influential parameter as per their impact on the decision-making output. In this aspect, the charting methodology considers the real life dataset of the alternatives for a real life situation and rates the parameter as per their significance to the output. Interference from either experts or stakeholder or literature is completely avoided. The resources (experts, stakeholders, and literature) are used only for the initial selection of the related parameters for performance analysis of HPPs (Solution to Problem ii).

The priority value or weight of importance for all the considered alternatives is estimated based on their role to generate optimal performance efficiency for the power plant. In this regards, the PV of each of the parameters was treated as the design factor of an optimization problem, whereas, the output function is selected as the objective function. The design parameters have constraints like the values will be within the minimum and maximum priority proposed by the ANP and MACBETH method. The Group Method of Data Handling (GMDH) algorithm was used as the programming technique to maximize the function which will be directly proportional to the performance efficiency of HPPs. The magnitude of the priority of the optimal condition of the output function is treated as the weight of importance for the parameters of the optimal condition (Solution to Problem iii). The Sect. 2 will describe the methods selected for application in the present investigation.

2 Methods Adopted

In this study two MCDM techniques, ANP and MACBETH were used to determine the boundary of the search space for the priority of the parameters. Here Group Method of Data Handling (GMDH) was used to maximize the performance function of efficiency. The MCDM method ANP was selected for their features like comparing alternatives with respect to criteria and vice versa. The impact of both criteria and alternatives to each other is considered before estimating the priority value of the alternatives. In the present study also as the decision making is required to be in both directions this method was considered in the determination of priority value. The MACBETH method is specializes in the identification of important based on its dissociation from the objective. The procedure of comparing the alternatives based on its inverse impact on the output will solve the problems of overlapping in the interdependence of criteria and alternative.

2.1 Analytical Network Process

In 1996 Analytical Network Process (ANP) was first proposed by Thomas Saaty (Mobin 2015). Transformation of qualitative value to the numeric value for comparative analysis is one of the most important features of ANP (Kabak and Dag deviren 2014; Yu et al. 2006). As ANP handles both quantitative and qualitative alternatives with respect to criteria and it can define the network structure it was chosen as one of the appropriate MCDM methods for this study. ANP has been applied to evaluate the performance indicators of reverse logistics in the renewable energy risk assessment (Wu et al. 2020; Hashemizadeh et al. 2020; Shaktawat and Vadhera 2020), determination of EIA report of Hydro-Power Plant (Sarmah et al. 2020), footwear industry (Guimaraes and Salomon 2015), wastewater treatment alternatives (Senante et al. 2015) and supplier selection (Hashemi et al. 2015).

2.2 Measuring Attractiveness by a Categorical Based Evaluation Technique

Measuring Attractiveness by a Categorical Based Evaluation Technique (MACBETH) was first proposed by Bana and Vansnick (1997). MACBETH is an MCDM approach which has the advantage of represents the subjective judgments to into quantitative scores (Karande and Karande 2014). One of the drawbacks of this method is related to linear programming (LP); it is well known that several optimal solutions (i.e. rankings) can be obtained with the LP method. These different ranks can be confusing for the decision maker (DM). As MACBETH handles subjective judgments about different alternatives with respect to criteria, it was applied in the current study. MACBETH has been used to determine the utility of governments to parties in coalition formation (Roubens et al. 2006), for health value measurement (Roubens et al. 2006), and for on-board hydrogen storage technologies (Montignac et al. 2009).

2.3 Group Method of Data Handling (GMDH)

Group method of data handling (GMDH) is a family of inductive algorithms for computer-based mathematical modeling of multi-parametric datasets that features fully automatic structural and parametric optimization of models. GMDH is used in such fields as data mining, knowledge discovery, prediction, complex systems modeling, optimization and pattern recognition. GMDH algorithms are characterized by inductive procedure that performs sorting-out of gradually complicated polynomial models and selecting the best solution by means of the external criterion. GMDH method has been applied in energy forecasting studies (Zor et al. 2020; Shakir and Biletskiy 2020), various renewable energy development studies (Lak Kamari et al. 2020).

3 Detailed Methodology

The methodology of the present objective can be divided into three steps:

i. Application of the new Convergent Point MCDM methods and GMDH to determine the P.V. of the selected parameters.
ii. Application of GMDH to find vulnerability index compared to the optimal condition.
iii. Sensitivity analysis, scenario analysis, case study analysis to validate the model.

3.1 Application of the New Convergent Point MCDM Methods and GMDH to Determine the P.V. of the Selected Parameters

The application of the selected MCDM method follows a basic methodology of criteria and alternative selection and use of the aggregation method which depend on the type of the method applied. In the next sections, the method followed in the selection of criteria, alternative and the way aggregation methods were used is described.

3.1.1 Selection of Criteria

In the present investigation, two levels of criteria were selected. All the methods used for ranking the selected parameters as per their contribution in the optimization of the decision function was taken as the Super Criteria.

One of the major drawbacks of the Multi Criteria Decision Making method is for rating the parameters it depends on the significance as determined by the group of experts and stakeholders employed for this specific decision making goal. The problem is change in the expert group may change the significance of the parameters. That is why the present study tried to implement a new unbiased, completely free of human influence method to determine the rank of the parameters as per their contribution in the optimization of the objective or decision function. In this aspect, the statistical control charts (SCC) were used to monitor the performance or quality of the samples produced. Here also the three different SCC was used to identify the contributions of the selected features in maximization of the decision function.

In this regard, at first five random samples of weights and parameter magnitude were generated following a normal distribution pattern for all the variables. The 0 values were taken as minimum value and 1 is assigned the maximum value possible for any parameter. The values were generated within the minimum and maximum values. Five sets of decision function were calculated from the five sets of samples for every iteration. Both the weights and values were used to find the decision function and are constantly iterated to identify a maximum value of the decision function within the upper and lower control limits determined following different SCC methods. That means with the help of this bounded optimization the values of the weights at the convergent point were used as an indicator to rate significance of the parameter in the optimization of the decision function.

The description of the method and how the SCC was used to in the present study is explained in Table 1.

At the end of this step, the absolute rank of the variables was determined in terms of its contribution in maximization of the decision function.

It is to be noted that, for each sub-criterion the decision function is developed in different way, keeping in view of the impact of the sub-criteria on the decision

Table 1 Description of the method used to achieve the present objective

Name of criteria	Description
Two-sample t-test	*Original Method*: A confidence interval is calculated and a hypothesis test of the difference between two means is done. Applied to test if a new process of treatment is superior to the current process or treatment. (Patwal and Narang 2020; Moser et al. 1989) *Analogy*: The confidence interval was determined within the five set of samples and from the maximum and minimum average value derived within a specific set of iterations. The maximum value of the decision function was found out when the magnitude lies within the confidence interval
x-bar	*Original Method*: x-bar can be used to identify whether measurement procedure has gone out of statistical control or not. In this aspect five or ten set of samples were generated and tested for quality. The control limit or the mean point was determined from the mean of the samples and a number which depends on the number of samples considered for the test. The upper and lower control limit is determined respectively by adding and subtracting this number. Whenever the sample gets out of the difference of upper and lower control limits it is assumed that an uncertainty has taken place. The method is used to monitor change in the average value of the monitored phenomena (Majumder et al. 2020b) *Analogy*: In the present method five set of samples were arbitrarily generated for weights as well as parameter values. The index was calculated from each set of samples and averaged to find the mean value. The maximum value of the index or decision function was identified when the values of the function lie within the upper and lower control limit
μ-chart	*Original Method*: The μ-Chart is a procedure which depicts the way of providing a denotational semantics to State charts and where problems occur with the original description. It is assumed that for a normally distributed variable there will be no uncertainty if it remains within the summation of mean and three times its standard deviation. Any value which remains outside this summation will indicate uncertainty. (Aslam et al. 2020; Reeve and Reeves 2000) *Analogy*: In the present investigation, the function values were generated within the summation of its mean and three times its standard deviation and the optimal value was identified within the values which satisfy the condition

Table 2 Description about the selected criteria

Name of criteria	Why used?
Productivity (L) (Economic Review of Tripura, 2013–14)	The productivity of the reservoir can be affected by the selected parameters as depicted in Table 3. The degree of impact on performance of the reservoir will vary with respect of the parameters
Quality (QM) (Flavin and Aeck 2005)	The change in the selected parameters can also imbibe change in the quality and the longevity of the reservoir
Energy requirement at the time of installation (ER)	The energy used at the time of installation of the reservoir has a sustainable impact on the financial liability of the reservoirs which in turn will impact on the parameters selected for representing efficiency of the reservoir
Change in location (LI)	The change in location will change the effect of selected features on the performance

function. Each parameter will also behave in its own way with respect to the sub-criteria. So, keeping both of this condition all the parameters were ranked for each of the criteria and by each of the methods. So, at the end of this stage there are 12 sets of ranks assigned to the parameter indicating its contribution towards the decision function and with respect to the sub-criteria.

3.1.2 Selection of Sub-criteria

The second level of criteria is depicted in Table 2. As discussed in the previous section the decision function was calculated considering the role of the selected features, i.e., whether it increase or decrease the performance of the power plant. If the variable increases the performance then it is taken as Incremental Variable (I) and if a change in the parameter decreases the performance then it is categorized as the Decremental Variable (D). Now, for some criteria some variables will be incremental and for some other it will be detrimental. Similarly, each of the second level criteria may have an incremental or detrimental effect on the objective function. Table 2 also depicts the way the second level criteria or sub-criteria influence the objective function in the present study.

3.1.3 Selection of Alternatives

The present investigation identified and selected parameters as per their citation frequency in specific kinds of literatures. In total four types of literatures were searched. The manuscripts which deal with loss in performance, related to the performance of different machines used in the power plant, energy requirement and locational interference on suitability for installation of hydropower plants.

After the five most cited parameters were identified for each type of literatures, log books of few power plants were checked to find common parameters. From the log book the ten most common parameters among the fifteen parameters (Five factors were common for two or more types of literatures) selected from literatures were used as the alternative to the present decision making problem. All the parameters and their impact on the decision function is shown in Table 3.

At the end of this phase the ten most significant alternatives will be identified. After the alternatives were selected the SCC method is used to determine the rank of the factors as per their contribution in maximization of the decision function.

3.1.4 Application of Convergent Point MCDM with MACBETH and ANP and GMDH Technique

In the present investigation a new procedure in determination of the priority value of the selected alternatives were proposed. In this new method, two MCDM techniques (in the present study ANP and MACBETH was used) were applied to determine the priority value or weight of importance of the alternatives which are then used as the input variables of the vulnerability index.

At first with the help of the selected SCC techniques the parameters were arranged in the descending order of significance separately for all the four selected criteria. Then based on the position of the parameters with respect to selected parameters a pairwise comparison was made with the help of two selected MCDM techniques having completely different procedure of pairwise comparison.

Now using the two different PV as determined by the two different techniques as the two bounds for each of the selected alternatives an objective function was determined which is directly proportional to the performance and inversely proportional with respect to the vulnerability of the reservoir where each of the alternatives were used as the design variable. The magnitude of the parameters was used as the constant. Here the average value of all the alternatives were utilized. The objective function was developed in such a manner that the I-parameters along with their PV becomes the numerator and the D-parameters along with there weight of importance was used as denominator. Now the GMDH algorithm was used as the optimization technique. The trade off will approximate the PV of each parameter at which the reservoir can yield optimal performance.

Now in the next phase of determination of the vulnerability index the magnitude of the parameters were used as the design variable and the maximum and minimum value observed for these parameters in one water year for the last five years were used as the higher and lower limit.

The function was the same function just the design variables were changed. The value of PV for each parameter at the trade off as determined in the previous phase was used as the magnitude of the PV for the selected parameters.

Table 3 Description of the factors

Name of factor	Why used?	Impact on decision function
Hydraulic features (H_1F)	Represents the self weight of dam, reaction on the foundation, water pressure (both vertical and horizontal component), uplift pressure due to subsoil flow and silt pressure	I
Hydrologic features (H_2F)	Represents the impact of wind and wave forces on the foundation	I
Design stability (DS)	Represent the safety factor to avoid shear failure	I
Efficiency of manpower (LE)	Depicts the availability and performance of the labours with the help of manhour per month that can be extracted from the available manpower (WER 2015; Wu et al. 2020)	I
Discharge (AD)	Represent the amount and frequency of discharge (Rahi and Kumar 2016)	I
Uncertainty factor (TW)	Represent the probability of turbulence in the reservoir storage, force generated due to earthquake or ice formation etc. uncertain phenomena	D
Deficiency in demand and supply (DSD)	Represent the difference between demand and supply of water with respect to the power plant (Yu and Tzeng 2006)	D
Difference between inflow and outflow (PIP)	The difference between inflow and outflow from the reservoir	I
Difference between the upstream and downstream water head (PH)	Represent the difference between the water head at the upstream and downstream	D
Distance from nearest powerhouse (DG)	Represents the distance between the reservoir and the location of powerhouse where the prime mover and generators are installed. More the distance less will be the efficiency of the reservoir in supplying energy to the powerhouse for rotating the turbines due to the conveyance loss	D

The result will give the parameter at which the performance of the reservoir will be optimal and vulnerability minimum. The reservoir is required to be developed based on these values as derived at the point of trade off. This technique will also ensure that the index value will be calculated with respect to the optimal status of the reservoir which makes the index relative not an absolute indicator.

3.2 Determination of Vulnerability Index

In the present phase the GMDH technique was applied to estimate the vulnerability index where the selected parameters, which are deemed as alternatives in Sect. 3.1, was considered as the input variables and the objective function is designated as the output variable. In this step the objective function, determined in Sect. 3.1, was modified by dividing the entire index by the optimal value of the index as determined in the Sect. 3.1. This value will change with change in the location and thus making the index flexible and universally applicable.

Section 3.3.1 describes about the validation techniques used in the present investigation.

3.3 Validation of the Model

The developed model and its results were validated by analysis of the sensitivity of the parameters with PV as determined from the method and by application of the indicator for estimation of performance for hydropower plant in operation. The Sects. 3.3.1 and 3.3.3 details the validation method.

3.3.1 Sensitivity Analysis

Multiple Input Single Output (MISO) Tornado strategy created by SenseIt Limited (Chakraborty 2018) has been used for the sensitivity analysis. The input variables range was varied between 0 and 1. The effect of each input has been observed on the output and the results were compared with the weights of the variables found from the new MCDM approach.

3.3.2 Scenario Analysis

As the method also aimed to include the impact of extreme events on the priority value of the control variables (C.V.), ten different scenarios were devised to represent the impact of extreme events in the priority values of the control variables following the IPCC Climate Change Scenarios A2 and B2 (Fig. 1).

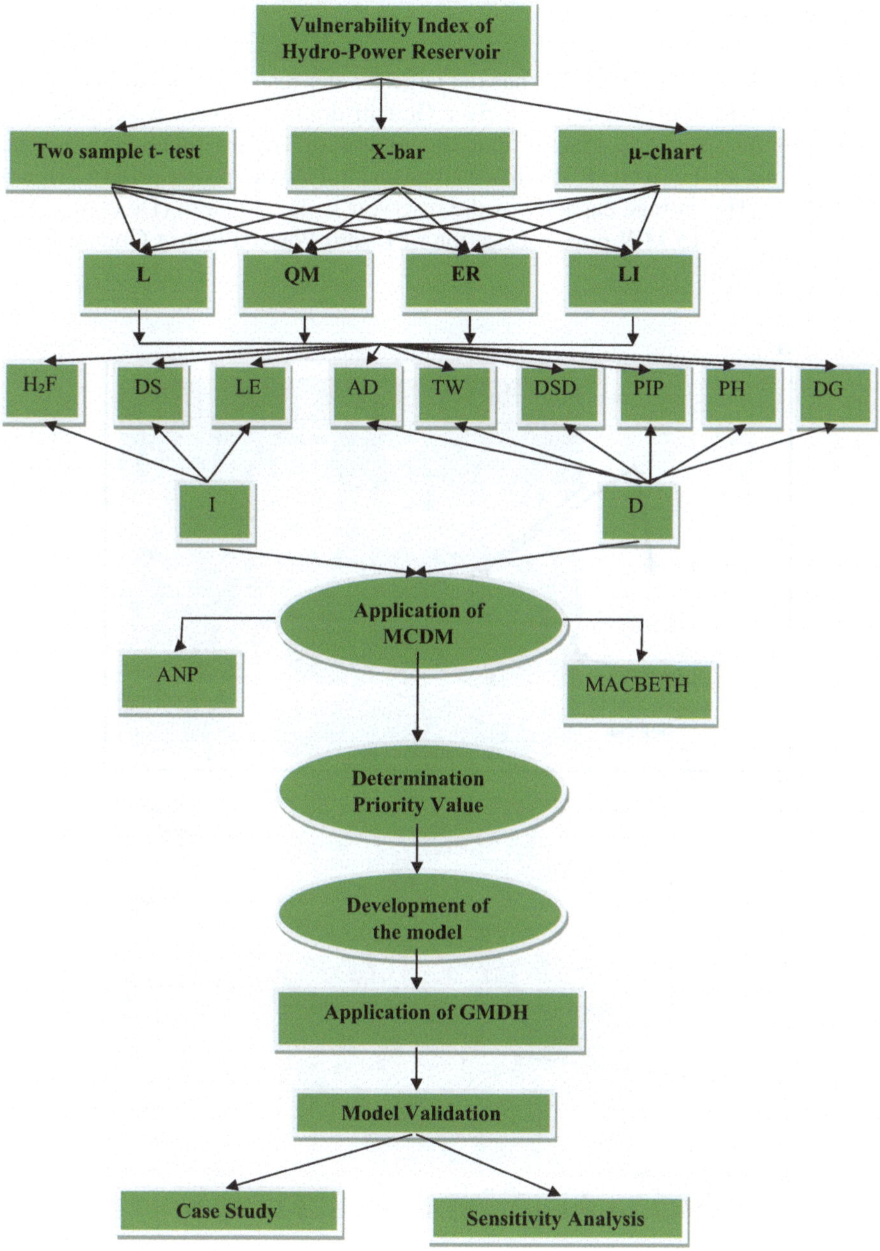

Fig. 1 Schematic diagram of convergent point decision making method

3.3.3 Case Study

The Gomati River basin (Fig. 2a) has the largest water resource availability among all the river basins of Tripura. The river Gomati drains so much water that it is responsible for around 31.45% of total annual flow of Tripura which is by far the largest among all the rivers. It also has the one and only hydro power project Tripura at Dumboor. The storage capacity of the reservoir is 23,570 ha m (De et al.). The project generates 15 MW power. Location of Hydro Power Plant in Gomati River Basin in shown in Fig. 2b.

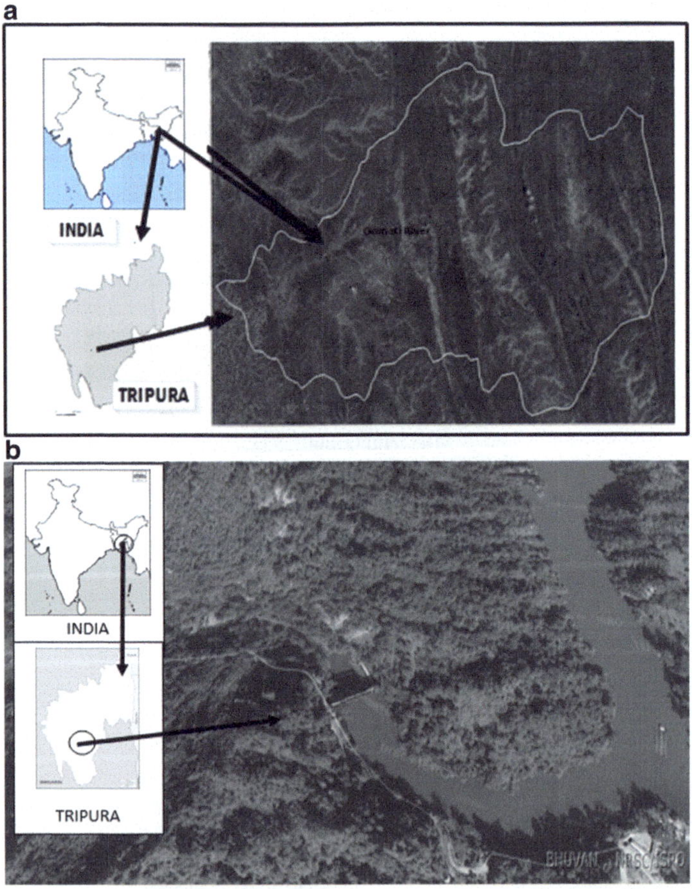

Fig. 2 **a** Location of Gomati River (*Source* Google Earth™). **b** Location of hydro power plant in Gomati River Basin (*Source* Bhuvan)

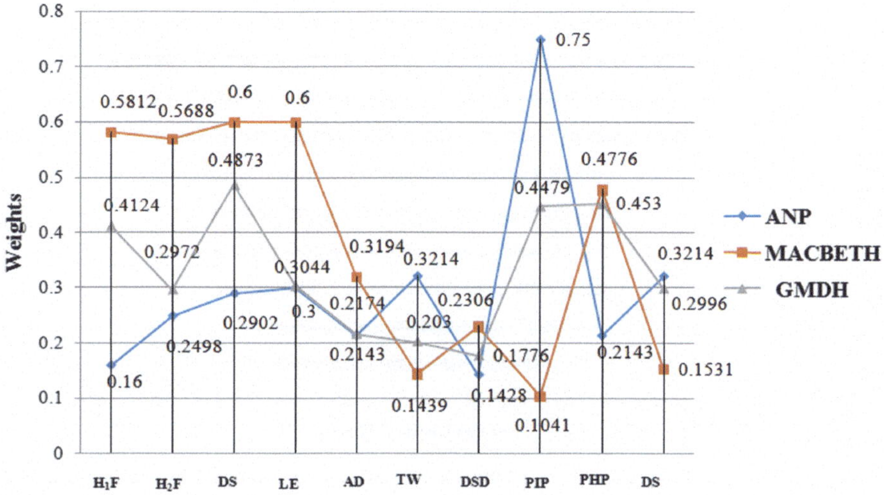

Fig. 3 Comparison of P.V. as estimated by ANP, MACBETH and convergent point MCDM represented as GMDH

4 Results and Discussion

Figure 3 shows the P.V. of each factor by MCDM techniques and convergent point decision making techniques. At the trade off point, most important parameter (MIP) was found to be DS and PV of the other parameters at the optimal pint was found to be 0.4124, 0.2972, 0.4873, 0.3044, 0.2174, 0.2030, 0.2876, 0.4479, 0.4530, and 0.2996 respectively for H_1F, H_2F, DS, LE, AD, TW, DSD, PIP, PHP and DG.

The validation of the model was conducted by sensitivity analysis and the application of the model. From Fig. 4, the most sensitive parameter found at the tradeoff point is DS which is also the MIP as found by the GMDH technique. Table 4 depicts the MIP as selected by the three different MCDM techniques. The MIP of GMDH and MACBETH was found to be same whereas the significance for the same parameter, DS, by ANP was found to be in 5th position with respect to the other parameters. In case of MACBETH the MIP selection was not pronounced and it has also identified LE as the MIP along with DS. As can be seen from the Fig. 3 and Table 4 the decision of from the three techniques are not same and as a result the trade off can give a clear indication considering the decision from both ANP and MACBETH techniques.

5 Scenario Analysis

The PV of each of the parameters was also identified under likely and unlikely conditions. The conditions were formulated by increasing and decreasing parameters simultaneously (H_1F, H_2F, DS, LE, AD, TW, DSD, PIP, PH, DG) with 5, 10, 15, 20, 50

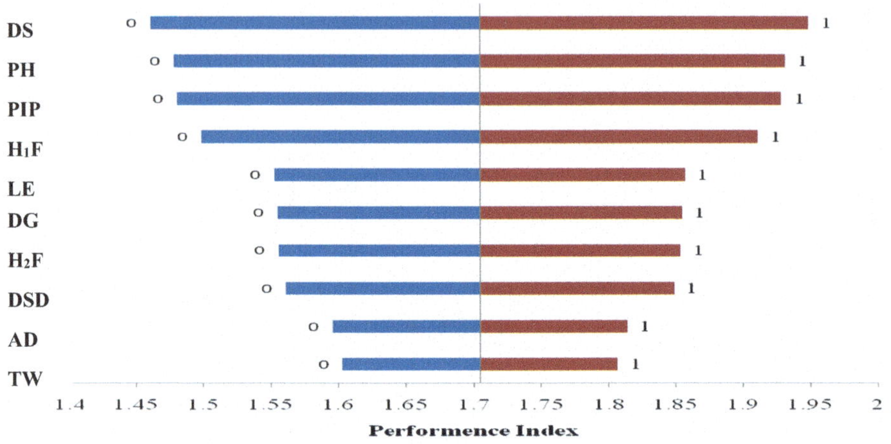

Fig. 4 Sensitivity of the parameter or in this case alternative as estimated by the Sense It tool (performance index is the objective function)

Table 4 Table showing the MIP with respect to the selected MCDM techniques and the proposed GMDH method

Technique	MIP
ANP	PIP
MACBETH	DS, LE
GMDH	DS

and 100%. If all these parameters increased or decreased simultaneously by 5–15% and 20–100% from normal conditions, these scenarios are called likely and unlikely scenarios. This means that the value of the Profit Function (P.F.) (see Table 5) was maximized at a scenario where all these parameters were 50% less than the cost at normal conditions. It was also assumed that all these parameters will be modified at the same time.

From the Table 5, scenario-6 is best and scenario-3 is worst with respect to the index compared to the value of the index and among the likely scenario as well as scenario-5 is best and scenario-3 is worst with respect to the index compared to the value of the index and among the unlikely scenario.

6 Conclusion

In this present study a new method of MCDM was proposed for estimation of Vulnerability Index for HPRs. The novelty of the new technique is the application of optimization technique in determination of the PV of the parameters and the utilization of two different MCDM techniques for estimation of the limit of feasible region of the optimization problem. As a result, the need of validation of the decision minimized.

Table 5 Scenario analysis in likely and unlikely conditions

Scenario		Increasing/Decreasing all the parameters simultaneously (%)	Index value
Likely scenario	Scenario 1	5	1.150162
	Scenario 2	10	1.040099
	Scenario 3	15	0.939606
	Scenario 4	−5	1.405045
	Scenario 5	−10	1.553728
	Scenario 6	−15	1.719902
Normal scenario			1.271232
Unlikely scenario	Scenario 1	20	0.847488
	Scenario 2	50	0.423744
	Scenario 3	100	0
	Scenario 4	−20	1.906847
	Scenario 5	−50	3.813695
	Scenario 6	−100	–

Although the new method was validated with the help of Sensitivity, Scenario and Case Study Analysis all of which were found to be represented the reliability of the new method. Another novelty of this investigation is the proposal of vulnerability index for Hydro Power Reservoirs or HPRs. In the study the Gumati Reservoir, which is a part of the Gumati HPP in Tripura, North East India was selected as the case study area. ANP, MACBET and GMDH was selected respectively as the MCDM techniques and optimization techniques. According to the results, DS was found to be the most important parameter and the optimal values of the design parameter at the trade off point was found to be 12.09, 8.72, 14.29, 8.93, 6.38, 5.95, 8.43, 13.14, 13.29 and 8.79% respectively for H1F, H2F, DS, LE, AD, TW, DSD, PIP, PHP and DG. A scenario analysis was also conducted with the P.V. of all the parameters under likely and unlikely scenarios. The study results show that in normal conditions DS will be the most important parameter to regulate the performance of the HPR. The change in the index value compared to the index value in the normal condition is 1.712 for the likely scenarios. The method can be utilized for monitoring the performance of the HPR in real time throughout the year. The index value can be used for comparing the performance of various HPRs on a common time scale so that proper mitigation measures can be adopted the improve the performance of the HPR which have a low value of the indicator. But the index has some limitation and among them the applicability of the index is required to be experimented with wide variation of case studies and on different systems such that the universality of the index can be determined.

References

Aslam, M., Saghir, A., & Ahmad, L. (2020). *Introduction to statistical process control.*

Assessment of wastewater treatment alternatives for small communities: An analytic network process approach. *Science of the Total Environment, 532*, 676–687.

Bana, C. A., & Vansnick, J. C. (1997). Applications of the MACBETH approach in the framework of an additive aggregation model. *Journal of Multi-Criteria Decision Analysis, 6*(2), 107–114.

Chakraborty, T. (2018). A MCDM-NBO approach for selection of installation location for wave energy power plants. In *Application of geographical information systems and soft computation techniques in water and water based renewable energy problems* (pp. 121–140). Singapore: Springer.

De, S. K. Geo-environmental status of the river Gumti, Tripura. Tripura State Pollution Control Board. ISSN: 2278-0181.

Economic Review of Tripura, 2013–14.

Flavin, C., & Aeck, M. H. (2005). *Energy for development: The potential role of renewable energy in meeting the Millennium Development Goals.* New York: World Watch Institute.

GEA. (2012). *Global energy assessment-toward a sustainable future.* Cambridge and New York, NY: Cambridge University Press and Laxenburg, Austria: The International Institute for Applied Systems Analysis.

Guimaraes, J. L. S., & Salomon, V. A. P. (2015). ANP applied to the evaluation of performance indicators of reverse logistics in footwear industry. *Procedia Computer Science, 55,* 139–148.

Hashemi, S. H., Karimi, A., & Tavana, M. (2015). An integrated green supplier selection approach with analytic network process and improved: Grey relational analysis. *International Journal of Production Economics, 159,* 178–191.

Hashemizadeh, A., Ju, Y., Bamakan, S. M. H., & Le, H. P. (2020). Renewable energy investment risk assessment in belt and road initiative countries under uncertainty conditions. *Energy*, 118923.

Hatata, A. Y., El-Saadawi, M. M., & Saad, S. (2019). A feasibility study of small hydro power for selected locations in Egypt. *Energy Strategy Reviews, 24,* 300–313.

IEA. (2012). World Energy Outlook 2012, Paris, France.

Jadoon, T. R., Ali, M. K., Hussain, S., Wasim, A., & Jahanzaib, M. (2020). Sustaining power production in hydropower stations of developing countries. *Sustainable Energy Technologies and Assessments, 37,* 100637.

Johnson, N., Kang, J., & Hathway, E. A. (2014). Acoustics of weirs: Potential implications for micro-hydropower noise. *Renewable Energy, 71,* 351–360.

Kabak, M., & Dag deviren, M. (2014). Prioritization of renewable energy sources for Turkey by using a hybrid MCDM methodology. *Energy Conversion and Management, 79,* 25–33.

Karande, P., & Karande, P. (2014). A facility layout selection model using MACBETH method. In *Proceedings of the 2014 International Conference on Industrial Engineering and Operations Management*, Bali, Indonesia (pp. 7–9).

Kling, H., Stanzel, P., & Preishuber, M. (2014). Impact modelling of water resources development and climate scenarios on Zambezi River discharge. *Journal of Hydrology: Regional Studies, 1,* 17–43.

Lak Kamari, M., Isvand, H., & Alhuyi Nazari, M. (2020). Applications of multi-criteria decision-making (MCDM) methods in renewable energy development: A review. *Renewable Energy Research and Application, 1*(1), 47–54.

Locatelli, G., Palerma, E., & Mancini, M. (2015). Assessing the economics of large Energy Storage Plants with an optimisation methodology. *Energy, 83,* 15–28.

Majumder, P., Majumder, M., Saha, A. K., & Nath, S. (2020a). Selection of features for analysis of reliability of performance in hydropower plants: A multi-criteria decision making approach. *Environment, Development and Sustainability, 22*(4), 3239–3265.

Majumder, P., Majumder, M., & Saha, A. K. (2020b). Real-time monitoring of power production in modular hydropower plant: Most significant parameter approach. *Environment, Development and Sustainability, 22*(5), 4025–4042.

Mobin, M. (2015). Multi-objective X-bar control chart design by integrating NSGA-II and data envelopment analysis. In *Conference: 2015 Industrial and Systems Engineering Research Conference,* at Nashville, Tennessee, USA.

Montignac, F., Noirot, I., & Chaudourne, S. (2009). Multi-criteria evaluation of on-board hydrogen storage technologies using the MACBETH approach'. *International Journal of Hydrogen Energy, 34*(10), 4561–4563.

Moriarty, P., & Honnery, D. (2012). What is the global potential for renewable energy? *Renewable and Sustainable Energy Reviews, 16,* 244–252.

Moser, B. K., Stevens, G. R., & Watts, C. L. (1989). The two-sample t test versus Satterthwaite's approximate F test. *Communications in Statistics-Theory and Methods, 18*(11), 3963–3975.

Nautiyal, H., & Goel, V. (2020). Sustainability assessment of hydropower projects. *Journal of Cleaner Production,* 121661.

Paish, O. (2002). Small hydro power: Technology and current status. *Renewable and Sustainable Energy Reviews, 6,* 537–556.

Pasalli, Y. R., & Rehiara, A. B. (2014). Design planning of micro-hydro power plant in Hink River. *Procedia Environmental Sciences, 20,* 55–63.

Patwal, R. S., & Narang, N. (2020). Optimal generation scheduling of pumped storage hydro-thermal system with wind energy sources. *Applied Soft Computing,* 106345.

Rahi, O. P., & Kumar, A. (2016). Economic analysis for refurbishment and uprating of hydro power plants. *Renewable Energy, 86,* 1197–1204.

Reeve, G., & Reeves, S. (2000, November). μ-Charts and Z: Hows, whys, and wherefores. In *International Conference on Integrated Formal Methods* (pp. 255–276). Berlin, Heidelberg: Springer.

Roubens, M., Rusinowska, A., & Swart, H. (2006). Using MACBETH to determine utilities of governments to parties in coalition formation. *European Journal of Operational Research, 172*(2), 588–603.

Sarmah, P., Nema, A. K., & Sarmah, R. (2020). An approach to determine the quality of EIA reports of hydropower plants using analytic network process and fuzzy logic toolbox. *Environmental Impact Assessment Review, 85,* 106462.

Shakir, M., & Biletskiy, Y. (2020). Forecasting and optimisation for microgrid in home energy management systems. *IET Generation, Transmission and Distribution, 14*(17), 3458–3468.

Shaktawat, A., & Vadhera, S. (2020). Risk management of hydropower projects for sustainable development: A review. *Environment, Development and Sustainability,* 1–32.

Wang, B., Nistor, I., Murty, T., & Wei, Y. M. (2014). Efficiency assessment of hydroelectric power plants in Canada: A multi criteria decision making an approach. *Energy Economics, 46,* 112–121.

World Energy Resources. (2015). *Charting the upsurge in hydropower development 2015.* Hydropower World Energy Council.

Wu, Y., Wang, J., Ji, S., & Song, Z. (2020). Renewable energy investment risk assessment for nations along China's belt & road initiative: An ANP-cloud model method. *Energy, 190,* 116381.

Yang, X., & Liu, H. (2019). Diphenylphosphine-Substituted Ferrocene/Silsesquioxane-Based Hybrid Porous Polymers as Highly Efficient Adsorbents for Water Treatment. *ACS Applied Materials Interfaces, 11*(29), 26474–26482.

Yu, R., & Tzeng, G. H. (2006). A soft computing method for multi-criteria decision making with dependence and feedback. *Applied Mathematics and Computation, 180,* 63–75.

Zor, K., Çelik, Ö., Timur, O., & Teke, A. (2020). Short-term building electrical energy consumption forecasting by employing gene expression programming and GMDH networks. *Energies, 13*(5), 1102.

Yu, L., Yuan, K., Phuong, H. T. A., Park, B. M., & Kim, S. H. (2016). Angiotensin-(1-5), an active mediator of renin-angiotensin system, stimulates ANP secretion via Mas receptor. *Peptides, 86,* 33–41

Recognition of Fatigue Failure in Wave Energy Converter Using Statistical Control Chart, Multi-criteria Decision Making Tools and Polynomial Neural Network Model

Satyabrata Saha

Abstract The current work exhibits the pattern of Fatigue influenced to the Wave Energy Converters. The fatigue failure happened because of different reasons in genuine field of use of Wave Energy Conversion. Every single conceivable factor are considered to discover the fatigue trending. To break down the exhaustion likelihood fatigue probability Statistical Control Process and Multi-Criteria Decision making tools such as Weighted Product Method and Weighted Summation Method are used, from the result a model is prepared through Polynomial Neural Network software named GMDH. There seven fatigue influenced parameter were considered as input from previous literature review. These parameter were applied theatrically in the analysis, and output results were trained in the software, then final outcome of the model predicted index was given by the frame work. However considering various mechanical design oriented hypotheses, it was found that the fatigue influenced factors are all non-beneficiary to Fatigue trend. After all the figuring it can foresee the closeness of Fatigue failure in a Wave Energy Converter irrespective of different location.

Keywords Wave energy converter · Statistical control method · Fatigue factors · MCDM · WSM · WPM · Group Method of Data Handling (GMDH)

List of Abbreviation

MCDM	Multi Criteria Decision Making
WEC	Wave Energy Converter
WSM	Weighted Summation Method
WPM	Weighted Product Method
PTO	Power Take off
GMDH	Group Method of Data Handling

S. Saha (✉)
School of Hydro Informatics Engineering, National Institute of Technology Agartala, Agartala, India
e-mail: antusaha84@gmail.com; ssaha.me@gmail.com

M. Majumder and G. D. Kale (eds.), *Water and Energy Management in India*,
https://doi.org/10.1007/978-3-030-66683-5_13

1 Introduction

The present demand of fossil fuel and burgeoning concentration of pollutants enforced scientists worldwide to look for alternatives to substitute conventional energy sources. Among the most suitable renewable, energy from ocean waves was found to be reliable and enough to satisfy global energy demand. But due to some location dependent factors this type of source is expensive and thus unpopular among the masses (Veritas 2005). One among the factors is the mechanical fatigue of the converters. This factor can be defined as weakness in machine parts or structures due to repeated variation of working load and widely (Ambühl et al. 2015).

The fatigue of converters affects both the efficiency and operational life time and is considered whenever a design related problems are solved. As the factor can reduce the operating efficiency as well as life time of a converter estimation of the same can ensure prevention of economical liability and the life time of the converters. Different factors can influence fatigue in a converter. Some of the factors are Wave impact on the converter, bearing life maintenance etc. But not all the parameters are equally significant in influencing mechanical fatigue in a converter. The Wave Energy Converters (WECs) are used in ocean in various position like Offshore, Onshore, near shore, partially submerged i.e. floating, totally under water or established in the open atmosphere in shoreline of sea i.e. fixed structure (Yang et al. 2010). Different types of difficulties may come in WECs. Fatigue failure in WECs is significant important area to analyze specially for offshore and near shore structures.

The objective of this research is to predict the trend of the fatigue probabilities for any WECs in all probable wave power generation situations by using Statistical Control Chart (X, P, and R Chart) (Tunas Bangsa Pematangsiantar 2017), AHP Multi Criteria Decision Making (MCDM) (Anastasakis and Mort 2001) and Neural Network (GMDH Software) Model (Hagan et al. 1996).

2 Objective of the Study

The study aims to develop of a new methodology to identify the fatigue probability in any wave energy converter. This aim includes the identification of the characteristics of most suitable factors which can make fatigue in the converters. This involves the use of Statistical Control chart and Multi Criteria Decision Making theory (AHP, WSM and WPM) combined with the GMDH algorithm to predict the probability of fatigue chances in the converter and to maximize the efficiency of the overall plant.

3 Fatigue Influenced Factors

In a wave energy converter it consists of different parts as well other components, in any kind of change or deformations occurs in those areas the fatigue can cause in the WECs. From literature review it was observed that seven most significant factors were considered for fatigue in a WEC. Details of the factors are discussed below as in Sects. 3.1–3.7.

3.1 Internal Structure

Internal areas of the WECs are considered which are accessible for regular maintenance. These areas mainly submerged into the sea portion. Metal joints are welded or nut bolt system. Ultimate Stress and Allowable stress are to defined properly and Safety Factor to be considered $\geq$ 1.0 (Veritas 2005). Time to time overview of these sections is required.

3.2 External Structures

External areas of WECs those are easily accessible for maintenance repairing and not accessible for regular inspection and repair in dry and clean conditions (Ambühl et al. 2015). Safety Factor to be considered $\geq$1.0 (Veritas 2005).

3.3 Non-accessible Areas

These particular areas not planned to be accessible for inspection and repair during operation (Veritas 2005). Because positions 150 m below water level, it should be assumed inaccessible for service and inspection for maintenance (Veritas 2005).

3.4 Wave Impact, Wave Climate and Weather Condition

Incremental change in the heaviness of steel expected to confront higher outrageous waves, and changes in exhaustion life of WEC parts from activity in more unpleasant climate (Thies et al. 2012). For weakness life significantly more than for basic costs, given the wide assortment of proposed WECs and their segments it is difficult to get results that could be summed up to every one of them. It was picked to represent the use of standard exhaustion computations to a specific WEC (Veritas 2010; Thies et al. 2014).

3.5 Replacement Schedule for Bearings System

The substitution plan for bearing will be as far as possible to the adjusting span. This part bolsters the gyroscopic system of this WEC, thus turn speeds are higher than those of direction in numerous other force take off instruments. Furthermore the bearing is housed inside the structure and bearing life will be very unique for outside moving parts presented to erosion, lubricant contamination (Babarit et al. 2012). It should be referenced regardless that the figuring were rehashed for various kinds of bearing under totally different burdens, and the relative changes in administration life between various ocean states and atmospheres were fairly comparative. Bearing exhaustion is an imperative on support that might be relied upon to be shared by numerous WECs (Nkurlu et al. 2020). These parts will be picked so stacks are inside their evaluated working reach, inside the exhaustion life.

3.6 The Manufacturer Defined Range Rating

It indicates the Hydraulic load on the WECs, Hydraulic Motor rotation, gyroscopic velocity, Types of lubricant used, Optimum operating temperature. If the WECs are operated beyond the manufacturer defined range rating of the instruments the fatigue trend will be more (Kuznecovs et al. 2019).

3.7 Foundation Design for Wave Energy Converters

Structure of establishments for wave vitality converters will be founded nearby and area explicit data. The determination of site examinations and the decision of these examination techniques will be considered the sort and size of the wave vitality gadgets, the consistency of land and seabed conditions. For use of stays the dirt qualities and scope of soil or land quality properties will be dissected. Site choice examinations ought to give adequate data about the land qualities to a profundity required to check impact of conceivable disappointment conditions (Wu et al. 2018).

4 Methodology

Strategy of this examination is portrayed in the flowchart of Fig. 1. At first the issues identified with fatigue failure of WECs are considered in light of better execution or improved yield of tremendous speculation in regards to Wave Energy Conversion process. Main considerations that impacted Fatigue are considered from the mechanical activity. Those boundaries are put appropriately by utilizing Statistical Control Charts like X, R and P Charts (Allen 2006). Control charts are used to

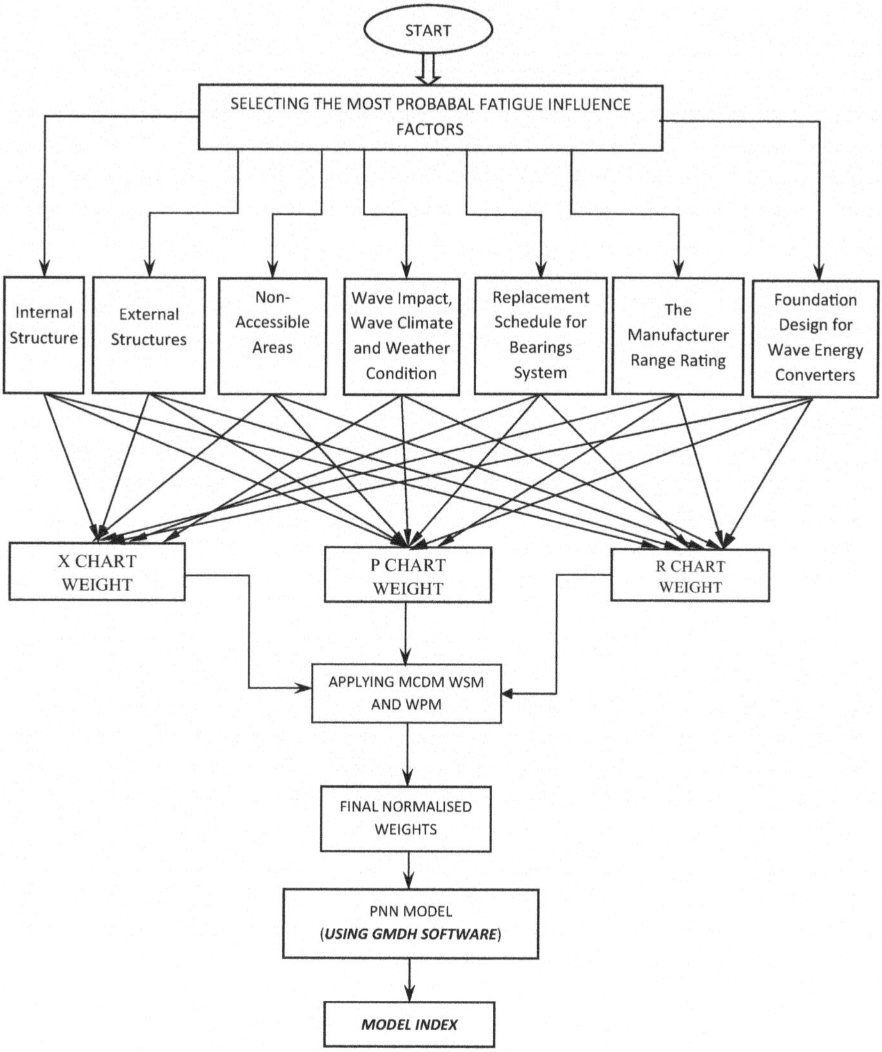

Fig. 1 Flowchart of the total process

routinely monitor quality. Depending on the number of process characteristics to be monitored. A control chart is used to monitor a process variable over time. That variable can be in any type of data. These attribute charts are considered as criteria and the Fatigue influenced parameters are considered as alternatives in the process of identifying the weights of parameters by pairwise comparison matrices which to be used in MCDM, WSM and WPM (Anastasakis and Mort 2001). An approach to ranking alternatives against a set of n criteria is the weighted sum method (WSM). It commonly takes the form.

$$R_j = \sum_{i=1}^{n} w_i q_{ij} \tag{1}$$

(Hazelrigg 2019). Where the rating of the jth alternative, R_j, is the sum of criteria (or attribute) measures, q_{ij}, times weights, $w_i > 0$, applied to each measure. The Weighted Product method is similar to the Weighted Sum method. The main difference is that instead of addition in this model there is multiplication. It commonly takes the form:

$$P\left(\frac{A_K}{A_L}\right) = \prod_{j=1}^{n} \left(\frac{a_{Kj}}{a_{Lj}}\right)^{wj} \tag{2}$$

.

for K, L = 1, 2, 3 … *m*.

If the ratio $P(A_K/A_L)$ is greater than or equal to the value 1, then it indicates that alternative A_K is more desirable than alternative A_L (in the maximization case) (Chang and Yeh 2001).

After weights are found, the PNN software was trained by those weights and the model was generated to predict the Index value of Fatigue influence (Nkurlu et al. 2020).

Initially from the literature review the most significant fatigue influenced factors were considered. Then the statistical control chart methods were used to make the proper ranking of those parameters by upper control limit as well as lower control limit of the charts. The control charts are taken as criteria and factors were taken as alternative in the multi criteria decision making theory. Then a pairwise comparison metric were form to identify the weight value of the factors. A scale value of 1–9 were considered. Eign vectors were archived and normalized weight vectors were fetched. Others two MCDM method WPM and WSM were also used to find the weight vectors of the fatigue influenced factors. Then the average weight values were taken. Using this Weight vectors a Polynomial Neural Network model (GMDH) was trained to predict the output results in different cases.

5 Results and Discussion

Statistical Control Chart (X, P and R Chart) were used to ranking the seven fatigue influenced parameters as per their weightage value. A pairwise comparison Matrices was constructed for finding importance of Criteria which was taken as the control charts. WPM and WSM were used to identify the relative importance of the alternative firstly then applied criteria weight to find out the normalized weight for each of the MCDM, then final weight calculated by averaging both results (Tables 1, 2, 3 and 4).

Polynomial Neural Network Software predicted algorithm:

$$F_I = N_C + \sum_{n=1}^{n} (w_n x_n) \qquad (i) \text{ (Nkurlu et al. 2020)}$$

$\boldsymbol{F_I}$ = Fatigue Influenced Index, $\boldsymbol{N_C}$ = Model Constant = $\mathbf{2.07784 \times 10^{-8}}$ [Obtained by GMDH Software].

$\boldsymbol{W_n}$ = Weight of the parameter, $\boldsymbol{X_n}$ = Parameter data (Scale to 1).

Table 1 Criteria versus criteria pairwise comparison matrices importance and their normalized values

Criteria	Score	X Chart	R Chart	P Chart	Foundation design for wave energy converters normalized weight
X Chart	5	1.00	1.25	1.67	0.4167
R Chart	4	0.8	1.00	1.34	0.3333
P Chart	3	0.6	0.75	1.00	0.2500

Table 2 Normalized weight of Seven Fatigue influenced parameters in WPM MCDM method

WPM	X chart	R chart	P chart	Sum	Normalized value
Criteria Wt.	*0.4167*	*0.3333*	*0.2500*		
Internal structure	0.1443	0.1456	0.1603	0.14859	0.14911
External structures	0.1519	0.1304	0.1636	0.14976	0.15028
Non-accessible areas	0.1502	0.152	0.1197	0.14023	0.14072
Wave impact, wave climate and weather condition	0.1393	0.1342	0.1478	0.14060	0.14109
Replacement schedule for bearings system	0.12	0.1547	0.1315	0.13140	0.13186
The manufacturer range rating	0.1508	0.15	0.1434	0.14821	0.14873
Foundation design for wave energy converters	0.1435	0.133	0.1336	0.13770	0.13818

Table 3 Normalized weight of seven fatigue influenced parameters in WSM MCDM method

WSM	X chart	R chart	P chart	Sum	Normalized value
Criteria Wt.	*0.4167*	*0.3333*	*0.2500*		
Internal structure	0.1443	0.1456	0.1603	0.14873	0.14887
External structures	0.1519	0.1304	0.1636	0.15033	0.15047
Non-accessible areas	0.1502	0.152	0.1197	0.14106	0.14118
Wave impact, wave climate and weather condition	0.1393	0.1342	0.1478	0.14074	0.14087
Replacement schedule for bearings system	0.12	0.1547	0.1315	0.13213	0.13224
The manufacturer range rating	0.1508	0.15	0.1434	0.14829	0.14842
Foundation design for wave energy converters	0.1435	0.133	0.1336	0.13783	0.13795

Table 4 Final mean weight values of seven parameters by WSM & WPM

Fatigue influenced parameter	Final weight calculated by MCDM	Rank
Internal structure	0.14911	2
External structures	0.15028	1
Non-accessible areas	0.14072	5
Wave impact, wave climate and weather condition	0.14109	4
Replacement schedule for bearings system	0.13186	7
The manufacturer range rating	0.14873	3
Foundation design for wave energy converters	0.13818	6

5.1 Application of Polynomial Neural Network

The current investigation is to identify and appraise the relationship between the factors as info and the yield as model file. Accordingly, the chose weakness affected boundaries are utilized information sources and the attainability list (model record) was considered as the yield or the forecast (Hagan et al. 1996) (Figs. 2, 3, 4, 5 and 6).

5.2 Sample Model Data Generated from GMDH Software

In GMDH software iteration was taken on around five hundred results, a sample of consecutive ten datasets from anywhere from the results are taken. From the Table 5

Plot | Residuals | Accuracy | Raw model

Error measure: Absolute — Target: INDEX of New Microsoft Office Excel Worksheet

Postprocessed results	Model fit	Predictions
Number of observations	278	185
Max. negative error	-9.88098E-15	-9.21485E-15
Max. positive error	9.15934E-15	9.40914E-15
Mean absolute error (MAE)	3.10902E-15	3.251E-15
Root mean square error (RMSE)	3.83787E-15	4.01692E-15
Residual sum	-5.71765E-14	-1.55292E-13
Standard deviation of residuals	3.83236E-15	3.92823E-15
Coefficient of determination (R²)	1	1
Correlation	1	1

Fig. 2 Accuracy of the model fit and prediction

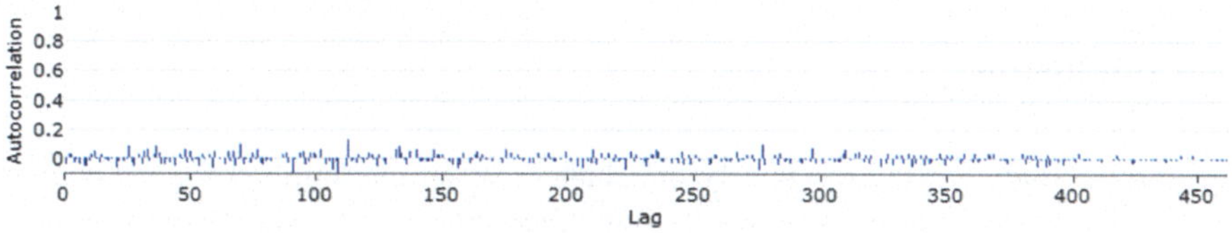

Fig. 3 Plot of autocorrelation

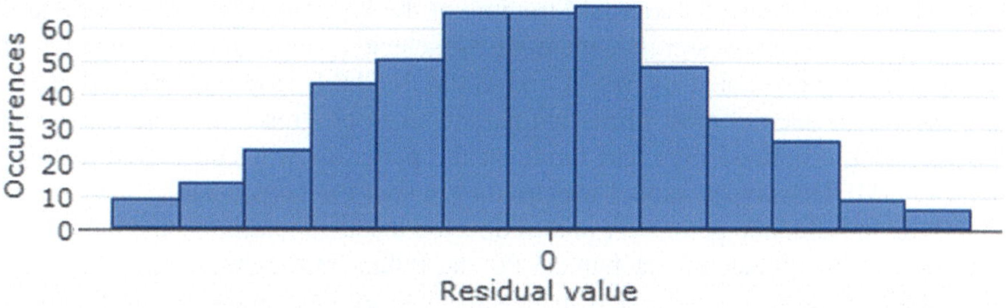

Fig. 4 Occurrence of residual values

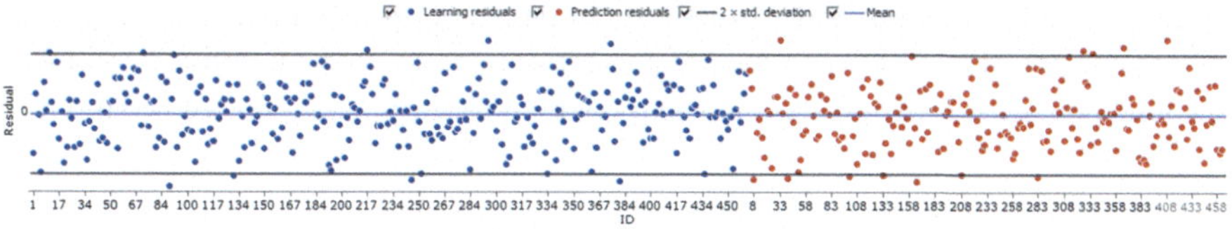

Fig. 5 Distribution of residual

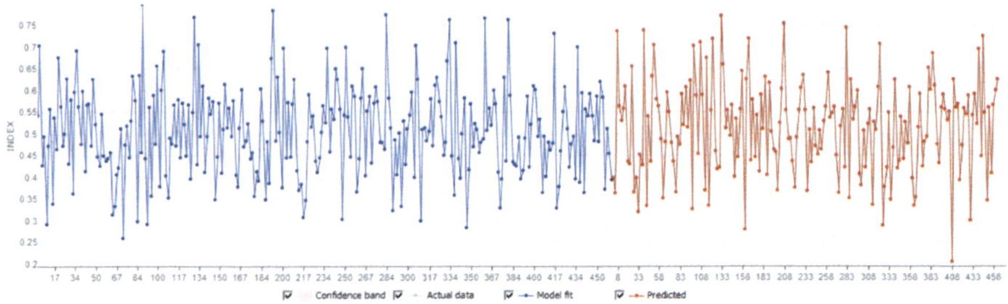

Fig. 6 Comparison between actual and predicted values

second reading has a more tendency of fatigue failure whereas forth one has a lesser chance of fatigue failure on WEC.

6 Conclusion

The current study is tried to identify the chances of fatigue failure of wave energy converter with the help of an index by implementation of statistical control charts X, R, P, Multi-criteria Decision Making tools such as WSM and WPM also a model software Group Method of Data Handling.

According to the results External Structure of the Wave energy converter parameter was found to be most significant among the selected factors and have the highest priority value as estimated by WSM and WPM MCDM. However internal structures was as the second most important parameter in progression of the index and manufacture pre-defined rage and rating of the materials was considered as third one. The GMDH software model was used to design the considerable input factor as the fatigue influence parameter and to develop a self-regulating substructure for estimation of the chance of mechanical fatigue failure in the system. In this angle the model execution was good and its exactness level was depending on the input variables given to the data sets. The analytical values of the seven factors can be applied in the real field, so, the achievement of the study can be applicable to real life practice. Although the lack of real field practical application may raise question about the practical feasibility of the indicator. But this can be dealt in future scope of the studies or some location wise test can be done for more accuracy of the model.

Table 5 Sample Model.Index data

Sl No	Internal structure	External structure	Non-accessible areas	Wave impact, climate	Bearing life	Manufacturer range rating	Foundation design	Index	Model.Index
1	0.68681	0.29240	0.82037	0.26004	0.62735	0.82497	0.43636	0.56421	0.56421
2	0.30874	0.38827	0.61557	0.95327	0.83773	0.68861	0.85045	0.65592	0.65591
3	0.00420	0.46709	0.61579	0.50804	0.70748	0.34593	0.69572	0.47004	0.47004
4	0.66439	0.17262	0.11698	0.31910	0.16138	0.85948	0.23192	0.36765	0.36765
5	0.60028	0.12669	0.27954	0.92533	0.05511	0.35980	0.93348	0.46820	0.46821
6	0.12783	0.63239	0.65053	0.98691	0.39330	0.32297	0.23053	0.47664	0.47664
7	0.70004	0.79652	0.24327	0.08312	0.70654	0.06092	0.95723	0.50453	0.50454
8	0.86464	0.53214	0.13274	0.04505	0.63748	0.95640	0.74312	0.56293	0.56293
9	0.11544	0.39705	0.72372	0.46413	0.35151	0.93625	0.85783	0.54836	0.54835
10	0.30574	0.15158	0.48638	0.34947	0.55409	0.14074	0.80014	0.39068	0.39068

References

Allen, T. T. (2006). *Introduction to engineering statistics and six sigma: Statistical quality control and design of experiments and systems*. Springer Science & Business Media.

Ambühl, S., Ferri, F., Kofoed, J. P., & Sørensen, J. D. (2015). Fatigue reliability and calibration of fatigue design factors of wave energy converters. *International Journal of Marine Energy, 10*, 17–38.

Anastasakis, L., & Mort, N. (2001). *The development of self-organization techniques in modeling: A review of the group method of data handling (GMDH)*. Research report-University of Sheffield Department of Automatic Control And Systems Engineering.

Babarit, A., Hals, J., Muliawan, M. J., Kurniawan, A, Moan, T., & Krokstad, J. (2012). Numerical benchmarking study of a selection of wave energy converters. *Renewable Energy, 41*, 44–63.

Chang, Y. H., & Yeh, C. H. (2001). Evaluating airline competitiveness using multi-attribute decision making. *Omega, 29*(5), 405–415.

Hagan, M. T., Demuth, H. B., & Beale, M. H. (1996). *Neural network design*. Boston: Pws Pub.

Hazelrigg, G. A. (2019). A note on the weighted sum method. *The Journal of Mechanical Design, 141*(10), 100301.

Kuznecovs, A., Ringsberg, J. W., Yang, S.-H., Johnson, E., & Anderson, A. (2019). A methodology for design and fatigue analysis of power cables for wave energy converters. *International Journal of Fatigue, 122*, 61–71.

Nkurlu, B. M., Shen, C., Asante-Okyere, S., Mulashani, A. K., Chungu, J., & Wang, L. (2020). Prediction of permeability using Group Method of Data Handling (GMDH) neural network from well log data. *Energies, 13*(3), 551.

Thies, P. R., Johanning, L., & Smith, G. H. (2012). Lifecycle fatigue load spectrum estimation for mooring lines of a floating marine energy converter. In ASME *2012 31st International Conference on Ocean, Offshore and Arctic Engineering* (pp. 667–676). American Society of Mechanical Engineers.

Thies, P. R., Johanning, L., Harnois, V., Smith, H. C. M., & Parish, D. N. (2014) Mooring line fatigue damage evaluation for floating marine energy converters: Field measurements and prediction. *Renewable Energy, 63*, 133–144.

Tunas Bangsa Pematangsiantar, S. T. I. K. O. M. (2017). Comparison of weighted sum model and multi attribute decision making weighted product methods in selecting the best elementary school in Indonesia. *International Journal of Software Engineering and Its Applications, 11*(4), 69–90.

Veritas, D. N. (2005). Guidelines on design and operation of wave energy converters. *Carbon Trust*.

Veritas, D. N. (2010). Fatigue design of offshore steel structures. *No. DNV-RP-C203*.

Wu, J., Yao, Y., Zhou, L., & Göteman, M. (2018). Real-time latching control strategies for the solo Duck wave energy converter in irregular waves. *Applied Energy, 222*, 717–728.

Yang, L., Hals, J., & Moan, T. (2010). Analysis of dynamic effects relevant for the wear damage in hydraulic machines for wave energy conversion. *Ocean Engineering, 37*(13), 1089–1102.

Conclusion

Ganesh D. Kale

Chapter "A Review of Multiple Criteria Decision-Making Methods in Reference to Water Resources and Climate Science Applications" provided a review of the literature on the topic of applications of MCDM in climate science and water resources problems. There were seventy references having 11 different approaches of MCDM and these were assessed in this chapter. This chapter provided an integrated source of references, which could be beneficial for practitioners and researchers. On the basis of review, some latest trends and directions for the future research were also highlighted. This chapter provided a comprehensive review, which will be helpful in selection of the suitable MCDM technique for the problem of general material selection. In this chapter, it was also suggested that, combination of multiple techniques addresses deficiencies which may be observed in certain methods. The inferences drawn from the literature review can be useful for the future studies on application of MCDM techniques in water resources problems and climate science.

Study in the chapter "Development of Spatial Cognitive Model for Estimation of Ungauged Runoff for Mesoscale Rivers" developed a model by using MCDM methods and neural networks, which was used for flow amount estimation in the Dhalai district of Tripura, India. The model had input variates, which were based on change in land use characteristics. Methodology of the study included data collection, image processing, development of the HEC-HMS model for runoff estimation, deciding parameter priority by using MCDM, development of an ANN predictive model, indicator development and model validation. From the study, it was found that, there was a 49% decline in runoff with an increment in 'forest cover' by 75%. Also, 9% decrease in runoff was found corresponding to a 25% increase in the forest cover. Also, corresponding to 50% decline in 'water body,' the model predicted a 25.5% increment in the runoff, which illustrated the model's sensitivity to each input. The significance of input parameter can change with change in number of

G. D. Kale (✉)
Civil Engineering Department, SV NIT, Surat 395007, Gujarat, India
e-mail: gdk@ced.svnit.ac.in

M. Majumder and G. D. Kale (eds.), *Water and Energy Management in India*,
https://doi.org/10.1007/978-3-030-66683-5_14

input parameters or change in MCDM methods. This problem can be resolved, if different methods and various combinations of inputs and output are compared to identify the method, which computes the significance utmost prominently and had minimal calculation steps for deciding importance. This type of study can be carried out in future to resolve the aforesaid problem.

In chapter "Indicator Based Impact Analysis of Urbanization with Respect to Evapo-Transpiration", by implementing multi-criteria decision-making, namely AHP and FLDM, evapotranspiration vulnerability was predicted from the chosen attributes that affect PET and were also affected by urbanization. The VI of four cities Agartala (India), Brisbane (Australia), Kolkata (India) and New York (USA) was estimated in the study. For aforesaid four cities having different levels of population density, a neuro-genetic model was applied to show the vulnerability of the evapo-transpiration to urbanization. Vulnerability was found to be higher in cities with denser populations and higher levels of urbanization. Similar study can be performed in future by using a number of MCMD approaches in addition to the approaches used in this study.

Study in chapter "Trend Analyses in Groundwater Levels of the Bikaner District, Rajasthan" analysed trends in GWLs of six blocks in the Bikaner district of the Rajasthan, India, for the period of 1994–2018. This study was carried out for the temporal scales of monsoon, POMKH, POMRB and PREMON. The SS test, MK test/MK test with correction factor-2, ITA plot and smoothing curve were utilized in the trend analyses. The results indicated presence of statistically significant declining trends in monsoon and pre-monsoon GWL time series at Khajuwal and Dungargarh blocks, respectively. All statistically significant trends were found to be declining, thus these trends indicated an improvement in GWLs at given blocks in corresponding seasons. Similar study can be carried out in future at other districts also.

In chapter "Climate Change Impact on Virtual Water Availability: A Categorized Polynomial Neural Network Approach", study developed predictive model for virtual water availability by using PNN based algorithms and evaluated the impacts of climate change on the availability of virtual water for the four metro cities namely New Delhi, Mumbai, Kolkata and Chennai. To assess the impact of climate change on the virtual water, output from the climate prediction model HadCM3 corresponding to two distinct scenarios of IPCC SRES namely A2 and B2 and three time periods viz. 2020–2030, 2031–2050 and 2051–2070 were utilized in the study. From the study, it was found that, New Delhi was worst affected and Kolkata was least affected in case of B2 scenario. For A2 scenario, New Delhi was again found to be worst affected and Mumbai was found to be least affected. The classifier was observed to be more proficient than the techniques viz. stepwise forward regression, logistic regression and decision forest used in the present investigation. In future, similar study can be carried out by using other IPCC SRES scenarios.

In chapter "Development of ANN Model for Simulation of the Runoff as Affected by Climatic Factors on the Jamuna River, Assam, India", authors have developed an ANN model for the identification of effects of climate change on the runoff process for the Jamuna river, Assam by utilizing six predictors namely specific humidity, surface upward latent heat flux, precipitation, maximum air temperature, wind speed and minimum air temperature as input variables. The runoff of the Jamuna catchment was observed to be increased with surface upward latent heat flux, precipitation, specific humidity, maximum and minimum temperature and runoff was found to be decreased with wind speed. The outcomes of the study showed that, the network was optimized with 5-1-1 network structure. Also, the R^2 and RMSE values were found to be 0.88 and 0.352, respectively. In future, similar study can be carried out by using an ANN model formulated by using sophisticated techniques of input selection, data division etc.

In chapter "Modelling of Reference Evapotranspiration for Semi-arid Climates Using Artificial Neural Network", daily ET_0 was modeled using data driven and empirical models for the Hyderabad, which is situated in Telangana, India, which is having semi-arid climatic conditions. Dataset used for the study comprised of daily meteorological data corresponding to the period of 1965–2015. Estimates of ET_0 from the Penman–Monteith model were considered as standard reference for various radiation and temperature based empirical models and data driven models also. The rates of daily ET_0 were estimated by using the technique of an ANN modeling, which was performed corresponding to two, three and five input variables. The results obtained from an ANN modeling were compared with other methods of ET_0 estimation viz. Penman Monteith method, Priestly-Taylor method, Hargreaves method and Turc method. When all climatic variables were given as input to an ANN model, it was found to simulate ET_0 values as estimated by the Penman Monteith method. Development of such data driven algorithms on the basis of empirical models can be used for prediction when the availability of climate variate data is limited. In future, ANN model formulated based on empirical methods can be used for estimation of other variates also.

In chapter "Verifying Storm Water Drainage System Capacity for Vadodara Airport", study attempted to assess the discharge carrying capacity of the prevailing stormwater drains of the Vadodara Airport in the context of probable flood scenarios. SWMM model was developed and storm water drainage system of the Vadodara airport was assessed for the precipitation corresponding to the return period of one year and time durations of 15, 30, 45 and 60 min as per guidelines of CPHEEO manual. For all rainfall durations and return periods corresponding to two-hour time of concentration, the nodes j6, j17, j31, j37, j41, j46, j48, j49, j51 and j53 were observed to get flooded. The simulation results can be utilized to modify the capacity of airport's prevailing storm water drainage system and to decide about necessary changes in the drainage network to prevent airport flooding. The similar study can be carried out in future by using other proficient model like MIKE Urban.

Chapter "Optimal Trade-Off Between the Energy—Economy of a Hydropower Plant for Better Management of the Renewable Energy Resources" attempted to identify the optimum trade-off amongst the financial liability and utilization efficiency with the help of two optimization techniques, DE and FFA. The trade-off analysis was conducted on HPP of the Tripura. Financial liability and utilization efficiency of the power plant was determined by numerous parameters and decision making methods were used to produce an index demonstrating the financial and utilization requirements of the power plant. It was found that, turbine efficiency and operational and maintenance costs were the most crucial indicators of the utilization and economic liability of the HPP, respectively. Also, DE was found to be superior optimization technique as compared to FFA in the present trade off study. For trade-off amongst the benefits of carbon reduction and ecological performance of addition of carbon capture and storage to the power plant, this model can be applied.

Study in chapter "Impact Analysis of Water, Energy, and Climatic Variables on Performance of Surface Water Treatment Plants" applied tools of cognitive and objective decision-making to establish a framework for identifying the most significant factor which will be largely affected by the uncertainty and its effect on the performance efficiency of the Barjala SWTP. The ANP-MCDM method was used for the selection of features throughout the investigation and the PNN model was employed to estimate the indicator from the chosen parameters. Data acquired from water treatment plants in the Tripura, which is located in North Eastern India were used to demonstrate the reliability of the method proposed. The Barjala SWTP performance for the year 2014 was observed to be 1.04 times lesser as compared to that in 2010. The ANP analysis application indicated that, it is appropriate to deal with complex decision-making problems as it permits the consideration and comparison of several criteria in a systemic way. The study suggested that, greater complexity entails greater overlap in terms of assessment criteria, which further strengthens the ANP model's suitability. Similar study can be carried out in future by using the objective method of assigning weights in ANP, so that, the ranking of criteria and alternatives are independent of thinking of decision makers.

In chapter "Power Allocation in an Educational Institute in India: A Fuzzy-GMDH Approach", a new technique was proposed for allocating available energy amongst the various types of consumers. In this technique, importance of consumers is decided at first by using fuzzy logic and after the ranking of consumers on the basis of their importance, the available energy was distributed amongst the various types of consumers by using various cognitive algorithms namely PSO, DE, Neuro-Genetic optimization and GMDH. National Institute of Technology, Agartala, Tripura, India was considered as the study area in this study. The study area had various types of consumers having various demands, patterns of consumption and impact on socio-economic features of the area. The results of the study clearly indicated that, PSO was the better algorithm for the optimization. The results also suggested an option for optimal use of available energy on the basis of consumers' socio-economic importance. The similar study can be performed in future for the larger area having large number of various types of consumers to identify the allocation procedure's sensitivity.

Chapter "Application of New Convergent Point Decision Making Method in Estimation of Vulnerability Index for Hydro Power Reservoirs" attempted to identify most crucial parameters which can maximize the Gumti HPP performance with the aid of optimization methods used as MCMD techniques. The Gumti HPP is situated in Tripura, India. Here two MCDM methods, MACBETH and ANP were utilized to decide the search space boundary, which was required for deciding the priority of the parameter. In this study, a nature based technique of optimization and TLBO were utilized to maximize the efficiency performance function. The results shown that, under normal conditions, most crucial parameter for regulating the HPP performance was efficiency of generator. The technique utilized in the study can also be used to determine the financial suitability of the project where the highest value of the profit function can produce the best financially suitable HPP projects. So, technique utilized in the present study can be used in future for determining the financial suitability of the project.

Chapter "Recognition of Fatigue Failure in Wave Energy Converter Using Statistical Control Chart, Multi-criteria Decision Making Tools and Polynomial Neural Network Model" developed a new methodology to recognize the fatigue probability in any wave energy converter. Every conceivable factor was contemplated to identify fatigue trending. Probability statistical control process and MCDM tools like weighted summation method and weighted product method were used to break down the exhaustion likelihood fatigue. From the results, a model is developed by using PNN software named GMDH. Seven fatigue influenced parameters were contemplated as input and these were identified through the literature review. These parameters were employed in the analysis and yielded results were trained in the software and then model predicted index was estimated by means of framework. On the other hand, contemplating various mechanical design oriented suppositions, it was observed that, fatigue influenced factors were all non-recipient to fatigue trend. The analytical values of seven factors can be employed in the actual field.

The manufacturer's authorised representative in the EU is Springer Nature Customer Service Centre GmbH, Europaplatz 3, 69115 Heidelberg, Germany. If you have any concerns regarding our products, please contact ProductSafety@springernature.com

Printed and bound by CPI Group (UK) Ltd, Croydon, CR0 4YY
30/09/2026
02225691-0006